AQUATIC FOOD SECURITY

AQUATIC FOOD SECURITY

Edited by Margaret Crumlish and Rachel Norman

5m Books

First published 2024

Published by
5M Books Ltd,
Lings, Great Easton,
Essex CM6 2HH, UK,
Tel: +44 (0)330 1333 580
www.5mbooks.com

Follow us on
Twitter @5m_Books
Instagram 5m_books
Facebook @5mBooks
LinkedIn @5mbooks

A Catalogue record for this book is available from the British Library

ISBN 9781789181326
eISBN 9781789183368
DOI 10.52517/9781789183368

Book layout by Cheshire Typesetting Ltd, Cuddington, Cheshire
Printed by CPI Antony Rowe Ltd, UK
Photos and figures by the authors unless otherwise indicated
Cover photo: Bedis ElAcheche: https://www.pexels.com/photo/fish-net-on-gray-surface-891118/

Contents

Contributors

Asadatun Abdullah – Department of Aquatic Product Technology, Faculty of Fisheries and Marine Sciences, IPB University (Bogor Agricultural University), Bogor, West Java, Indonesia
Amaya Albalat – Institute of Aquaculture, University of Stirling, Stirling, Scotland, UK
Brian Austin – Institute of Aquaculture, University of Stirling, Stirling, Scotland, UK
Eirin Marie S. Bar – Norwegian University of Science and Technology (NTNU), Norway
Joy Becker – School of Environmental and Life Sciences, Faculty of Science, University of Sydney, Australia
Malcolm Beveridge – Institute of Aquaculture, University of Stirling, Scotland, UK
Ram C. Bhujel – Aqua-Centre, Asian Institute of Technology (AIT), School of Environment, Resources and Development (SERD), Thailand
Ioannis S. Boziaris – School of Agricultural Sciences, University of Thessaly, Volos, Greece
Margaret Crumlish – Institute of Aquaculture, University of Stirling, Stirling, Scotland, UK
Lynne Falconer – Institute of Aquaculture, University of Stirling, Stirling, Scotland, UK
Marine Furhmann – Sydney School of Veterinary Science, Faculty of Science, University of Sydney, Australia
Fiona Harris – University of the West of Scotland, Paisley, Scotland, UK
Ross D. Houston – Director of Innovation, Benchmark Genetics, United Kingdom
Anita N. Jakobsen – Norwegian University of Science and Technology (NTNU), Norway
Ioannis T. Karapanagiotidis – Department of Ichthyology & Aquatic Environment, School of Agricultural Sciences, University of Thessaly, Greece
Jørgen Lerfall – Norwegian University of Science and Technology (NTNU), Norway
Rachel Norman – University of Stirling, Stirling, Scotland, UK
Roni Nugraha – Department of Aquatic Product Technology, Faculty of Fisheries and Marine Sciences, IPB University (Bogor Agricultural University), Bogor, West Java, Indonesia
Tati Nurhayati – Department of Aquatic Product Technology, Faculty of Fisheries and Marine Sciences, IPB University (Bogor Agricultural University), Bogor, West Java, Indonesia
Mala Nurilmala – Department of Aquatic Product Technology, Faculty of Fisheries and Marine Sciences, IPB University (Bogor Agricultural University), Bogor, West Java, Indonesia
Nurjanah Nurjanah – Department of Aquatic Product Technology, Faculty of Fisheries and Marine Sciences, IPB University (Bogor Agricultural University), Bogor, West Java 16680, Indonesia
Yoshihiro Ochiai – Graduate School of Agricultural Science, Tohoku University, Aramaki, Aoba, Sendai, Miyagi, Japan
Rohana Subasinghe – FUTUREFISH, London, UK
Ruddy Suwandi – Department of Aquatic Product Technology, Faculty of Fisheries and Marine Sciences, IPB University (Bogor Agricultural University), Bogor, West Java, Indonesia

Trevor Telfer – Institute of Aquaculture, University of Stirling, Stirling, Scotland, UK
Smaragda Tsairidou – The Roslin Institute and Royal (Dick) School of Veterinary Studies, University of Edinburgh, Edinburgh, UK
Ruth N. Zadoks – Sydney School of Veterinary Science, Faculty of Science, University of Sydney, Australia

Abbreviations

AA	arachidonic acid
ACT	Agriculture Certification Thailand
AFAS	Asian Fisheries and Aquaculture
AGD	amoebic gill disease
AHPND	acute hepatopancreatic necrosis disease
AHPNS	acute hepatopancreatic necrosis syndrome
AIT	Asian Institute of Technology
AMR	antimicrobial resistance
ASC	Aquaculture Stewardship Council
ASP	amnesic shellfish poisoning
BAP	best aquaculture practices
BKD	bacterial kidney disease
bt	billion tonne
BV	breeding values
CAGR	compound average growth rate
CCP	critical control points
CCRF	Code of Conduct of Responsible Fisheries
CCS	closed containment systems
CDC	Centers for Disease Control
CE	carp erythrodermatitis
CFP	ciguatera food poisoning
CP	crude protein
DF	degree of filling
DHA	docosahexaenoic acid
DNA	deoxyribonucleic acid
DSP	diarrhetic shellfish poisoning
DWTC	Dubai World Trade Centre
EC	encephalopathy crisis
EFA	essential fatty acids
EMS	early mortality syndrome
EPA	eicosapentaenoic acid
EU	European Union
EUS	epizootic ulcerative syndrome
FA	fatty acids
FAO	Food and Agriculture Organization

FCR	feed conversion rate
FPI	fishery performance indicator
FRSS	Fisheries Resources Survey System
GAA	Global Aquaculture Alliance
GAP	good aquaculture practice
GBS	Group B *Streptococcus*
GEBV	genomic estimated breeding values
GFSI	Global Food Safety Initiative
GHG	greenhouse gas
GHP	good hygiene practice
GIFT	genetically improved farmed tilapia
GMP	good manufacturing practices
GWAS	genome-wide association studies
HAB	harmful algae bloom
HACCP	hazard analysis and critical control point
HCG	human chorionic gonadotropin
HGT	horizontal gene transfer
HOG	head-on-gutted
HPP	high-pressure processing
IBD	identity by descent
IBS	identity by state
IFPRI	International Food Policy Research Institute
IFS	International Featured Standards
IHHN	infectious hypodermal and haematopoietic necrosis
IHN	infectious haematopoietic necrosis
IHNV	infectious hematopoietic necrosis virus
IMNV	infectious myonecrosis virus
IMTA	integrated multi-trophic aquaculture
IoT	internet of things
IPIFF	International Platform of Insects for Food and Feed
IPN	infectious pancreatic necrosis
IPNV	infectious pancreatic necrosis virus
ISA	infectious salmon anaemia
IUU	illegal, unreported and unregulated
JICA	Japanese International Cooperation Agency
KHVD	KHV disease
kt	kilotonne
LA	linoleic acid
LAB	lactic acid bacteria
LD	linkage disequilibrium
LNA	alpha-linolenic acid
MA	modified atmosphere
MAP	modified atmospheric packaging
MAR	multiple antibiotic resistance
MAS	marker assisted selection

MRL	maximum residue limits
MSC	Marine Stewardship Council
mt	million tonnes
NASA	National Aeronautics and Space Administration
NGS	National Geographic Society
NSP	neurotoxic shellfish poisoning
OIE	Office International des Epizooties (now WOAH)
PAH	polycyclic aromatic hydrocarbons
PAP	processed animal proteins
PCB	polychlorinated biphenyls
PCR	polymerase chain reaction
PD	pancreas disease
PG	pituitary gland
POP	persistent organic pollutants
PSP	paralytic shellfish poisoning
PUFA	polyunsaturated fatty acids
QIM	quality index method
QTL	quantitative trait loci
RAS	recirculating aquaculture systems
REF	Research Excellence Framework
RF	radio-frequency
RNA	ribonucleic acid
RPS	relative percent survival
RSBID	Red Sea bream iridovirus disease
RSIVD	Red Sea bream iridoviral disease
RTE	ready-to-eat
RTFS	rainbow trout fry syndrome
SDG	Sustainable Development Goals
SGR	specific growth rate
SGS	soluble gas stabilization
SHPN	septic hepatopancreatic necrosis
SIMP	seafood import monitoring program
SNP	single nucleotide polymorphisms
SPF	specific pathogen free
SQF	Safe Quality Food
SSO	specific spoilage organisms
SSOP	sanitation standard operating procedure
SVC	spring viraemia of carp
SVCV	spring viremia of carp virus
t	tonne
TSP	triple super-phosphate
TSV	taura syndrome virus
TVC	total viable counts
UN	United Nations
UNCLOS	United Nations Convention on the Law of the Sea

VAP	value-added product
VHS	viral haemorrhagic septicaemia
VHSV	viral haemorrhagic septicaemia virus
VPC	viremia primaveral de la carpa
WAS	World Aquaculture Society
WHC	water holding capacity
WHO	World Health Organization
WOAH	World Organisation for Animal Health (formerly OIE)
WSD	white spot disease
WSS	white spot syndrome
WSSV	white spot syndrome virus

1

Introduction to global aquatic food security

Margaret Crumlish and Rachel Norman

Food security is a complex and multifactorial issue. While there are many definitions, most agree with the Food and Agriculture Organization (FAO) description that food security 'exists when all people, at all times, have physical and economic access to sufficient, safe and nutritious food to meet their dietary needs and food preferences for an active and healthy life' (FAO, 1996). Achieving this is not a simple task and technical solutions alone will do little to tackle food inequality on a global scale. While international recognition of the contribution aquatic food makes to global food security has been slow compared with that of terrestrial farming, it is now very firmly established as the main food growth production area (Garlock et al., 2022), with a larger contribution originating from the aquatic farming sector than the capture fisheries, and is forecast to continue (Kobayashi et al., 2015). In this book, we consider aquatic food security to be any food type that is produced for human consumption in an aquatic environment, however, most of the examples given in the chapters will focus on fish or animal production. Unless otherwise specified, throughout this publication, the term 'fish' indicates fish, crustaceans, molluscs and other aquatic animals, but excludes aquatic mammals, reptiles, seaweeds and other aquatic plants. It is not the intention of this book to provide an overview of the marine fisheries sector nor the diversity in aquatic farming production systems, nor do we focus on reactional or subsistence fisheries but instead our approach is inclusive of all seafood: capture fisheries and farmed food. Numerous factors influence aquatic food consumption and production, including product availability, political stability, economic viability and growth, not to mention the cultural and ethical beliefs of consumers. However, the primary driver that inspired this book was the rapid expansion of the global seafood sector and the scientific and commercial development supporting aquatic food security as we currently understand it.

In the natural sciences, diversity is often considered a strength as it provides resilience within the ecological or biological system. Multiple fish and shrimp species are being farmed globally within a diverse range of production systems and environments. If we consider the availability and accessibility of food being significant contributors to achieving food security globally, then surely a key strength is the diversity of production/fishing systems, the range of species being farmed/fished for human consumption combined with the scope of products available. This biological diversity ensures

Margaret Crumlish and Rachel Norman (eds) **Aquatic Food Security**
DOI: 10.52517/9781789183368.001, © 5m Books Ltd, 2023

seafood supply chains remain buoyant, meeting demand from consumers and facilitates access to essential micro and macro nutrients (Belton et al., 2018). It is this range of products that makes aquatic food truly accessible on a global scale, satisfying the spectrum of consumers: from wealthy consumers in developed countries to disadvantaged or marginalized communities that have few other food choices.

On a global scale, products traded internationally are more commonly raised in intensive farming systems that are run as commercial businesses. Sustainable intensification is challenging for any food production system and the practice of monoculture farming is often described as 'high risk but high reward'. The reader is directed to Chapter 2, **The role of intensification in global aquaculture production**, which identifies both the benefits of and challenges to delivering sustainably intensive aquaculture systems/sectors. The growth of the aquaculture sector has supported development of associated industries such as the aquafeed and aquatic pharmaceutical industries. One of the biggest financial investments in any in-feed aquaculture system is the cost of the animal feed. Correct nutrition is essential to support the growth and health of the animals in the production system. Now, commercially produced feed has many advantages over traditional feed types, but the reliance on marine resources to supply essential fish meal and fish oil in aquafeeds continue to be contentious issues (Tacon, 2004). Providing an alternative source to the existing supply of marine oils and fish meal added to the commercial pelleted diets can help reduce the environmental impact of aquaculture. The reader is directed to Chapter 3, **The importance of animal nutrition and selective breeding in aquaculture production**, which provides a historical context, explores the existing nutritional requirements and sources as well as offers a comprehensive overview of the animal genetics. Nutrition and genetics are key contributors to the production of animal health and welfare within any farming system, particularly applicable to aquatic animals given the large diversity of species and their individual dietary requirements.

Aquatic food is produced and sourced from freshwater, brackish water and marine environments, depending on the species being farmed, and is available in both temperate and tropical locations around the globe. Chapter 4, **Aquaculture production now and in the future – an ecosystems perspective**, introduces the reader to the concept of 'environmental goods and services' of direct relevance to existing and future seafood production systems. Chapter 4 provides insight into the impact of waste (primarily from animal excreted products and feed) on the local, regional and global environments. Farm-level waste management is complicated, as many of the production systems found globally are practised in open systems, susceptible to environmental changes, particularly those driven by climate (e.g. fluctuating temperatures, unpredictable weather patterns and changing salinity levels). By adopting environmental modelling as described in Chapter 4, critical local and regional evidence can be gathered and mapping of the aquaculture sector can reduce the risk of environmental harm from poor farming practises or mismanagement. The adoption of good practises in animal genetics, nutrition and environmental monitoring all support the sustainable intensification of global aquaculture and go hand in hand with the previously discussed resilience to contribute to aquatic food security. However, poor biosecurity combined with lack of knowledge and understanding of animal husbandry can exacerbate the onset of infectious diseases, leading to lower production levels and contributing to food insecurity. The lack of efficacious disease prevention tools (e.g. vaccines) combined with the mismanagement of chemotherapeutants in some production systems contribute to the on-going disease outbreaks from emerging and re-emerging infections. All aquatic animals are susceptible to infectious diseases and can suffer from outbreaks of viruses, bacteria, fungi and parasites. As the aquaculture industry has expanded so too has the list of recognized diseases and the reader is directed to Chapter 5, **Production level diseases and**

public health considerations in aquaculture. While tremendous efforts are made to promote the highest level of hygiene, sanitation and welfare in all farming systems, infectious diseases do not remain static and instead evolve. The recent global pandemic from COVID-19 has raised awareness by the general public of the concept of 'disease variants' and 'transmission routes/times'. Research on aquatic diseases often concentrates on a single pathogen at one time. Unfortunately, infectious disease outbreaks are rarely that linear and it is not unusual to experience co-infections or multiple infectious threats during production cycles. A further complication is the impact of climate change on animal physiology and pathogenic capacity leading to novel disease outbreaks and resurgence of previous diseases. The information provided in Chapters 3, 4 and 5 show the complexity in recognizing and addressing the multifactorial challenges in meeting the global demand for aquatic food production. This would be akin to meeting the 'availability' pillar from the FAO food security description.

Chapter 6, **Global aquatic food production**, contextualizes aquatic food security and introduces the complexity of socio-economic-political challenges to achieving the Blue Food agenda. Technical solutions implemented in isolation without the broader context of the drivers behind food insecurity are unlikely to solve the issues being experienced, and in some circumstances, may exacerbate food insecurity particularly in vulnerable communities reliant on seafood as their primary source of protein. Chapter 6 highlights the relationship between seafood availability, accessibility and safety in achieving the UN Sustainable Development Goals. Seafood is traded on the global market and aquatic food supply chains have grown in size and length in parallel with the aquaculture sector development. It is in Chapter 7, **The role of markets in global aquatic food security**, that we introduce the importance of post-harvest markets and product placement. Historically, seafood has been sold in markets supplying coastal communities. Many consumers in low- to middle-income countries prefer purchasing from fresh or traditional wet markets. Whereas consumers in higher-income countries rely heavily on the retail sectors and supermarkets to supply their seafood, often as pre-packaged products. Chapter 7 describes how the seafood markets have evolved and developed over time and provides details at country-specific levels. It is clear that seafood plays an important part in individual and global food consumption and Chapter 8, **The role of processing and retail sectors in aquatic food security**, illustrates the varied preservation methods available and practises applied globally to preserve and process these highly perishable products. Chapter 8 introduces the role and value of using traditional food preservation methods to promote long-term shelf-life and provide rural families with a source of protein in their diet. Novel methods in secondary processing and packaging of seafood not only promotes accessibility but also plays a vital role in reducing the risk of microbial spoilage. Chapter 8 describes the range of seafood preservation and processing methods applied globally, which directly contribute towards the accessibility and safety of the seafood being marketed and traded.

To achieve food security, we must not only consider the availability and accessibility of the food, these are important, but the food must also remain fit for human consumption. This introduces the concept of food safety, which relies on science and policy to identify food-borne hazards to human health and implementation of prevention strategies to reduce the risk of food-borne infections in human populations. When considering food safety, we must include all stages from the post-harvest onwards, e.g. transportation, handling, preparation and, of course, storage. Chapter 9, **Raw seafood practises, risk and public health in Southeast Asia**, introduces the concept of 'food practices', highlighting the cultural aspects surrounding food that we are not always consciously aware of. This concept explores the social, political and economic context of how food is produced and consumed – it brings a social dynamic by understanding the human behaviour behind food choices.

In many parts of the world, seafood is consumed raw or lightly preserved/cooked, where the preparation (or lack of preparation) may be culturally influenced. To promote public health and support behavioural changes in food preparation practices and reduce the risk of human ill health we need to understand the drivers behind food preparation. All food is susceptible to poor food safety, and contamination during production and/or food preparation may exacerbate these risks leading to human ill health. The burden of food-borne infections in disadvantaged communities can be especially high, particularly where access to medical treatment may not be readily available. In any food production system or any food manufacturing process there are recognized chemical, physical and biological risks. Chapter 10, **Aquatic food safety**, provides the reader with a comprehensive overview on the range of risk factors unique to seafood production and highlights mitigation strategies to reduce the hazard or lessen the risk of human ill health from seafood consumption. This chapter focuses on the public health risks associated with seafood production, processing, and consumption. Chapter 11, **Quality issues in our global seafood**, distinguishes unique seafood hazards compared with generic food risks from any perishable product. A sound understanding of these is critical to support the production of safe and nutritious seafood fit for human consumption. Chapter 11 uses specific examples from the seafood value chain to explore food-quality issues and risks leading to food spoilage. The reader is introduced to novel technologies to reduce spoilage on a wide range of seafood products and their contribution to prolonged product shelf-life.

Finally, in Chapter 12, **Horizon scanning for aquatic food security**, we discuss the uncertainties surrounding aquatic food security. We explore the challenges and opportunities facing global aquatic food production and the drivers of instability within the global seafood system. Here we bring together the biological, social, economic and political dimensions influencing stability in this relatively new food production sector. The reader will be familiar with many of these challenges as they are not necessarily unique to aquatic food production or consumption, however, in this chapter, we highlight the direct impact these have on the availability, accessibility and safety of seafood compared with terrestrial farming produce. The opportunities open to the seafood industry in addressing aquatic food insecurity lie in a combined and collaborative approach, one requiring interdisciplinary research. Our overarching goal with this book is to provide a 'one stop shop' for anyone interested in aquatic food security.

References

Belton, B., Bush, S.R. & Little, D.C. (2018) Not just for the wealthy: Rethinking farmed fish consumption in the Global South. *Global Food Security*, 16, 85–92. https://doi.org/10.1016/j.gfs.2017.10.005.

FAO (1996) Rome Declaration on World Food Security. Available at https://www.fao.org/3/w3613e/w3613e00.htm (accessed 1 May 2023).

Garlock, T., Asche, F., Anderson, J., Ceballos-Concha, A., Love, D.C., Osmundsen, T.C. & Pincinato, R.B.M. (2022) Aquaculture: The missing contributor in the food security agenda. *Global Food Security*, 32. https://doi.org/10.1016/j.gfs.2022.100620.

Kobayashi, M., Msangi, S., Batka, M., Vannuccini, S., Dey, M.M. & Anderson, J.L. (2015) Fish to 2030: The role and opportunity for aquaculture. *Aquaculture Economics and Management*, 19, 282–300. https://doi.org/10.1080/13657305.2015.994240.

Tacon, A.J.G. (2004) The use of fish meal and fish oil in aquaculture: A global perspective. *Aquatic Resources, Culture and Development*, 1, 3–14. https://doi.org/10.1079/cabireviews20063050506.

2

The role of intensification in global aquaculture production

Rohana Subasinghe and Rachel Norman

Fish is an important, nutritious food commodity with a high consumer demand that requires a continuously increasing volume of supplies. Global food fish supplies were recorded at about 171 million tonnes (mt) in 2016 (FAO, 2018), with aquaculture representing 47% (80 mt). Aquaculture is one of the fastest growing sectors among all the food production systems and has been growing at an average annual rate of approximately 6% in the past three decades. Inland water capture fisheries also produced 11.6 mt of fish in 2016, which represented 12.8% of total marine and inland catches. With marine fish catches relatively static since the late 1980s, aquaculture has been responsible for the continuing impressive growth in the supply of fish for human consumption (Figure 2.1). The average global per capita fish consumption in 2015 was 20.2 kg, this increased to 20.5 kg in 2019 but then declined again to 20.2kg in 2020. It is predicted to increase to 21.4 kg by 2030 (FAO, 2022).

The UN predicts that the global population will reach 8.5 billion in 2030 (United Nations, 2015). This will inevitably increase the pressure on food production sectors to maximize production and reduce waste. Production increase must occur in a sustainable way and in a context in which key resources, such as land and water, are likely to be scarcer and in which climatic change impact will intensify. The fish-production sector is no exception. In this context two key questions emerge.

1. Will fisheries and aquaculture be able to maintain the current global fish consumption rate of 20 kg per capita per year (with equivalent regional values and variations), and if not, how will society address the shortfall and effects of this (Barange et al., 2014)?
2. Are sustainable fisheries and aquaculture able to help address the bigger food security issue that will affect the world in the coming decades?

According to a recent Food and Agriculture Organization (FAO) assessment (Cai and Leung, 2017) if fish production and trade continue at current 'business-as-usual' levels (supply based on continued recent growth trends), there will be a demand–supply gap of 48 mt by mid-2020 (i.e. only 40% of the demand will be met by the supply based on a business-as-usual scenario). Since the trend of aquaculture growth would only cover 40% of the projected world fish demand, growth and

Margaret Crumlish and Rachel Norman (eds) **Aquatic Food Security**
DOI: 10.52517/9781789183368.002, © 5m Books Ltd, 2023

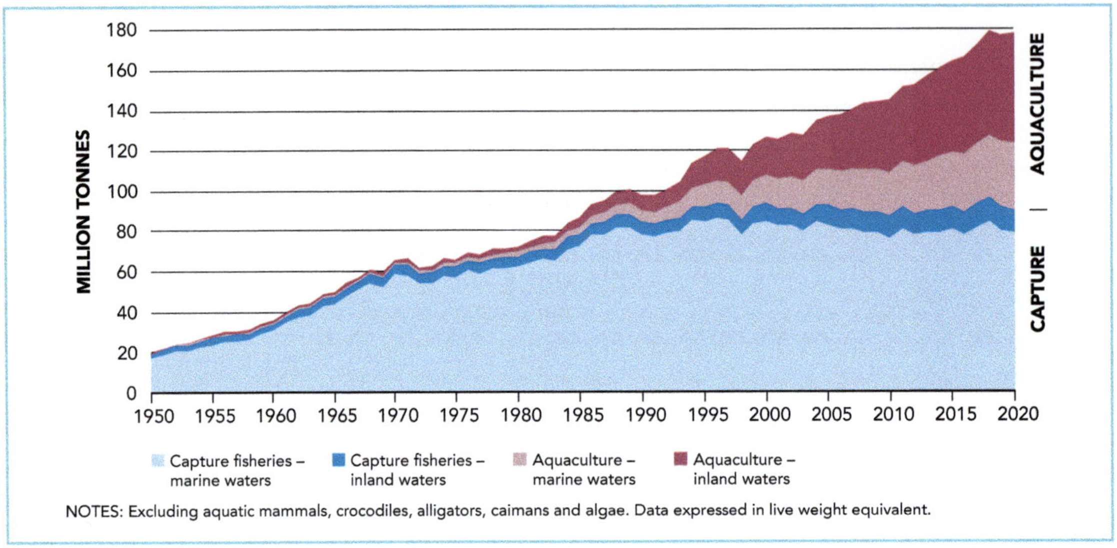

Figure 2.1 World capture fisheries and aquaculture production, 1950–2020. Source: Food and Agriculture Organization of the United Nations. Reproduced with permission (FAO, 2022).

capture fisheries production is expected to show little growth, the resulting excess demand would tend to drive up fish prices, which would dampen the demand growth. Therefore, per capita fish consumption (determined by fish demand) in the mid-2020s would likely be less than 25 kg/year if global aquaculture follows its recent trend, or what we consider business as usual. However, the 25 kg/year consumption target could only be sustained if aquaculture growth could be doubled from 4.5% business-as-usual growth to 9.9% a year. Will this be possible through sustainable intensification?

What is sustainable intensification?

Sustainable intensification is a concept that challenges global agriculture to achieve a doubling in world food production while sustaining the environment in which we live. Sustainable intensification of aquaculture aims to increase aquaculture output from the same available land and water area, while reducing the negative environmental impacts. This means increasing production, while maximizing resource use efficiency, environmental performance and social equity through improved governance, management practices and the adoption of innovative technologies.

Godfray and Garnett (2014) argued sustainable intensification is a new, evolving concept, its meaning, and objectives subject to debate and contest. They also stated sustainable intensification is only one part of what is needed to improve food system sustainability and is by no means synonymous with food security. Both sustainability and food security have multiple social and ethical, as well as environmental, dimensions. Achieving a sustainable, health-enhancing food system for all will require more than just changes in agricultural production, essential though these are. Equally radical agendas will need to be pursued to reduce resource-intensive consumption and waste and to improve governance, efficiency and resilience. They presented four premises underlying sustainable intensification: (1) the need to increase production; (2) the importance of achieving increased production through higher yields; (3) improving environmental sustainability; and (4) the importance of understanding that sustainable intensification is a goal but does not specify a priori how it should be attained or which techniques to deploy. The aquaculture value chain consists of several key components: (1) inputs and services; (2) production; (3) processing; and (4) marketing. Our discussion in this chapter focuses on intensification aspects relevant to aquaculture production.

The continued increase in global aquaculture production of recent decades has been the result of diversification, expansion and intensification. Unfortunately, we do not have accurate estimates of the relative contribution of intensification to the overall increase of global aquaculture production and rate of growth. Globally, there is experience and evidence of unsustainable intensification practices which lead to serious and significant production losses and economic impacts for example in shrimp (Joffre et al., 2018). Evidence from different production systems and practices used for certain high-value species, such as salmon, reveal sustainable intensification (including technological advancement) has been responsible for significant increases in production and will continue to do so in the coming years (Ellis et al., 2016; Little et al., 2018).

What drives intensification?

Increasing profit is the biggest driver for intensification. Aquaculture production, at all levels, is practised as business. It has a cost and it yields profit. Inputs and services are becoming costly, the market price of both inputs (e.g. feed) and the market value of the final products regularly fluctuates resulting in reduced profit. This phenomenon has been pushing many small-scale farmers out of the aquaculture business (Mulokozi et al., 2020). Survival for all operators requires reducing production costs, while increasing productivity, which is generally achieved through making productivity and production optimal. Provision of an optimal culture environment, better farm inputs and applying better farm management practice are critical aspects of intensification. The challenge is to intensify production system while ensuring long-term overall sustainability.

Little et al. (2018) proposed that the key tenets to adopting sustainable intensification, as opposed to a business-as-normal approach to intensification, are based on the greater regularization and integration of aquaculture within global food systems. They elaborated that aquaculture value chains need to evolve towards meeting the nutritional and health demands of more informed, increasingly urbanized consumers who need assurance that their food has been produced with the welfare of all key stakeholders involved in mind, including the farmed animals themselves. The key aspects of aquaculture intensification in many parts of the world so far have been the use of improved nutrition through formulated diets for hatchery-produced animals and, increasingly, the deployment of genetically improved breeds. Such developments have resulted in rapid increases in yield compared to traditional systems. 'Closing the yield gap', whereby commercial yields are increased to meet the yields possible under controlled conditions has, in general, lagged behind the achievements with arable crops and livestock.

Nutrition, feeds and feed management

Intensification enhances productivity. Increased production requires optimization of fish health and growth rates, while minimizing mortalities and controlling the costs associated with capital investment and on-going production (Little et al., 2016). Provision of quality feed is one of the key requirements in sustainable intensification. Feed constitutes between 50% and 70% of aquaculture operational costs. Although feed is a costly input, the criteria for selecting and/or formulating a feed should not only be the price per unit. The level of performance should also be given equal consideration. Final unit cost of production will be determined by the unit price of feed, increased growth rate and reduced feed conversion ratios. Feed performance is based on multiple characteristics, which include: (1) feed efficiency; (2) growth rate; (3) feed quality; (4) digestibility; (5) health benefits; and (6) effects on water quality. Research reducing the use of animal proteins (especially marine resources such as fishmeal) in fish feed by supplementation with plant resources have produced successful results. Turchini et al. (2019) claimed that reliance on marine resources remains an on-going constraint, and the progress yielded by continued unidimensional research into alternative raw materials is becoming increasingly marginal. They suggested that greater emphasis on nutrients, including those not considered strictly nutritionally essential, is required to encourage further evolution of the industry and to efficiently move aquaculture nutrition beyond the incremental advances achieved in recent years. They expressed the need for rethinking or becoming reacquainted with the nutrient-based approach to aquaculture nutrition science, which might spur further innovation within the fish nutrition field and the aquaculture industry and, ultimately, help transform the use of marine-origin resources in aquaculture.

Disease control and health management

Farm-level health management is also vital for sustainable production. Addressing health issues with both proactive and reactive programmes has become a primary requirement for sustainable aquaculture intensification (Bondad-Reantaso et al., 2005). Epizootic-level incursion and spread of disease is not a novel phenomenon in aquaculture. One of the earliest recorded epizootics which occurred in Asia in 1971 was epizootic ulcerative syndrome (EUS) in freshwater fish (Chong, 2022). Asia, the region leading global aquaculture production, has been facing health challenges for several decades. A significant addition to the long list of aquatic diseases/pathogens severely affecting the aquaculture sector since 2009 is acute hepatopancreatic necrosis disease (AHPND), which devastated shrimp aquaculture in several Asian countries (e.g. China, Malaysia, the Philippines, Thailand) (Flegel, 2012). The rapid and timely investment in research that helped to develop early detection and diagnostic tools and effective management measures confirms that investment in applied health research is an integral part of sustainable intensification of aquaculture.

Despite investment in finding remedies for diseases in commercial species, rural small farmers contributing the lion's share of production often have little or no knowledge of aquaculture health management and therefore have inadequate opportunities to improve management skills and respond effectively to disease problems, preventing them from achieving sustainable intensification. Earthen ponds remain the dominant culture unit but within that classification there is a wide range of variation in terms of design and management for water quality and control of pests, parasites and pathogens. It is therefore of critical importance to focus efforts not only on the prevalence of diseases and pathogens but also on the development of smallholder, farmer-oriented health management programmes.

Genetics and breeding

Aquaculture has seen significant success in the improvement of productivity and production through the application of better and more efficient technology. Selective breeding has produced faster growing fish. Genetically improved farmed tilapia (GIFT) developed by WorldFish is an example of a successful selective breeding programme (WorldFish, 2020). GIFT now account for nearly 50% of the global tilapia production from over 20 countries worldwide.

The development of specific pathogen free (SPF) shrimp is another successful technological development that supported sustainable intensification of shrimp aquaculture. Although the concept of SPF animals was well defined for terrestrial animals that could be grown out in isolated installations, it was relatively new in aquaculture where it is difficult to isolate the animals in the aquatic environment. A major impetus for eventual wide adoption of the SPF shrimp concept was the emergence and spread of white spot disease (WSD) of shrimp caused by white spot syndrome virus (WSSV) in the mid-1990s (Flegel and Alday-Sanz, 1998). The availability of SPF stocks of *Penaeus vannamei* together with a pathogen exclusion biosecurity strategy was highly effective and rapidly led to it becoming the dominant cultivated shrimp species in Asia (Wyban, 2007). The benefits of using domesticated and genetically improved SPF stocks of *P. vannamei* to produce healthy post-larvae stages for pond stocking is documented (Alday-Sanz et al., 2020).

Intensification and environment

Intensification does not happen without environmental impacts, both at local and global levels. That is the very reason intensification should be sustainable with minimal negative impacts (Cao et al., 2007; Iwama, 1991). Intensification efforts can cause both localized and wider ranging impacts. For example, receiving waters may not have the natural capacity to remove rich nutrients from the effluents originating from high-density, fed production systems, thus requiring 'effluent treatment' systems to avoid overloading receiving waters (Ross et al., 2013). 'Escapees' from aquaculture production systems can cause negative environmental impacts, including potential genetic pollution, in the wider aquatic environment (Arechavala-Lopez et al., 2013).

Intensification of aquatic production requires more feed sources. Reliance on marine ingredients in aquaculture feeds, especially in high-protein diets targeting shrimp and carnivorous and/or marine species, has been a major point of criticism by environmentalists (Naylor et al., 1998). Years of research on fishmeal replacement with plant ingredients, more so for economic reasons, than environmental, has been largely recognized and successful (Hua et al., 2019; Pratoomyot et al., 2010). However, increased use of plant material for protein substitution and as a source of carbohydrate, has increased pressures on other sensitive ecosystems, notably from soybean production in South America (Schmidt, 2010; Newton and Little, 2018).

Recirculating aquaculture systems (RAS) is another technological development contributing to sustainable intensification of aquaculture. Closed systems potentially lead to a much lower risk of genetic contamination and pathogen-related health problems. Moreover, nutrients can be almost completely retained. These benefits come with significant energy consumption and costs. Mechanical failure and poor fail-safes, rather than disease, appear to be the main risks. The fallibility of 'completely secure' systems is, however, well known and the impacts of events such as mechanical failures are known to be particularly damaging (Badiola et al., 2012).

Little et al. (2018) claimed that super-intensification can reduce on-going local environmental costs through (1) advanced feed formulation, which improves feed efficiency, and (2) better water quality management, but that these come at a cost of larger global impacts related to the use of high levels of fossil fuels. These factors can reduce waste in the production systems and effluents, making intensification more sustainable. The use of RAS can not only improve water quality but also reduce disease and pathogen risks as recirculation minimizes the incursion of pathogens, if good biosecurity is maintained. There is a tendency for increased use of renewable energy and more efficient filtration technology in high-tech commercial production systems, which has the potential to further reduce the environmental impacts of intensified systems. Making RAS or systems reflecting similar principles available to smallholders will help in their efforts to sustainably intensify smallholder systems.

Geographical variation

Like all other food production systems, aquaculture is carried out at many different scales. Indeed, over 80% of the global aquaculture production is from small-scale farms that are usually owned and managed by families (Mulokozi et al., 2020, and references therein). The extent of aquaculture production varies significantly globally: China leads with 67.3% of production and 22.3% comes from the rest of the Asia-Pacific region (Subasinghe et al., 2009). The EU depends on imported seafood with more than one-half of the seafood it consumes coming from imported sources (Little et al., 2018), but even within the EU there is a huge amount of trading between countries, for example, much of the seafood produced or caught in the UK is exported to mainland Europe (0.445 mt in 2021) while much of the seafood eaten in the UK is imported (1 mt [Seafish, 2021]). Given the diversity of species produced or caught, their geographical ranges and which species are in demand in different regions then the potential for sustainable intensification is complex.

Mulokozi et al. (2020) studied aquaculture in rural Tanzania and found that it contributes, on average 13% to household incomes. For these farms it provides extra income, food and a strategy for diversification. However, they also found that, despite the potential opportunities, production in Tanzania was increasing slowly because of the high costs of import and production, and a lack of enabling policies.

Training has been identified as a factor that is likely to increase profit in several countries (Dey et al., 2010; Dickson et al., 2016; Kassam and Dorward, 2017; Mulokozi et al., 2020). Other actions that have been identified as ways to facilitate aquaculture growth in Tanzania include development of suitable production systems, and the availability and accessibility of food quality seeds and feeds (Kaliba et al., 2006; Mulokozi et al., 2020; Rutaisire et al., 2009).

Little et al. (2018), analysing the timeline of intensification in Asia, found there was a major wave of expansion in shrimp production in the 1980s. The timing of this varied from country to country because of a range of factors. For example, Thai shrimp farming was uniformly intensive by 2002 while Vietnam had a wider range of production intensities (Thanh, 2014). Little et al. (2018) identified that policy options to stimulate sustainable intensification of export orientated aquaculture in Asia included supporting trade into the EU through a system of health and safety and quality assurance and ensuring that sustainable intensification is better understood by consumers. They suggest that sustainable intensification cannot be assessed on the farm alone but requires a value-chain approach integrated and informed by life cycle assessment to consider environmental and social impacts.

The future

As has been discussed above, although there are many drivers behind sustainable intensification, the seafood system is so complex and variable geographically that the route to successful implementation is not straightforward. In the early stages of aquaculture development then much could be achieved through training and support to access seed and feed. In countries where aquaculture is more developed then it is support for trade and clear communication with consumers that will support further expansion. In addition, the availability and adoption of appropriate technologies, such as more sustainable feeds, are required in order to achieve the environmental sustainability which is required to distinguish sustainable intensification from just intensification.

As will be discussed in the rest of this book, sustainable intensification of aquaculture is only one way in which seafood can contribute to global food security. Sustainable intensification must happen alongside reduction in waste, changes in consumer behaviour and adaptation to climate change. However, as we have seen there is significant potential for sustainable intensification to increase levels of production and hence support both local and global food security at all scales and for large parts of the world.

References

Alday-Sanz, V., Brock, J., Flegel, T.W., McIntosh, R., Bondad-Reantaso, M.G., Salazar, M. & Subasinghe, R. (2020) Facts, truths and myths about SPF shrimp in Aquaculture. *Reviews in Aquaculture*, 12, 76–84. https://doi.org/10.1111/raq.12305.

Arechavala-Lopez, P., Sanchez-Jerez, P., Bayle-Sempere, J., Uglem, I. & Mladineo, I. (2013) Reared fish, farmed escapees and wild fish stocks: A triangle of pathogen transmission of concern to Mediterranean aquaculture management. *Aquaculture Environment Interactions*, 3, 153–161. https://doi.org/10.3354/aei00060.

Badiola, M., Mendiola, D. & Bostock, J. (2012) Recirculating Aquaculture Systems (RAS) analysis: Main issues on management and future challenges. *Aquacultural Engineering*, 51, 26–35, https://doi.org/10.1016/j.aquaeng.2012.07.004.

Barange, M., Merino, G., Blanchard, J.L., Scholtens, J., Harle, J., Allison, E.H., Allen, J.I., Holt, J. & Jennings, S. (2014) Impacts of climate change on marine ecosystem production in societies dependent on fisheries. *Nature Climate Change*, 4, 211–216. https://doi.org/10.1038/nclimate2119.

Bondad-Reantaso, M.G., Subasinghe, R.P., Arthur, J.R., Ogawa, K., Chinabut, S., Adlard, R., Tan, Z. & Shariff, M. (2005) Disease and health management in Asian aquaculture. *Veterinary Parasitology*, 132, 249–272. https://doi.org/10.1016/j.vetpar.2005.07.005.

Cai, J. & Leung, P.S. (2017) Short-term projection of global fish demand and supply gaps. FAO fisheries and aquaculture technical Paper no. 607. Available at www.fao.org/3/a-i7623e.pdf. Food and Agriculture Organization: Rome.

Cao, L., Wang, W., Yang, Y., Yang, C., Yuan, Z., Xiong, S. & Diana, J. (2007) Environmental impact of aquaculture and countermeasures to aquaculture pollution in China. *Environmental Science and Pollution Research International*, 14, 452–462. https://doi.org/10.1065/espr2007.05.426.

Chong, R.S., (2022) Chapter 50 – Epizootic ulcerative syndrome. In *Aquaculture Pathophysiology*. Editor(s): Kibenge, F.S.B, Baldisserotto, B, Chong, R.S. Academic Press. Pages 621–627 https://doi.org/10.1016/B978-0-12-812211-2.00050-0.

Dey, M.M., Paraguas, F.J., Kambewa, P. & Pemsl, D.E. (2010) The impact of integrated aquaculture–agriculture on small-scale farms in Southern Malawi. *Agricultural Economics*, 4, 67–79. https://doi.org/10.1111/j.1574–0862.2009.00426.x].

Dickson, M.D.A., Nasr-Allah, A., Kenawy, D. & Kruijssen, F. (2016) Increasing fish farm profitability through aquaculture best management practice training in Egypt. *Aquaculture*, 465, 172–178. https://doi.org/10.1016/j.aquaculture.2016.09.015.

Ellis,T., Turnbull, J.T., Knowles,T.G., Lines, J.A., Auchterlonie, N.A. (2016) Trends during development of Scottish salmon farming: An example of sustainable intensification? *Aquaculture*, 458, 82–99, https://doi.org/10.1016/j.aquaculture.2016.02.012.

FAO (2018) The state of world fisheries and aquaculture 2018 – Meeting the sustainable development goals. Rome. Licence: CC BY-NC-SA 3.0 IGO. Available at www.fao.org/3/I9540EN/i9540en.pdf.

FAO (2022) In Brief to The State of World Fisheries and Aquaculture 2022. Towards Blue Transformation. Rome, FAO. https://doi.org/10.4060/cc0463en.

Flegel, T.W. & Alday-Sanz, V. (1998) The crisis in Asian shrimp aquaculture: Current status and future needs. *Journal of Applied Ichthyology*, 14, 269–273. https://doi.org/10.1111/j.1439-0426.1998.tb00654.x.

Flegel, T.W. (2012) Historic emergence, impact and current status of shrimp pathogens in Asia. *Journal of Invertebrate Pathology*, 110, 166–173. http://dx.doi.org/10.1016/j.jip.2012.03.004.

Godfray, H.C.J. & Garnett, T. (2014) Food security and sustainable intensification. *Philosophical Transactions of the Royal Society of London. Series B, Biological Sciences*, 369, 20120273. https://doi.org/10.1098/rstb.2012.0273.

Hua, K., Cobcroft, J.M., Cole, A., Condon, K., Jerry, D.R., Mangott, A., Praeger, C., Vucko, M.J., Zeng, C., Zenger, K. & Strugnell, J.M. (2019) The future of aquatic protein: Implications for protein sources in aquaculture diets. *One Earth*, 1, 316–329. https://doi.org/10.1016/j.oneear.2019.10.018.

Iwama, G.K. (1991) Interactions between aquaculture and the environment. *Critical Reviews in Environmental Control*, 21, 177–216. https://doi.org/10.1080/10643389109388413.

Joffre, O.M., Klerkx, L. & Khoa, T.N.D. (2018) Aquaculture innovation system analysis of transition to sustainable intensification in shrimp farming. *Agronomy for Sustainable Development*, 38, 34. https://doi.org/10.1007/s13593-018-0511-9.

Kaliba, A.R., Osewe, K.O., Senkondo, E.M., Mnembuka, B.V. & Quagrainie, K.K. (2006) Economic analysis of Nile tilapia (Oreochromis niloticus) production in Tanzania. *Journal of the World Aquaculture Society*, 37, 464–473. https://doi.org/10.1111/j.1749-7345.2006.00059.x.

Kassam, L. & Dorward, A. (2017) A comparative assessment of the poverty impacts of pond and cage aquaculture in Ghana. *Aquaculture*, 470, 110–122. https://doi.org/10.1016/j.aquaculture.2016.12.017.

Little, D.C., Newton, R.W. & Beveridge, M.C. (2016) Aquaculture: A rapidly growing and significant source of sustainable food? Status, transitions and potential. *Proceedings of the Nutrition Society*, 75, 274–286. https://doi.org/10.1017/S0029665116000665] [PubMed: 27476856].

Little, D.C., Young, J.A., Zhang, W., Newton, R.W., Al Mamun, A. & Murray, F.J. (2018) Sustainable intensification of aquaculture value chains between Asia and Europe: A framework for understanding impacts and challenges. *Aquaculture*, 493, 338–354. https://doi.org/10.1016/j.aquaculture.2017.12.033.

Mulokozi, D.P., Mmanda, F.P., Onyango, P., Lundh, T., Tamatamah, R. & Berg, H. (2020) Rural aquaculture: Assessment of its contribution to household income and farmers' perception in selected districts, Tanzania. *Aquaculture Economics and Management*, 24, 387–405. https://doi.org/10.1080/13657305.2020.1725687.

Naylor, R.L., Goldburg, R.J., Mooney, H., Beveridge, M., Clay, J., Folke, C., Kautsky, N., Lubchenco, J., Primavera, J. & Williams, M. (1998) Nature's subsidies to shrimp and salmon farming. *Science*, 883–886. https://doi.org/10.1126/science.282.5390.883.

Newton, R.W. & Little, D.C. (2018) Mapping the impacts of farmed Scottish salmon from a life cycle perspective. *International Journal of Life Cycle Assessment*, 23, 1018–1029. https://doi.org/10.1007/s11367-017-1386-8.

Pratoomyot, J., Bendiksen, E.Å, Bell, J.G. & Tocher, D.R. (2010) Effects of increasing replacement of dietary fishmeal with plant protein sources on growth performance and body lipid composition of Atlantic salmon (Salmo salar L.). *Aquaculture*, 305, 124–132. www.sciencedirect.com/science/journal/00448486. https://doi.org/10.1016/j.aquaculture.2010.04.019.

Ross, L.G., Telfer, T.C., Falconer, L., Soto, D. & Aguilar-Manjarrez, J., editors (2013) Site selection and carrying capacities for inland and coastal aquaculture. FAO/institute of aquaculture, University of Stirling, expert workshop, 6–8 December 2010. Available at www.fao.org/3/i3322e/i3322e.pdf. Food and Agriculture Organization: Stirling, UK Fisheries and Aquaculture Proceedings No. 21. Food and Agriculture Organization: Rome.

Rutaisire, J., Char-Karisa, C., Shoko, A.P. & Nyandat, B. (2009) Aquaculture for increased fish production in East Africa. *African Journal of Tropical Hydrobiology and Fisheries*, 12, 74–77. https://doi.org/10.4314/ajthf.v12i1.57379.

Schmidt, J.H. (2010) Comparative life cycle assessment of rapeseed oil and palm oil. *International Journal of Life Cycle Assessment*, 15, 183–197. https://doi.org/10.1007/s11367-009-0142-0.

Seafish (2021) Available at: https://www.seafish.org/insight-and-research/seafood-trade-data/ (accessed 2nd May 2023).

Subasinghe, R., Soto, D. & Jia, J. (2009) *Global Aquaculture and Is Role in Sustainable Development. Reviews in Aquaculture 1 p2–9*. https://doi.org/10.1111/j.1753-5131.2008.01002.x.

Thanh, L.P. (2014) Sustainable development of export-orientated farmed seafood in Mekong Delta. PhD thesis. University of Stirling.

Turchini, G.M., Trushenski, J.T. & Glencross, B.D. (2019) Thoughts for the future of aquaculture nutrition: Realigning perspectives to reflect contemporary issues related to judicious use of marine resources in aquafeeds. *North American Journal of Aquaculture*, 81, 13–39. https://doi.org/10.1002/naaq.10067.

United Nations (2015) Population 2030: Demographic challenges and opportunities for sustainable development planning (ST/ESA/SER.A/389). Available at https://www.un.org/en/development/desa/population/publications/pdf/trends/Population2030.pdf.

Worldfish (2020) https://www.worldfishcenter.org/pages/gift/Accessed, Vol. 1.

Wyban, J. (2007) Thailand's white shrimp revolution. *Global Aquaculture Advocate*, May–June, 56–58.

3

The importance of nutrition and selective breeding in aquaculture production

Smaragda Tsairidou, Ross D. Houston and Ioannis T. Karapanagiotidis

Introduction

Seafood production via capture fisheries and aquaculture is a major contributor to food and nutrition security worldwide (Béné et al., 2016). Seafood (including fish, shellfish and crustaceans) is an excellent source of high-quality animal protein, healthy fats (e.g. omega-3 fatty acids), essential nutrients and micronutrients. Fish is highly efficient in converting feed to protein, for example, Atlantic salmon has a feed conversion ratio in the range of 1.3 and tilapia in the range of 2 (Fry et al., 2018). According to World Health Organization data, fish is the main source of animal protein for about 1 billion people worldwide (WHO, 2019). Consumption of micronutrient-rich foods, such as seafood, can help prevent and control micronutrient deficiencies especially in the developing world (Béné et al., 2016). Despite a more complex relationship between seafood production and poverty reduction, capture fisheries and aquaculture contribute to both local and national economies (Béné et al., 2016). Fish and related products are currently among the most traded food items worldwide (FAO, 2018), therefore bringing significant national revenue from exports and international fish trade (Béné et al., 2016). At the household economy level, fish-related activities provide household livelihoods for many local communities (Garaway, 2005; CFS, 2014; Béné et al., 2016).

Aquaculture has been rapidly increasing and is now approximately equal to capture fishery production (FAO, 2018). Farmed species of major economic interest include Whiteleg shrimp (10.72% of world production value of all species in 2017), Atlantic salmon (6.69%), species of carp (15.57%), Chinese mitten crab (3.82%) and Nile tilapia (3.05%) (Cai et al., 2017). Aquaculture helps to increase seafood availability and hence prevents price inflation, bridging the gap between demand and supply in current and future markets (Merino et al., 2012; Troell et al., 2014; Béné et al., 2016). Future markets and the farming industry are confronted with emerging global challenges including population growth and climate change, and similar to all agriculture sectors, aquaculture is vulnerable to those. For the reasons discussed above, aquaculture products can substantially contribute to meeting the increasing demand by the growing human population for animal protein. Global fish consumption has been rapidly increasing, and with a rate higher than that of population growth

Margaret Crumlish and Rachel Norman (eds) **Aquatic Food Security**
DOI: 10.52517/9781789183368.003, © 5m Books Ltd, 2023

(Figure 3.1, [FAO, 2018]). However, reliance on inland freshwater systems and feeds derived from land crops, brings aquaculture into conflict with land use for the production of food used directly for human consumption. The most noted disadvantages of aquaculture involve its environmental impact, its effects on marine ecosystems and the potential spread of diseases. In addition, the continuous interaction with environmental factors makes aquaculture particularly vulnerable to climate-change-induced fluctuations. Meeting all these challenges is crucial for ensuring sustainable aquaculture production and remaining a valuable contributor to aquatic food security. Through optimization of the use of resources and effective fisheries management, it is possible for aquaculture to use diverse feed sources and feeding strategies (Troell et al., 2014), and to minimize its environmental impact. Controlling infectious diseases remains a key challenge for aquaculture, and as discussed in this chapter, genetics can offer an alternative approach for the design of effective disease control strategies with long-term benefits. This chapter will describe the challenges facing aquatic animal nutrition and genetics and provide mitigation strategies to support aquatic food security.

Importance of nutrition and feeding in aquaculture production

Nutrition and feeding play a significant role in aquaculture production as they affect many aspects of aquatic animal biology, such as survival and health, appetite, growth and reproduction, and nutrient body composition and thus the nutritional value of the aquatic animal end products. Moreover, nutrition directly affects the environmental impact of farming as nutrient leaching, uneaten and undigested parts of feed can all increase the nutrient loading in the aquatic ecosystem. Thus, better feeding management as well as improvements in feed digestibility are needed. Of most importance to the farmer is the fact that nutrition and feeding represents the major operational cost of intensive aquaculture production and therefore improvement in the cost effectiveness of animal diets

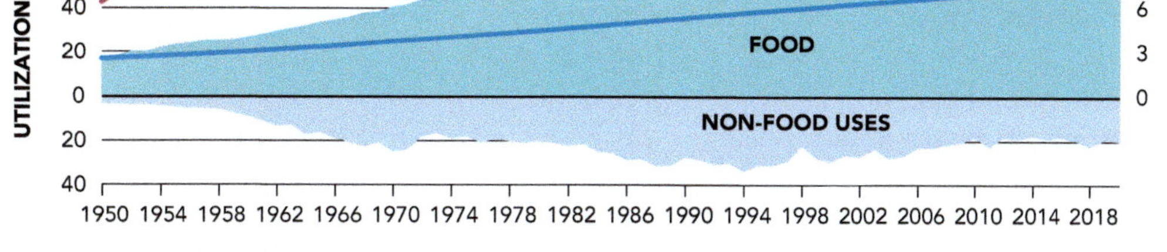

NOTES: Excluding aquatic mammals, crocodiles, alligators, caimans and algae. Data expressed in live weight equivalent. For algae and apparent consumption, see Glossary in the main report, including Context of SOFIA 2022. Source of population figures: United Nations. 2019. 2019 Revision of World Population Prospects. In: *UN*. New York. Cited 22 April 2022. https://population.un.org/wpp

Figure 3.1 World fish utilization and apparent consumption. Source: Food and Agriculture Organization of the United Nations. Reproduced with permission (FAO, 2022).

is constantly sought. As aquaculture develops and intensifies all around the globe, the industry demands more industrially compound aquafeeds. Apart from finding suitable feed ingredients the sector should also ensure that these feedstuffs have a low environmental footprint. It is well known that although the sector has reduced its dependency on fishmeal and fish oil over the recent decades, it still relies on a significant supply of these limited natural resources. Thus, making aquafeeds more sustainable is a priority in order to enhance aquaculture sustainability. Consequently, it is imperative to have a basic understanding of the general principles of fish nutrition in relation to these issues and how these can affect the stability of aquatic food security.

Understanding nutrition in aquatic production

In this chapter, the term 'fish' is used as a generic term to include all aquatic animals that are currently or have the potential of being farmed, despite the pronounced differences in nutrition and feeding needs of fish and crustaceans. The science of fish nutrition is the branch of physiology that studies all the biochemical processes that take place in the organism of farmed aquatic animals in order to supply the nutrients needed to perform their vital functions, such as body maintenance, growth, health and reproduction. Specifically, fish nutrition studies the following, among others:

- feeding behaviour and the regulation of feed intake
- digestive physiology: feed digestion, absorption of nutrients, nutrient metabolism, digestive system function, etc.
- nutritional energetics: the transfer and distribution of dietary energy within the body to perform its various physiological functions
- nutritional requirements: for proteins, lipids, carbohydrates, vitamins and minerals
- applied nutrition in aquaculture: the feeding of cultured species according to their aquaculture farming system, estimation of feed rations and their distribution, etc.
- feed formulation and feed technology: methods for development of formulae and manufacturing process of compound feeds
- effects of feeds and feeding on the environmental impact of aquaculture
- effects of feeds and feeding on the nutritional value and on the end product quality
- effects of nutrients on genome expression
- effects of nutrients on health and nutritional diseases.

The term 'feed' encompasses any matter of vegetable or animal origin consisting of nutrients and which can be utilized by the animal organism. Feeds of farmed aquatic animals are mainly of two types.

1 Natural food, which consists of various plant and animal organisms living on the culture medium such as phytoplankton, zooplankton, small invertebrates (e.g. worms, molluscs, arthropods, insect larvae), vertebrates (e.g. amphibians, fish larvae and eggs, fish). Natural food supports fish nutrition exclusively in extensive and majorly in semi-intensive culture systems.
2 Exogenous feed, which can range from simple-unprocessed feedstuffs of terrestrial plant and animal origin to industrial compound feeds in the form of pellets. The first type of feed is generally

produced on-farm (farm-made feeds) in semi-intensive systems and is used both to supplement fish nutrition and to act as indirect organic fertilizers of water. Still, a common practice in many semi-intensive systems, particularly in the tropics, is the use of trash fish. In contrast, compound feeds are nutritionally balanced diets requiring a higher cost and technology investment and which support entirely the nutrition of fish in intensive systems.

Nutritional requirements of fish

Fish nutritional requirements are species-specific and are differentiated to the various growth stages. The nutritional requirements of fish and crustaceans has been extensively covered by many authors (Cowey, 1988; Halver and Hardy, 2002; NRC, 1993, 2011; Sargent et al., 2002; Steffens, 1989; Tacon and Tran, 2022), however, key points will be explored in this chapter.

A nutrient balanced diet is important for aquaculture production because this satisfies all the nutritional requirements and energy needs of fish, promoting growth, health and welfare, and thus affecting the time taken and volumes of fish produced. There are five groups of nutrients that are required by all living organisms, including fish and crustaceans. Proteins, lipids and carbohydrates, commonly called macronutrients, are mainly body structural nutrients that provide metabolic energy to the organism. Vitamins and minerals are dynamic nutrients that do not provide energy but are essential for the digestive physiology of fish. Therefore, the nutritional requirements of fish encompasses the qualitative and quantitative requirements for proteins, lipids, carbohydrates, minerals and vitamins. Although there have been major improvements in diet formulations over the years, our knowledge on the nutritional requirements of many aquaculture species is still limited (Hamre et al., 2013).

Proteins are considered the most important nutrients that affect fish growth and the cost of the diet. Dietary proteins play significant physiological roles in the fish body from simply building up and repairing cells and tissues to supplying enzymes and producing hormones. They are present in all cells, tissues and organs of the fish body and therefore, fish have constant protein requirements for maintenance (regeneration of body proteins), catabolism (proteins as substrates for energy production) and anabolism (synthesis of new body protein – growth). For applied fish nutrition, there is an optimum protein level in the diet that will yield the maximum growth, and this has to be defined for each cultured species, often expressed as a percentage of the diet. A higher than optimum protein level will result in reduced growth rates and higher feed costs. Additionally, excess dietary protein is less digestible producing more nitrogenous wastes to the production unit. Protein sparing using increased dietary levels of carbohydrates, is not as beneficial to the animals, as most cultured species are carnivorous and cannot utilize high levels of carbohydrates in the feed. However, in diets of omnivorous species, such as carps and tilapia, dietary carbohydrates can have a significant protein sparing effect (Azaza et al., 2015; Shimeno et al., 1995). Furthermore, the digestibility of a protein determines the actual amount of the protein to be used for anabolic and catabolic purposes by fish. Thus, the different digestibilities of dietary protein sources could affect the optimum protein level. For example, the lower digestibility of a plant protein compared with fishmeal implies that a higher level of protein will be required in the diet in order to meet the protein requirements of a carnivorous fish.

In fact, fish do not have a specific requirement for crude proteins but rather for specific amino acids that they cannot synthesize *de novo* and thus these must be provided by the diet.

All aquaculture species require 10 essential amino acids; namely arginine, histidine, isoleucine, leucine, lysine, methionine, phenylalanine, threonine, tryptophan and valine, while cysteine and tyrosine can be synthetized by methionine and phenylalanine, respectively. Proteins are also composed of non-essential amino acids, the importance of which has been overshadowed by that of essential amino acids because they are synthetized sufficiently in fish body. However, their significance in fish nutrition should not be ignored as non-essential amino acids play very important roles, such as regulating gene expression, cell signaling pathways, digestion and absorption of other nutrients, DNA and protein synthesis, proteolysis, metabolism of glucose and lipids among many other functions (Hou et al., 2015). Thus, a balanced mixture of essential and non-essential amino acids in the dietary protein is of significant importance (Peres and Oliva-Teles, 2006). The quantitative dietary requirements for essential amino acids vary between species, but only marginally. Moreover, essential amino acids quantitative dietary requirements within species vary according to its life stage/size class. As a general rule, younger individuals have higher quantitative essential amino acid requirements than adults of the same species, although the differences are minimal. For example, fingerlings of gilthead seabream (*Sparus aurata*) require arginine at 5.5% of dietary protein, while juveniles and growers require arginine at 5.4% (FAO, 2019). However, our knowledge of the quantitative requirements for essential amino acids at the various developmental stages of the various farmed species is still limited.

Dietary lipids, especially triglycerides, are the main source of metabolic energy for fish as most species cannot efficiently digest complex carbohydrates. Lipids, especially polar lipids, are also structural components of cell membranes controlling the permeability and fluidity of the intracellular space (Guillaume et al., 2001). In addition, these nutrients act as transporters of fat-soluble vitamins and carotenoids in intestinal absorption. One of the most important roles of dietary lipids is that they provide the essential fatty acids (EFA). Fatty acids are the structural units of most lipid classes and consist of saturated, monounsaturated and polyunsaturated fatty acids (PUFA) depending on the presence and the number of double bonds in their carbon atom chain. Fish, as all vertebrate animals, cannot synthesize *de novo* the PUFA of the n-3 and n-6 series that are necessary for cellular metabolism and which are therefore considered essential and must be supplied by the diet (Sargent et al., 1997, 1999a, 1999b, 2002). Both n-3 and n-6 series PUFA are essential in all fish species, but the EFA requirements for a given species are dependent on the ability of the species to convert endogenously PUFA with 18 carbon atoms in their chain (C18), such as 18:3n-3 (alpha-linolenic acid, LNA) and 18:2n-6 (linoleic acid, LA), into PUFA with 20 and 22 carbon atoms (C20 and C22, respectively), such as 20:5n-3 (eicosapentaenoic acid, EPA), 22:6n-3 (docosahexaenoic acid, DHA) and 20:4n-6 (arachidonic acid, AA). These PUFA biosynthetic pathways in fish include chain elongations and desaturations that are complex and still are not fully characterized. Most marine fish species have very low capacities, if none at all, for PUFA bioconversions (Sargent et al., 1989, 1995, 2002) and thus EPA, DHA and AA are considered the EFA that must be provided sufficiently by the diet, with dietary LNA and LA not satisfying their requirements. Whereas, the conversions are well established for many freshwater species of fish (Sargent et al., 1989, 1995), so that these species have a requirement for LNA and LA, though C20 and C22 PUFA are more effective nutritionally.

Defining the quantitative requirements of EFA is difficult and has been discussed in detail elsewhere (Sargent et al., 2002; Tocher, 2003). A complication is that competitive interactions exist between different series of fatty acids for the same biosynthetic enzymes in the formation of C20 and C22 PUFA. Such interactions also exist in the actions of the different eicosanoids produced from AA and EPA. In addition, competition exists in phospholipid biosynthesis, where an excessive dietary intake of a specific PUFA, e.g. EPA, can lead to an elevation of that PUFA in tissue phospholipids at the expense of another PUFA present in much lower concentration in the diet, e.g. DHA. Furthermore, the specificities of PUFA as substrates for β-oxidation generating metabolic energy can also determine the tissue fatty acid composition that in turn will affect the required levels of EFA in the diet. Therefore, defining the exact dietary requirements of EFA in fish requires consideration not only of the absolute as well as the relative amounts of individual fatty acids in fish diets, but also the fish's innate abilities to metabolize these fatty acids, whether anabolically or catabolically (Sargent et al., 2002). The quantitative requirements of fish for EFA mainly vary according to the species and life stage. In general, all species require total PUFA between 0.5–6% of their diet. As a rule of thumb, marine, cold-water and carnivorous fish have higher quantitative requirements for n-3 than for n-6 PUFA. Another generalized rule is that in the younger development stages (e.g. larvae, fry, fingerling, juvenile) the requirements for n-3 PUFA and especially for EPA and DHA are higher compared with their adult stages. For example, *S. aurata* larvae and fingerlings require 3–5.5% of EPA + DHA in the diet (Izquierdo, 2005), while in the adult stage quantitative requirements reduce below 2.5% (Ibeas et al., 1994; Izquierdo, 2005).

Regarding quantitative requirements of total lipids, it should be borne in mind that one of the basic functions of lipids in the organism is the production of metabolic energy. As previously mentioned, modern fish feeds are formulated to contain high levels of eupeptic total lipids making use of their protein sparing effect. Owing to the metabolic interactions between dietary proteins, lipids and carbohydrates it is therefore meaningless to determine the exact level of total lipids in feeds that will yield optimal fish growth. It is more the digestible protein to digestible energy ratio that is of concern in feed formulations. In one sense, extremely high lipid levels applied to the diet is more of a technological limitation rather than a nutritional one. However, care should be taken to ensure high dietary lipid levels do not result in undesirable body fat deposition and digestive organ malfunctions, which can then negatively affect fish growth and consumer's acceptance. It is generally accepted that a dietary percentage between 10% and 30%, depending on the fish species, is sufficient for protein sparing and maximum fish growth, as well as avoiding excessive amounts of lipids being deposited in fish tissues in most carnivorous species. So far, the maximum dietary lipids levels are unknown for most of the farmed species.

In fish, no defined dietary requirement for carbohydrates has been demonstrated. The main physiological role of these nutrients is to provide metabolic energy, organic carbon and hydrogen to fish as well as acting as structural components of fish tissues. In fish feeds, carbohydrates are the least expensive energy source in the diet, acting also as binding agents and thus enhancing pellet stability. Dietary carbohydrates are not highly digestible by fish (Cowey, 1988), but the hydrothermal processing of plant feedstuffs (e.g. milling, heating and extrusion) can improve the digestibility. Fibre, which consists of non-starch polysaccharides, is indigestible by fish but can serve other roles enhancing the digestion process. For example, fibre changes the nature of the contents of the gastrointestinal tract, and thus changes how other nutrients and chemicals are absorbed. Although there is not a strict dietary requirement, carbohydrates should be provided by the diet in appropriate amounts to assist in the protein sparing effect. In compound feeds, several plant feedstuffs, mainly cereals such as wheat, corn and rice, are used as carbohydrate sources. Particular attention is paid, both at the level of total carbohydrates and at the level of indigestible celluloses, to the diets of carnivorous fish to keep to low levels. For example, in gilthead seabream and European seabass grower feeds, the levels of carbohydrates in the diet are kept below 25% and those of fibre below 6%. On the contrary, in the feeds of omnivorous and herbivorous species, such as tilapia and carps, the level of total carbohydrates in the diet can reach 40%.

As far as vitamins and minerals are concerned, these are essential micronutrients that are required for optimum health and normal physiological functions such as growth, development and reproduction of fish. Most vitamins cannot be synthesized by fish and therefore they must be obtained from the diet. They can be classified into fat-soluble vitamins, such as A, D, E and K, that are absorbed by dietary lipids and for which excessive intake can cause several hypervitaminosis symptoms, and water-soluble vitamins, such as C, choline, inositol and those of B complex, whose excessive intake can easily be eliminated from fish body. Their main physiological role is their action as co-enzymes, assisting proton and electron transportation in the body, regulating hormone production, protecting cell membranes (Combs, 1992). Vitamin A is required mainly for the maintenance of the epithelial cells of tissues and organs and is particularly important for fish skin, where cells are lysed and renewed constantly. Vitamin A deficiency can cause skin wounds and sensitiveness to skin diseases, e.g. exoparasite infections. For the epithelial cells of fish eyes, a deficiency will lead to keratinized cells that can cause infection by bacteria and impaired fish vision. Vitamin D mainly assists the homeostasis of calcium and phosphorus, their absorbance by the intestines and their deposition in the bones, regulating also the action of the parathyroid hormone involved in bone formation (Lock et al., 2010). Vitamin E has a major antioxidant action against the auto-oxidation of PUFA, carotenoids and other vitamins. In feeds, the antioxidant action of vitamin E is often combined with the antioxidant action of selenium and vitamin C. Vitamin K is mainly responsible for normal blood coagulation by regulating prothrombin and thrombin formation, and bone minerilization (Krossøy et al., 2011). Vitamins of the B complex, such as B_1, B_2, B_3, B_5, B_6, B_7, B_9 and B_{12}, are mainly involved in protein, lipid and carbohydrate metabolisms regulating appetite, digestion, growth, fecundity, immunity etc. Vitamin C is involved in the formation of collagen in connective tissues and bone matrix, assists wound repair, facilitates the absorption of iron preventing anemia and acts as an antioxidant preventing lipid peroxidation (Darias et al., 2011). Inositol and choline are mainly involved in phospholipid synthesis regulating lipid metabolism homeostasis. Qualitative and quantitative vitamin requirements of fish have been defined for some species, but remain unknown for several others. In compound feeds, vitamins are commonly added as a ready multi-vitamin premix, usually at 0.2–1% of diet. In diets with high lipid and carotenoid content, vitamin E is often added in extra amounts in order to act as antioxidant.

Minerals are also essential nutrients for fish nutrition and can be classified as trace elements, such as iron, copper, zinc, selenium, fluorine, chromium, manganese and iodine (Fe, Cu, Zn, Se, F, Cr, Mn and I) that are required by fish in trace amounts and macrominerals, such as calcium, phosphorus, magnesium, potassium, sodium, chlorine and sulfur (Ca, P, Mg, K, Na, Cl and S) that are required in larger amounts. Minerals are structural components mainly of skeletal structures, serve as enzyme co-factors and act as ions of body fluids regulating osmosis, neural signals and endocrine functions (Guillaume et al., 2001; Lall, 2002). It is considered that fish, as all farmed animals, require about 29 minerals in their diet (Lall and Milley, 2008). However, there are only a few established quantitative dietary requirements for specific minerals and these only exist for specific fish species. Fish can absorb some minerals also from the culture water. As with vitamins, in compound feeds minerals are commonly added as a ready multi-mineral premix, usually at 0.2–1% of diet. Usually, mono-calcium phosphate or di-calcium phosphate and selenium are added as additional supplements.

Tailoring fish nutrition to intensity of culture

In aquaculture, the satisfaction of fish nutritional requirements depends on the type of culture system. In extensive systems, for example, no exogenous feeds are provided by the farmer and fish nutrition is based solely on the natural food available in the production system. Natural food is rich in all nutrients but its amount within an enclosed water volume can be limited and quite variable. Natural food is dependent on the productivity of autotrophic organisms, such as phytoplankton, anaerobic bacteria and blue-green algae that synthesize organic matter from sunlight and carbon dioxide, as well as that of heterotrophic organisms, such as detritus, that consume non-living organic matter (Jauncey, 1998). Thus, fish nutrition in extensive systems relies on the availability of nutrients in the water, especially nitrogen, phosphorus and carbon that are the main ingredients of natural food organisms. The various abiotic factors of water, mainly the light transmittance for photosynthesis and water temperature, affect the natural productivity and in turn fish nutrition in these systems.

In semi-intensive culture systems, fish nutrition can be supplemented by a broad variety of feedstuffs, depending on their availability and cost, including oilseeds and cereal grains residues, arable crop and brewery wastes, aquatic macrophytes, animal offal, terrestrial invertebrates and even kitchen waste (De Silva, 1995; Tacon, 1988). Supplementary farm-made feeds, usually in unprocessed form or simply processed as doughs and semi-moist pellets, are of low cost and nutritional quality and thus can only partially satisfy nutrient requirements. Pond fertilization via inorganic fertilizers, manures, composts and/or integration with terrestrial livestock are commonly practised in order to enhance the natural productivity of the water. In semi-intensive systems, the polyculture of species with different food habits yields a maximized utilization of the different trophic and spatial niches of the enclosed water body (Rahman et al., 1992).

In intensive systems, in which high stocking densities and growth rates are sought, fish nutrition is exclusively dependent on the exogenous feeds that are provided by the farmer. These feeds, in the form of pellets, are nutritionally complete diets that are formulated with a high expertise and are manufactured using high-tech equipment achieving a uniform product quality. Owing to their high nutrient specifications and specialized production, exogenous feed cost can account for more than 60% of the total production costs of a fish farming procedure.

Nutrition at early life stages

During the yolk-sac stage, larvae are fed on the nutrient reserves in the yolk sac, the quality of which is determined by the quality of broodstock nutrition. The period after mouth-opening is the most crucial stage in a fish's life as the newly hatched larvae have to search out their food. In some species, such as salmonids and carps, the mouth of first-feeding fry is relatively large and their digestive tract is functional and so they can accept compound feeds of small pellet size. However, first-feeding fry of most species, and mainly those of marine origin, cannot accept compound feeds as their digestive systems are not fully developed and functional. Additionally, fry also prefer to feed on the water surface seeking out swimming prey rather than a sinking pellet. Although, research and feed technology has advanced the knowledge and application of the so-called 'inert microdiets' that are targeted for the onset of exogenous feeding (Cahu and Zambonino-Infante, 2001; Teshima et al., 2000), 'live prey or live food' is still the preferable method for fish feeding at this life stage. Live food mainly consists of two distinct planktonic organisms: microalgae and zooplankton. Bacteria (Nevejan et al., 2018) and yeasts (Navarrete and Tovar-Ramírez, 2014) are also being tested but are not commonly used in practice. In earthen pond-raised fish, such as tilapias, feeding on natural food provides all the necessary live food for the fry nutrition. In contrast, in the larviculture of most marine and euryhaline species, such as gilthead seabream (*S. aurata*), European seabass (*Dicentrarchus labrax*), red seabream (*Pagrus major*), yellowtail (*Seriola quinqueradiata*), Asian seabass (*Lates calcarifer*) and turbot (*Scophthalmus maximus*), the nutrition of first-feeding fry is performed in a specialized in-house hatchery environment in which fry are being fed on live food that is produced in mass cultures parallel to the fry production (Lee, 2003; Shields, 2001).

Microalgae is used as direct food source either for fish larvae, in the case of bivalve molluscs and peneid shrimp, or for the zooplankton that in turn are fed to fish larvae (Conceição et al., 2010; Hemaiswarya et al., 2011; Muller-Feuga, 2000). From the numerous microalgal species that have been tested, only few are routinely grown in aquaculture hatcheries based on their nutritional and technical characteristics (Muller-Feuga et al., 2003; Shields and Lupatsch, 2012). The most frequently used genera are those of *Chlorella, Tetraselmis, Isochrysis, Pavlova, Phaeodactylum, Chaetoceros, Nannochloropsis, Skeletonema* and *Thalassiosira* (Hemaiswarya et al., 2011). Recent developments include the use of some species of microalgae in the form of live concentrates, frozen, freeze-dried or pastes, in order to reduce the high operating cost and variable productivity of microalgae production (Conceição et al., 2010). The nutritional value of microalgae is highly variable, both between and within species; critically affected by the culture conditions and the growth stage at harvest (Becker, 2013; Brown et al., 1997), and thus a mixture of microalgal species is commonly used in order to avoid nutritional deficiencies in larvae nutrition. In general, microalgae contain moderate to high amounts of protein, typically 20–55% of their dry matter (Becker, 2007; Brown et al., 1997), but in some classes, such as Cyanophyceae, their protein contents can be even higher. It is worth mentioning, however, that the reported values for microalgae protein content contain also a significant fraction of non-protein nitrogen, such as nucleic acids, amines, glucosamides (Becker, 2007). They have a favourable amino acid profile for use in fish nutrition; though low in methionine, and they are rich in antioxidants, carotenoids and vitamins. The lipid content of microalgae is typically 5–20% (Muller-Feuga et al., 2003) with significant amounts of EPA found in some species of Diatoms, Eustigmatophytes, Cryptomonads, Rhodophytes and Prymnesiophytes, and of DHA found in Cryptomonads, Prymnesiophytes and Thraustochytriidae, whereas Eustigmatophytes, Rhodophytes and Diatoms are rich in AA (Becker, 2013; Conceição et al., 2010).

Regarding zooplankton, the most commonly used are rotifers of the species *Brachionus plicatilis* and brine shrimp *Artemia salina* nauplii, while copepods (Calanoida, Harpacticoida, Cyclopoida) and Cladocera (*Moina macrocorpa, Daphnia* sp.) are being used to a lesser extent (Lavens and Sorgeloos, 1996; Lubzens and Zmora, 2003; Rasdi and Qin, 2016). Rotifers are cultured in large volumes in hatcheries at a high operational cost, while *Artemia* is mainly sourced as dried cysts from natural stocks; the viability of this practice to support the long-term growth of aquaculture is questionable. Fish larvae are initially fed on rotifers that are of smaller size (ranging 50–350 μm) than *Artemia* nauplii (400–800 μm), the ratio gradually moves from the former to the latter during larval rearing. Rotifers contain 28–63% protein and 9–28% lipids in their dry matter (Lubzens and Zmora, 2003), while the various life stages of *Artemia* used as live food contain 39–64% protein and 1–30% lipid (Dhont and Van Stappen, 2003). However, both rotifers and *Artemia* contain low amounts of the highly unsaturated fatty acids such as EPA, DHA and AA (Conceição et al., 2010) that are essential for marine fish larvae. Therefore, the 'enrichment' of these organisms with commercially available products (e.g. lipid emulsions, spray-dried microalgal products, etc.) containing specific fatty acids, as well as other nutrients, such as amino acids, minerals and vitamins, and dietary components, such as immune-stimulants and probiotics (Rainuzzo et al., 1997; Sargent et al., 1997; Sorgeloos et al., 2001; Rasdi and Qin, 2016), is a prerequisite step in order to provide a nutritionally balanced diet to the fish larvae. Copepods, in contrast, have a superior nutritional value to rotifers and *Artemia* as they are much richer in EPA and DHA (Rasdi and Qin, 2016), but their mass culture in hatcheries is currently restricted due to the lack of cost-effective mass production protocols (Conceição et al., 2010).

The so-called fish larvae 'weaning' stage feeding with compound diets usually starts with the co-feeding of live food at the final stages of larval rearing. These diets can be very high in protein, which can reach 70%, and are commonly called 'microdiets' or 'microparticulate diets' due to their very small particle size (50–600 μm). At early weaning, as soon as a few days after mouth-opening, microdiets of sophisticated manufacture, usually in the form of microencapsuled, microcoated or microbound, can be used to substitute or totally replace live food. At later stages of weaning, microdiets can be applied in the form of small pellet size or granules. Microdiets mostly are formulated empirically, as it is technically difficult to estimate fish nutritional requirements at larval stage (Cahu and Zambonino-Infante, 2001), while also their digestive physiology is still incomplete (Dabrowski, 1984). Most are commonly formulated based on a mixture of protein sources of marine origin, such as fishmeal, krillmeal, squid meal, hydrolysed fish protein, etc., and mixtures mainly of various fish oils. They may contain other dietary components such as algae, yeast extracts, probiotics, mixtures of amino acids, vitamins and minerals, lecithins, gelatins, attractants and cellulose in order to be nutritionally enriched (Baskerville-Bridges and Kling, 2000b; Izquierdo et al., 2019; Ruiz et al., 2019). The use of microdiets in the larvae nutrition of various fish species has had variable weaning success, but the most successful cases are when microdiets are co-fed with live food at a low substitution level of the latter (Holt, 1993; Kolkovski, 2001; Ma et al., 2014; Mata-Sotres et al., 2015; Rosenlund et al., 1997). Poor performance of inert diets in first-feeding larvae has been attributed to low palatability/intake, low digestibility, low residence time in the water and poor nutritional composition of the diet among reasons (Baskerville-Bridges and Kling, 2000a; Cahu and Zambonino-Infante, 2001). It seems that co-feeding with live food to stimulate feeding and digestion combined with nutritionally balanced microdiets of special design manufacture can significantly reduce weaning time and advance larval performance.

Fish feeding

Appropriate feeding is of great importance for the successful and sustainable management of aquaculture production. In intensive culture, given the high cost of the diets, the producer firstly must ensure an even distribution of the feed to all fish and that all fish are being fed. Simultaneously, the producer must pay special attention to ensure minimal feed losses in the aquatic environment. Uneaten feed will dissolve in the water resulting in nutrient enrichment and eutrophication of the culture medium, which in turn may result in oxygen depletion, increased turbidity and sediment organic matter, harmful algal blooms and a decreased benthic biodiversity (Boyd, 1982; Price et al., 2015).

In practice, the producer should determine:

1 the exact amount of feed given daily to the fish stock; the so-called 'feed ration or feeding level'
2 how often the fish must be fed; the so-called 'feeding frequency'
3 how to feed the fish stock; the so-called 'feeding method'.

The feed ration determines the required amount of feed to be delivered to the fish stock daily and most often is expressed as percentage of fish body weight/day. The appropriate ration size influences both fish growth and feed conversion, and is therefore critical in effective feed management (Goddard, 1996). The most frequently used measures of growth and feed conversion are the specific growth rate (SGR) and the feed conversion rate (FCR), respectively:

$$\text{SGR } (\%\ \text{day}^{-1}) = 100 \times (\text{LnW}_2 - \text{LnW}_1) \times (\text{t})^{-1}$$
$$\text{FCR} = \text{feed intake} \times (\text{W}_2 - \text{W}_1)^{-1}$$

where W_2 is the fish wet weight at time t, W_1 is the initial fish wet weight, LnW_2 and LnW_1 are the natural logarithms of W_2 and W_1, respectively.

For any given feed, fish growth increases with increasing ration level until voluntary feed intake reaches a maximum, at which point maximum growth rate occurs (Talbot and Hole, 1994). However, this does not mean that the more feed fish consumed the better feed efficiency will be. There is an optimum point of feed intake where the maximum feed efficiency is reached, and this occurs at a ration level below that of the maximum growth. In fact, the optimum ration level returns the lowest FCR value. This is because the rate of fish metabolism and the degree of feed utilization do not necessarily coincide with the saturation of fish appetite. Fish are voracious and continue to consume feed beyond the maximum point of feed efficiency. Thus, it is desirable that the fish are being fed to achieve their highest fish growth rates (highest SGR) with the highest feed efficiency (lowest FCR).

In general, feed can be given in excess (*at libitum*), to apparent satiation or in restricted rations. When feed is given in excess to their appetite, fish are overfed and thus a large proportion of nutrients might be excreted in the water column undigested. This practice, however, is rarely used in intensive systems where feed is costly, as the uneaten feed cannot be recollected and re-fed. A more common practice is to feed the fish to apparent satiation, in which the ration size is close to the saturation of fish appetite. This implies that the farmer should carefully monitor fish appetite as this can vary day by day and that feeding is typically performed more than once per day. A low feeding level can result in underfed fish, reduced growth and fish production, while a high feeding level will overfeed the fish increasing the dietary costs and cause degradation of water quality. Besides, high FCR values can be obtained by under-feeding as well as by over-feeding (Talbot and Hole, 1994). So, feeding at apparent satiation, if it is not properly managed, may not be a rational practice.

Restricted ration size is the most common practice in intensive aquaculture, in which the ration size is set below the saturation level and ideally this represents the optimum ration level. The accurate calculation of optimum ration level is practically difficult, as the farmer needs to know the exact total fish biomass. In order to estimate the fish biomass frequent sampling is needed, which again may be proved impractical. Feed producers provide species-specific feeding tables with indicative values for ration level according to the fish size and water temperature. These tables are compiled based on experimental records and estimations performed by the industry and academia. However, each production unit differs and fish appetite is influenced by farming conditions and the fluctuation of environmental parameters. Therefore, in order to achieve an effective feed management, the farmer can use the provided feeding tables as a guide but, simultaneously, should also record and monitor the amounts of feed consumed by fish daily. No one knows better the appetite of their fish than the producer, as long as there is a perceptive observation and a good record keeping. In general, the ration level is higher in younger fish and declines as fish grow in size due to the slower rate of metabolism of the latter. In addition, the ration level of both cold- and warm-water species increases with rising water temperature up to the point that temperature remains within the species optimal range and declines thereafter.

Feeding frequency indicates how often the fish must be fed, usually based on how many times per day. An optimal feeding frequency is known to reduce feed wastage, size variation and cannibalism (Dwyer et al., 2002; Folkvord and Ottera, 1993; Tucker et al., 2006), while it can affect feed intake, growth and feed efficiency (Biswas et al., 2010; Oh et al., 2019). The optimal feeding frequency may differ among species and growth stages, while it may be influenced by environmental factors and the diet itself. Again, feeding frequency should ensure the maximum SGR and the lowest FCR. In most fish species studied, feeding two to three times daily usually results in best fish performance (Biswas et al., 2010; Booth et al., 2008; Oh and Venmathi Maran, 2015; Silva et al., 2007), although there are cases where just one daily meal is enough (Lee et al., 2000) or that a more frequent feeding over a wider spread of time gives better results (Chiu et al., 1987; Hossain et al., 2001; Xie et al., 2011).

Working towards more sustainable aquafeeds

Today, the global community is challenged to meet the pressing food needs of a growing population using finite natural resources. Fish have always been considered as an important part of the human diet and it has long been recognized as a health-promoting food for human nutrition. Fish consumption is expected to continue rising due to an expected increased demand, which is driven by many factors including population growth, rising incomes, urbanization, advances in processing and expansion of distribution channels among others (FAO, 2018). Given the stagnating volumes from capture fisheries, aquaculture appears to have significant potential to meet the increasing demand for fish products. In 2016, world aquaculture production attained another all-time high at 80 million tonnes (mt) of food fish (FAO, 2018) providing almost one-half of all fish produced for human consumption.

Aquafeeds are an integral component of aquaculture with about 70% of total farmed fish produced requiring some form of additional feed to grow (Hua et al., 2019). In fact, the growth in production of fed species is faster than that of un-fed species, and as aquaculture production increases and intensifies, the sector relies increasingly on the use of industrially compound aquafeeds. Global compound aquafeed production is estimated at about 40 mt per year with an annual growth rate for 2018 of around 4% (ALLTECH, 2019). One of the most critical issues that threatens the sustainability and further growth of aquaculture and aquafeed production is their dependency on the fishmeal and fish oil that are included in industrially compound aquafeeds. Its high-protein content, excellent essential amino acid profile, rich vitamin and mineral content, high digestibility and palatability to fish, and its lack of anti-nutritional factors, make fishmeal the primary protein source of choice in diets for most farmed (especially carnivorous) fish species. Fish oil has been, and still is, the only high volume commercially available rich source of EPA and DHA that are essential to fish nutrition. Around 5 mt of fishmeal and 1 mt of fish oil are produced annually (Auchterlonie, 2019). The noted decline in production since mid-1990s, which has steadied over the past decade, reflects the overfishing of wild fish stocks. Most of the raw material used for fishmeal and fish oil production originates from wild capture fisheries. Fisheries management and monitoring activities require quotas, so the relative amounts of these natural products available for aquaculture are expected to remain stable at best or decline (Jackson, 2012). Thus, given the inelastic supply and increased demand not only from aquaculture but from the pharmaceutical industry and swine and poultry agriculture, prices are also generally rising (FAO, 2014), which threatens the economic sustainability of the aquaculture sector. Furthermore, aquaculture is criticized for contributing to the overexploitation of capture fisheries in order to produce fish meal and fish oil for farmed species (Naylor et al., 2000), with an on-going debate about the 'fish in–fish out' concept (Byelashov and Griffin, 2014; Tacon and Metian, 2008), which is the quantity of live fish from fisheries needed to produce the amount of fishmeal and fish oil required to produce a unit of farmed fish (IFFO, 2020).

Aquaculture has in the recent decades reduced the fishmeal and fish oil inclusion levels in aquafeeds, mainly using plant alternatives (Tacon and Metian, 2015). However, as aquaculture continues to grow, the demand for fishmeal and fish oil will intensify. Thus, it is generally accepted that in order to support sustainable, cost-effective and eco-efficient aquaculture production further reductions to the dietary inclusion levels of fishmeal and fish oil are needed. Improved feed formulations, advances in feed manufacture and a better on-farm feed management strategies may further reduce the amount of feed used, and thus of fishmeal and fish oil for a single production cycle (FAO, 2018). Moreover, fishmeal and fish oil produced from fish by-products, such as processing wastes, discards and farmed by-products, represent a growing share of total production, that nowadays is estimated at 33–35% of the volume produced (Auchterlonie, 2019; FAO, 2018). Aside from using plant alternatives, new feedstuffs using terrestrial animal proteins, insect proteins and microalgae among others are attracting increasing interest from both the aquafeed industry and academia (Alfiko et al., 2022; Gasco et al., 2021; Nagappan et al., 2021; Luthada-Raswiswi et al., 2021).

Terrestrial plant proteins

Terrestrial plant proteins and oils have been used successfully to reduce the fishmeal and fish oil used in aquafeeds in the few last decades (Bell and Waagbø, 2008; Turchini et al., 2009). However, terrestrial plant ingredients could not serve as a panacea as their nutritional value do not resemble that of fishmeal and fish oil and may lead to reduced fish growth rates when used in high levels. In addition, plant feedstuffs may result in reduced nutrient bioavailability (Bell and Waagbø, 2008) producing undesirable disturbances to the aquatic environment (Hardy, 2010), while also altering fish quality, especially with respect to the n-3 PUFA content that are considered as health-promoting nutrients in human nutrition (Matos et al., 2017). Furthermore, aquaculture competes in the international markets for the use of these plant alternatives, predominantly competing against the agriculture sector and the biofuel industry, thus further stressing the sustainability of aquatic food production. Besides, many plant ingredients also come with their own environmental health (Fry et al., 2016) and sustainability challenges (Matos et al., 2017). Optimization of the use of plant feedstuffs in fish diets is one fish oil/fishmeal alternative. In this context, several dietary strategies have been used successfully including the supplementation with crystalline essential amino acids (Monge-Ortiz et al., 2016), plant protein combinations (Zhang et al., 2012), the use of specific harvesting diets (Benedito-Palos et al., 2010), supplementation with additives (Gunathilaka et al., 2019) and application of exogenous enzymes (Castillo and Gatlin, 2015) among others.

Terrestrial animal proteins

Terrestrial animal proteins are suitable fishmeal replacers as they have higher protein content, more favourable amino acid profiles and fewer carbohydrates compared with plant feed ingredients, while also lacking anti-nutritional factors. Their use was banned in European aquafeeds from 2001 to 2013 during the bovine spongiform encephalopathy crisis (EC regulations 999/2001 and 1234/2003). They have been used successfully in fish and crustacean diets outside Europe. The approved reintroduction of non-ruminant processed animal proteins (PAPs) in European aquafeeds (EC Regulation 56/2013) further reduced the inclusion levels of fishmeal in a more cost-effective manner, as their price is lower than that of fishmeal. In addition, the use of terrestrial animal proteins could enhance aquaculture's sustainability and eco-efficiency, as these feedstuffs utilize animal by-products and have in general a more favourable carbon footprint when compared with fishmeal and plant alternatives (den Hartog and Sijtsma, 2013). Animal proteins, such as poultry by-product meal (Karapanagiotidis et al., 2019), feathermeal (Campos et al., 2017; Psofakis et al., 2020) and meat and bone meal (Moutinho et al., 2017) have been explored and proven suitable alternatives to fishmeal.

Insect proteins

Interest in the use of insect proteins to replace fishmeal was boosted during the last decade with the European Union giving the green light (EC Regulation 893/2017) to their inclusion in aquafeeds. Insects could contribute to further replacements of fishmeal and to the production of sustainable aquafeeds as they possess several suitable nutritional characteristics. Most insect species have a protein content between 40% and 70% that can be even higher in defatted insect meals (Makkar et al., 2014; Sánchez-Muros et al., 2014). Insects are also a good source of essential amino acids with some species being particularly rich in lysine and methionine, in which most plant proteins are usually deficient (Nogales- Mérida et al., 2019). Moreover, insects have been shown to have antimicrobial and anti-bacterial properties that could potentially serve as functional feed ingredients enhancing fish health (Zhao et al., 2010). Apart from their high nutritional value, insect farming itself seems to have a low ecological footprint as it has low requirements for culturing area and water, while insects can be fed on food wastes contributing to the nutrient recycling (Sánchez-Muros et al., 2014). Studies using insect meals as fishmeal replacers have showed that they can successfully fulfil this role and even total replacement can be achieved mainly for omnivorous species but also for some carnivorous species (Belghit et al., 2019; Henry et al., 2015; Sánchez-Muros et al., 2014; Tran et al., 2015). The aqua-feed industry has been slow to incorporate insect meals in diet formulations, possibly due to their high price (IPIFF, 2018) and limited availability (Arru et al., 2019), but also to their unstandardized nutrient profiles (Tran et al., 2015). There are also issues around food safety, potential public health concerns and consumer acceptability (Van Huis, 2020), which all contribute to the fact that insect meals are not yet widely exploited by the aquafeed industry. Most industrial applications involve meals from black soldier fly (*Hermetia illucens)* and yellow mealworm (*Tenebrio molitor*) in a kind of a defatted form.

Microalgae

Microalgae perhaps has the highest potential as a fish oil replacement as some species contain remarkable levels of n-3 highly unsaturated fatty acids that are even higher than those of fish oils. Prymnesiophytes such as *Pavlova* sp. and *Isochrysis* sp. are relatively rich in DHA that may account up to 11% of total fatty acids (Hemaiswarya et al., 2011), while *Schizochytrium* sp. can even reach 45% (Kamlangdee and Fan, 2003). Many microalgae families such as Diatoms, Eustigmatophytes, Cryptomonads and Rhodophytes are very good sources of EPA that may account for up to 34%

(Becker, 2007). The lipid content of microalgae can reach 50%, while their protein content has been reported to range from 30% to 75% with a suitable amino acid profile (Becker, 2013). Moreover, microalgae are very good sources of vitamins (particularly C and E), minerals, antioxidants and pigments (Brown et al., 1997). The nutrient content of microalgae is highly variable both between and within species as it is affected by numerous parameters such as culture conditions and the harvesting stage among others (Becker, 2013; Brown et al., 1997). Studies using microalgae as feed additives or fish meal replacers have showed growth promoting effects and enhanced fish quality (Guimarães et al., 2019; Metsoviti et al., 2018; Sarker et al., 2018; Vizcaíno et al., 2014). The main challenge with microalgae is to develop the expertise and technology for mass cultivation, as at present the cost of algal meals and oils is prohibitively high. The use of microalgae in aquafeeds could also enhance the environmental sustainability of the sector as microalgae culture has a low carbon footprint (Chung et al., 2017) and can utilize industrial by-products (Chi et al., 2017).

With the aquaculture sector becoming the major food fish provider to human diets, the demand for compound aquafeeds is expected to increase. Thus, the global aquafeed industry is facing many challenges and needs to produce (a) more aquafeeds to satisfy the increasing demand, (2) sustainable aquafeeds that will not rely on finite natural resources, such as wild-sourced fishmeal and fish oil, (3) safe aquafeeds to ensure the health of fish and human consumers and (4) aquafeeds of high nutritional value for the well-being of the consumer.

Selective breeding in aquaculture

Selective breeding programmes have the potential to generate cumulative and long-term improvement in traits of economic, welfare and environmental importance. Improving such traits is crucial to ensuring the economic and environmental sustainability of aquaculture production. The first family based breeding programme for salmon was the Norwegian national breeding programme in the 1970s (Gjedrem et al., 1991a, 2012). Since then, selective breeding has been increasingly undertaken by the aquaculture industry and breeding programmes have been developed for Atlantic salmon (*Salmo salar*) and other aquaculture species in several countries. Atlantic salmon is the species for which the most advanced breeding programs exist, and the technology and methods used in salmon tend to be transferred to other less advanced sectors (Lekang, 2022).

Principles of quantitative genetics and complex traits

When a single gene of major effect underlies the control of a trait, then individual phenotypes typically fall within distinct categories, where the phenotype refers to any observable and measurable characteristic that an individual carries. In contrast, quantitative traits with polygenic genetic architecture show continuous variation. Quantitative or complex traits are typically controlled by a large number of genes and the environment (Dekkers and Hospital, 2002). This implies that the individual effects of the underlying genes on the observed phenotype are small, and individuals form a continuously graded series rather than falling within distinct categories. Genomic loci affecting quantitative traits are called quantitative trait loci (QTL).

For example, the genetic architecture of disease resistance is known to range from single major genes to being highly polygenic (Bishop et al., 2010). When disease resistance is controlled by a major QTL, then the phenotypes in the population are often observed as binary (i.e. dead or alive). However, when disease resistance is under polygenic control, all the intermediate values, ranging from very susceptible to very resistant individuals, can be observed in the population, and variation is expressed on a continuous scale, which is often referred to as the underlying liability to disease.

Decomposing variation – the heritability

Observed phenotypic variation can be decomposed into genetic variation and variation due to environmental factors. Genetic variation can be further decomposed into the additive genetic variance (the variance of the breeding values [BV]), the dominance deviation due to interactions between alleles within the same locus, and the interaction (or epistatic) deviation due to interactions between genes at different loci (Falconer and Mackay, 1996). The BV or genetic merit of an individual is the contribution of an individual's genotype to the phenotype of the next generation and, as will be discussed below in detail, it is a key parameter in animal breeding programmes. Environmental factors include, for example, farm management and nutrition. It is well known that dietary nutrients can influence the genome, while the genome affects the ways an animal will respond to dietary nutrients (Müller and Kersten, 2003; Panserat and Kaushik, 2010).

The proportion of the total phenotypic variation that can be attributed to the genes transmitted from the parents is the trait heritability (in the narrow sense) (Falconer and Mackay, 1996). This reflects the proportion of observed variation that is due to additive genetic effects. In typical commercial breeding programmes, genetic selection operates on genetic variation already present in the population, hence, the amount of genetic variation present, or the trait heritability, determines the amount of progress that can be achieved through genetic selection.

Breeding programmes and breeding objectives in aquaculture

Initially in aquaculture selective breeding programmes the target production trait for improvement was growth rate. Through genetic selection, an average genetic gain for growth rate per generation of 17.8% has been reported for Atlantic salmon (Gjedrem and Rye, 2018). In Nile tilapia (*Oreochromis niloticus*), one of the most important aquaculture species globally, the 'genetically improved farmed tilapia' or GIFT strain resulted in 11.9% genetic gain per generation in body weight (Hamzah et al., 2014). In addition, in Atlantic and coho salmon it has been found that feed efficiency was further improved by selective breeding programmes for growth rate (Neely et al., 2008; Thodesen et al., 1999). Those data highlight the potential for rapid improvement of production traits via the use of well-managed selective breeding programmes.

In modern, advanced breeding programmes the breeding goals became wider to include traits such as age at sexual maturation, carcass yield, quality and colour, feed conversion efficiency, body composition and fat content, body shape and morphology, reduced deformities, survival and fecundity. In addition, disease resistance traits identified in salmonids, for example, resistance to furunculosis (Gjedrem et al., 1991b), resistance to infectious salmon anaemia (Ødegård et al., 2007a, 2007b), and resistance to infectious pancreatic necrosis (Houston et al., 2014a) have also been included in the breeding goal. In other species, for example, in Tilapia, traits such as environmental robustness, and specifically salinity and temperature tolerance, have been added in the breeding goal. However, aquaculture still lags behind plant and terrestrial livestock industries in exploiting genetic variation through genetic selection (Gjedrem and Rye, 2018), with much aquaculture production globally still being derived from wild or near-wild broodstocks.

Aquatic species have reproductive biological features that are advantageous for selective breeding and high-resolution genetic studies. Response to genetic selection (genetic gain) depends on the selection intensity, the genetic variance for the trait (the additive genetic standard deviation), and the prediction accuracy, while it is inversely related to the generation interval (Falconer and Mackay, 1996). In typical aquaculture breeding programmes, owing to the high fecundity of aquatic species, large full-sibling families are available that allows application of high selection intensities. The availability of large full-sibling families is particularly practical in performing genomic selection for disease resistance traits. Given that the breeding stock (i.e. the selection candidates) are not exposed to diseases for biosecurity reasons, performance testing can be conducted on close relatives (siblings) of the selection candidates, a process known as sib-testing. Moreover, the possibility of external fertilization adds great flexibility to the design of breeding programmes. Owing to their very recent domestication, marine species still have high levels of genetic variation, demonstrating moderate to high heritability in most traits of economic importance. This variation is available for genetic improvement via selective breeding (Table 3.1). Despite the relatively long generation intervals (e.g. 3–4 years in Atlantic salmon), selective breeding programmes can utilize this genetic variance to achieve impressive genetic gain in aquaculture.

In early breeding programmes improvement of the traits of interest was achieved through family based selection and through using pedigree data, that is, records of descent which capture identity by descent (IBD). However, with genetic selection using phenotype and pedigree data alone, it was possible to perform only between-family selection if traits were measured on full-siblings of candidates (such as disease resistance traits). In other words, the BV would correspond to the entire family and all full-siblings would be considered as genetically equal. This cannot capitalize on 50% of individual variation that is introduced by the Mendelian sampling process during meiosis. In contrast, as it will be discussed below, genomic selection using genetic markers can also capture within-family variation,

Table 3.1 Summary of example heritability for major farmed species and production traits of economic interest.

Species	Trait	Heritability	Source
Pacific white shrimp (*Litopenaeus vannamei*)	Body weight at harvest	0.24–0.45	Castillo-Juárez et al. (2007)
Atlantic salmon (*Salmo salar*)	Body weight and length	0.6, 0.61	Tsai et al. (2015)
	Resistance to infectious and parasitic diseases	0.1–0.6	Yáñez et al. (2014)
Common carp (*Cyprinus carpio*)	Body weight	0.49	Prchal et al. (2018)
Nile tilapia (*Oreochromis niloticus*)	Fillet yield and harvest weight	0.21, 0.36	Yoshida et al. (2019)
Pacific oyster (*Crassostrea gigas*)	Shell length and height	0.23, 0.26	Gutierrez et al. (2018)
	Wet weight	0.35	Gutierrez et al. (2018)

and genomic selection using marker genotype data has resulted in higher prediction accuracy of BVs than the use of pedigree information alone, improving the predictive ability of the BVs (Ødegård et al., 2014; Tsai et al., 2015).

Genetic markers

One type of genetic markers are the microsatellite markers. Those are stretches of repetitive DNA motifs, usually 1–5 bases long (e.g. a dinucleotide repeated 100 times). Their alleles are defined on the basis of variation in the number of repeats, and they are highly polymorphic with high mutation rates. Microsatellites are useful for tracing ancestry and kinship analysis, and for constructing linkage genetic maps.

Single nucleotide polymorphisms (SNPs) are the genetic markers commonly used in animal breeding. These are changes in a single DNA base and usually have only two alleles (bi-allelic). High density SNP genotyping platforms have been developed for several livestock species and are commercially available (e.g. the 777,000 SNPs high density chip for cattle, 50,000 SNP chip for sheep, etc.). Salmonids underwent a whole genome duplication event approximately 25–100 million years ago, and they are thought to be in the process of reverting to a diploid state. This has hindered the development of a salmon SNP array due to the difficulty in distinguishing paralogous loci from genuine bi-allelic SNP variation at unique genomic locations.

Genome-wide association studies

SNP marker genotypes are commonly used in genome-wide association studies (GWAS). This type of analysis allows us to shed light on the genetic architecture of the trait under study. More specifically,

Table 3.2 Summary of example SNP arrays for major farmed species.

Species	SNP array	Reference
Pacific white shrimp (*Litopenaeus vannamei*)	Illumina Infinium ShrimpLD-24 v1.0	Jones et al. (2017)
Atlantic salmon (*Salmo salar*)	Affymetrix Axion 132K Atlantic salmon SNP chip	Houston et al. (2014b)
	200K Affymetrix Axiom myDesign Custom Array	Yáñez et al. (2016)
	55K Affymetrix Axiom	Bangera et al. (2018)
	6K custom design Illumina iSelect SNP-array	Lien et al. (2011)
Common carp (*Cyprinus carpio*)	Affymetrix Axion 250K	Xu et al. (2014)
Nile tilapia (*Oreochromis niloticus*)	Onil50-array Affymetrix Axion	Joshi et al. (2018)
Pacific (*Crassostrea gigas*) and European oyster (*Ostrea edulis*)	Affymetrix Axiom Custom Array 40K	Gutierrez et al. (2017)
Rainbow trout (*Oncorhynchus mykiss*)	Affymetrix Axion 57K	Palti et al. (2015)

associations between a genetic variant and a phenotype are detected based on the marker *p*-value, and they are compared against pre-determined significance thresholds. In order for a marker to be declared significant, it has to explain a substantial amount of the variance, and that occurs in cases of major gene effects. Once a genomic region is identified through this process, then it can be further examined for candidate genes and causative mutations through fine-mapping, which helps us improve our understanding of the underlying biological pathways involved and discover the causative genes. However, for polygenic traits, detecting QTLs using this approach is challenging and large sample sizes of phenotyped and genotyped individuals are required.

GWAS approaches have been used for identification of traits of economic importance in several aquaculture species. For example in Atlantic salmon, GWAS revealed a major gene effect for resistance to the viral disease infectious pancreatic necrosis or IPN (Houston et al., 2010), while it has helped improve our understanding of the polygenic genetic architecture of other traits, such as resistance to sea lice (Tsai et al., 2016; Correa et al., 2017b), resistance to amoebic gill disease (Robledo et al., 2018a), or growth traits (Tsai et al., 2015). In coho salmon (*Oncorhynchus kisutch*) GWAS revealed a moderately polygenic genetic architecture for resistance to the bacterial infection from *Piscirickettsia salmonis* (Barria et al., 2018), in Nile tilapia GWAS showed the polygenic nature of fillet yield and harvest weight (Yoshida et al., 2019), while in common carp (*Cyprinus carpio*) GWAS led to the detection of candidate genes for flesh quality traits (Zheng et al., 2016).

Marker assisted selection

Marker assisted selection (MAS) allows selection of animals based on their genotype, where the genotypes have been obtained using dense SNP markers across the genome, linked with QTL. Once a QTL of major effect has been identified, MAS can be used to selectively breed individuals that carry the marker allele linked to the beneficial QTL allele. This can be considered as an indirect method, as selection is on the basis of the SNP marker genotype, where the marker is linked to the trait of interest, rather than directly on the locus controlling the trait. MAS has been applied to assist animal selection for improving traits controlled by major genes of large effects. For example, in Atlantic salmon, a single major gene effect was identified for resistance to IPN and has been successfully implicated into commercial breeding programs (Houston et al., 2010). IPN is caused by a birnavirus, which is responsible for high mortality in fry and post-smolts during infections. Using survival data from 'natural' IPN virus outbreaks in seawater, it was possible to identify a single locus that explained almost all genetic variation in resistance in freshwater and seawater, and utilize a SNP genetic marker to predict resistance to the IPN virus (Houston et al., 2010).

Genomic selection

Polygenic traits can be improved through genomic selection using SNP marker genotypes, under the assumption that the SNPs are either causal or in strong linkage disequilibrium (LD) with QTLs. In genomic selection animals are selected based on their SNP marker genotypes and their genomic estimated breeding values (GEBVs). Genome-wide SNP genotype data are used to calculate the genomic relationship matrix which captures identity by state (IBS). Then GEBVs are estimated using quantitative genetic models and specialized variance component analysis software (e.g. ASReml [Gilmour et al., 2009], DMU [Madsen et al., 2006], MiX99 [Vuori et al., 2006]). The GEBV of an individual is calculated as the genome-wide sum of the effects of SNP marker genotypes (Meuwissen et al., 2001). GEBVs can be calculated using best linear unbiased predictions methods (GBLUP [VanRaden, 2008]) or more sophisticated Bayesian (e.g. Bayes RC [MacLeod et al., 2016]), and weighted GBLUP (e.g. Tiezzi and Maltecca, 2015) methods that allow to assign different weight to SNPs in LD with QTLs affecting the trait of interest. Finally, the predictive ability of the BVs can be assessed through the correlation between the predicted breeding value and the trait value, adjusted for the trait heritability. That is known as the accuracy of prediction, and it increases with the amount of information available.

In typical aquaculture breeding programmes records of phenotypes are obtained on siblings of the selection candidates either through experimental disease challenge trial or from 'field' data from a natural challenge. Those siblings are also being genotyped (SNP marker genotype data), so that they form the 'training population' with both phenotypes and genotypes. This 'training population' can be used to estimate the marker effects, and then predict the GEBVs for the selection candidates for which only genotypic information is required.

Genome sequencing

Higher resolution genomic data can be obtained through whole genome sequencing rather than using genetic markers (Elshire et al., 2011; Robledo et al., 2018b). Genotyping by sequencing (GBS) techniques, for example restriction-site associated DNA sequencing (RAD-Seq) and variations, can be used to directly obtain genotype data, to build genetic linkage maps and improve reference genome assemblies, or to perform GWAS and genomic selection (Robledo et al., 2018b). Using genome sequence data sets the genomic selection process free of the requirement for LD between marker and QTL, and it has been hypothesized that this will benefit across-population predictions where the presence of different LD patterns dramatically reduces the prediction accuracy. However, incorporating such data into routine genomic selection is a challenge for the available computational and modelling approaches in order to remove noise and perform efficient analysis. Especially for aquaculture species with highly polymorphic genomes and paralogous sequences, thorough quality control is required in order to remove spurious data and genotyping errors (Zenger et al., 2019). Nevertheless, by using specialized software, stringent data filtering criteria can be applied, and paralogous loci can be identified, while improving coverage and read depth can help reducing genotyping errors.

Breeding for disease resistance in aquatic species

The family structure of most aquatic species allows great flexibility in the design of disease challenge experiments, to study the genetics underlying disease resistance. Infectious diseases impose a major and persisting limiting factor in aquaculture industry worldwide. They cause significant financial losses due to the cost of disease control (e.g. treatments, vaccinations, culling for biosecurity), impaired growth, and mortalities. Traditional disease control and prevention measures include medicines and vaccinations. These measures tend to be expensive, logistically impractical, only partially effective and often requiring regulatory approval. Moreover, commonly used chemical therapeutants have a significant environmental impact, while some pathogens, for example sea lice can develop high levels of resistance (Sevatdal and Horsberg, 2003; Tully and McFadden, 2000). Additional challenges for disease control arise given the continuous contact of farmed marine species in the open ocean with the wild environment and the potential for disease transmission from farmed hosts to wild hosts and vice versa.

Given the encouraging published evidence for host genetic variation in resistance to several major viral, bacterial and ectoparasitic infections (Yáñez et al., 2014), genetic selection presents an appealing and feasible complementary approach for disease control. In addition, a feature of modern breeding programmes is the simultaneous selection for multiple traits, therefore disease resistance traits can be targeted while still selecting for other economically important traits, for example, growth. This can be achieved through using selection indices to define a selection objective that includes the optimal combination of measurements of several traits (including indicator traits) on an individual, to improve target-traits (selection objective). Commonly used indicator traits for disease resistance in fish are survival (surviving days) or mortality, pathogen or parasite load (from cell culture assays or qPCR), biomarkers for host immune response, the number of parasites attached to the fish (e.g. sea lice count) or lesion scores (e.g. mean gill score for amoebic gill disease).

Potential target-traits for genetic disease control include *disease resistance* (considered the opposite of *susceptibility*, disease resistance can be defined as an individual's propensity of becoming infected when exposed to infectious material); *infectivity* (defined as the ability of an individual, once infected, to transmit infection (Lipschutz-Powell et al., 2014; Tsairidou et al., 2019)); and *tolerance* (the ability of an infected host to limit the impact of infection on fitness or health (Lough et al., 2015)). In many studies, a *broad-sense resistance* is often used, which represents the ultimate desirable trait in the general sense of disease avoidance. Disease resistance (broad-sense) has been a target trait for aquaculture breeders since the early 1990s as the salmon breeding programmes began to broaden their breeding goals (Gjøen and Bentsen, 1997). Moreover, 'umbrella-traits', such as *resilience* (comprising both resistance and tolerance), *immunocompetence* (the immunological ability to resist and recover from infection (Ask et al., 2007)), or general *robustness to disease*, are used to describe the attributes of quick recovery, high survivability and satisfactory performance in sub-optimal conditions and under pathogen challenge, which would be valuable in environments with endemic diseases and complex multi-factorial syndromes. As new global challenges arise due to emerging diseases and climate change, it may become pertinent to understand the genetics underlying genotype-by-environment interactions and the ability of farmed marine species to adapt and express genetic plasticity in their response to sub-optimal environments.

One example of a disease resistance trait suitable for genomic selection, is resistance to sea lice in farmed Atlantic salmon. Sea lice (*Lepeophtheirus salmonis* in Europe and *Caligus rogercresseyi* in Chile) is a parasitic marine copepod of the family of Caligidae, and it has been the largest disease-related problem in salmon industry, causing significant financial losses due to the cost of treatment, and increased morbidity and mortality (Gjerde et al., 2011; Treasurer et al., 2022). There is host genetic variation in resistance to sea lice, with a heritability of ~0.3, and the trait is known to be polygenic (Gjerde et al., 2011; Tsai et al., 2015, 2016). Therefore genomic selection has been proposed as an effective method to reduce sea lice prevalence in farmed Atlantic salmon, with reported genomic prediction accuracies in the range of 0.5–0.6 (Correa et al., 2017a; Tsai et al., 2016; Tsairidou et al., 2019b), and even when using medium SNP array densities (e.g. 5000 SNPs) obtained via genotype imputation (Tsairidou et al., 2019b). Salmon breeding companies have developed advanced breeding programmes that include genomic selection for sea lice resistance in salmon eggs and juvenile fish (e.g. Hendrix Genetics, n.d.).

New technologies and future trends

The fast adoption of new technological advances will allow aquaculture to adapt to emerging challenges and ensure long-term sustainability for the sector, thus supporting its role in aquatic food security. Commonly used chemical treatments have a considerable environmental impact, while pathogens, for example, sea lice in Atlantic salmon, increasingly develop high levels of drug resistance (Tully and McFadden, 2000). Therefore, genetic selection for disease resistance can be used as an alternative approach for more sustainable, long-term disease control.

In recent years, selective breeding has increasingly been incorporated in aquaculture production. Novel computational techniques such as genotype imputation and large-scale data with functional annotation information hold the promise of cost-effective genomic selection with enhanced prediction accuracy. Genotype imputation has been tested in several studies in simulated (Dufflocq et al., 2019; Li et al., 2010) and empirical data (Sanna et al., 2008; Tsai et al., 2017; Willer et al., 2008; Yoshida et al., 2018), and is one promising avenue for reducing the costs of genomic selection. Including the causal mutation itself rather than relying on linkage disequilibrium between the QTL and the SNP marker, and prioritizing functional SNPs into genomic prediction can assist in maintaining, and potentially enhance, prediction accuracy over generations and across-populations (Edwards et al., 2016; McLeod et al., 1995).

Selection using genomic data can also help us in the conservation of genetic diversity and managing inbreeding. High rates of inbreeding is associated with reduced genetic diversity and inbreeding depression, that is, the loss of heterozygosity and the disappearance of hybrid vigour (Woolliams, 2007). Inbreeding depression has negative impacts on fitness traits and causes a decline in performance for traits of economic importance (Wiener et al., 1992, 1994). Genomic selection exploits both between- and within-family genetic variation, which removes the emphasis from specific families and better control the reduction of the effective population size due to selection. This allows us to perform selection within families and distinguish between the genetic merit of full-sibs by predicting the Mendelian segregation term. Minimizing the rate of inbreeding can be achieved through the optimum genetic contributions theory and it may be possible to further enhance its performance by incorporating genomic data into this methodology (Woolliams et al., 2015).

The use of reproductive technologies can address species- or production system-specific logistical challenges. Artificial insemination and technologies for controlling gender can be used to maximize production efficiency and avoid overpopulation. Artificial induction of triploidy for sterilization can be used to prevent the negative impact of early maturation on growth and flesh quality, while it also protects the surrounding ecosystem from genetic contamination due to interbreeding between the wild populations and escapees of farmed salmonids (Shumway, 2021).

Genome editing and specifically CRISPR/Cas9 systems have been successfully applied in salmon to generate sterile fish (Wargelius et al., 2016), and have a huge potential for future applications to advance our understanding of salmon biology and improve important production and welfare traits. For example, identification of causative genes has been possible through the GeCKO technique (Shalem et al., 2014). These methods may offer solutions that are not otherwise possible, for example, complete disease resistance.

Interaction between genetics and nutrition: genotype-by-environment and genotype-by-diet interactions

As discussed above, the phenotype of an individual is a function of its genes and environmental effects. However, when selected individuals derived from the same population are required to perform under different production environments, their performance varies considerably, depending on their genotypes. In other words, the phenotype also depends on the interaction between genotypes and environment, and the environmental deviation has a different impact on different genotypes (Falconer and Mackay, 1996); those are known as genotype-by-environment interactions and result in re-ranking of genotypes across different environmental conditions, depending on the magnitude of the genetic correlation between traits measured in the different environments. This is pertinent particularly in aquaculture where improved stock from a few, core breeding programmes are distributed around the globe, and hence there is large variation in the geographical locations and production systems of the farms where they are reared (Sae-Lim et al., 2016). Furthermore, aquaculture species are exposed to different environments during their different life stages, and there is variation in the environmental conditions between years and different time of the year (Sae-Lim et al., 2016).

In addition, a correlation between genotype and environment may arise when superior genotypes are provided better environmental conditions, for example, if individuals selected for better performance are given more food, or food that is more suitable for their (genetic) requirements (Falconer and Mackay, 1996). In the latter, the feeding system is better adopted to allow those higher performing individuals to fully express their genetic potential. Such genotype-by-nutrition interactions may also result in re-ranking and different performance of animals across diverse feeding regimes. For example, low energy budget due to the limited food available in natural environments has been found to hinder farmed salmon from expressing their full genetic potential in growth when transferred to the wild (Glover et al., 2018).

It is possible to design fish feeding systems tailored to the breeding strategy in order to meet the needs of fish after genetic selection (e.g. for faster growth). For example, it has been reported that feed utilization and allocation of dietary lipids and protein for energy and growth differs between unselected and selected coho salmon (Neely et al., 2008) and Atlantic salmon (Thodesen et al., 1999). Alternatively, breeding programmes can be designed to accommodate specific feeding systems or feed availability. Given the environment (the feeding system) where the animals are expected to perform, selection can favour genotypes that will perform better in that specific environment. Environment-specific breeding values can guide the selection of individuals in the breeding nucleus to target higher performance in commercial production environments (Sae-Lim et al., 2016). Furthermore, increasing the size of the breeding nucleus and the number of families has also been suggested as a means to compensate for the re-ranking and subsequent loss of genetic gain due to genotype-by-environment interactions (Gjerde et al., 2014).

Several statistical models have been developed to quantify genotype-by-environment interactions, which either explicitly model the interaction between genotype and environment or consider the trait measured in different environments as being different traits (Sae-Lim et al., 2016). Furthermore, sensitivity to environmental changes can be described through the slope of reaction norms (i.e. the regression of performance on the environmental gradient), where a flat norm indicates less sensitivity (Sae-Lim et al., 2017). Experimental designs for quantifying genotype-by-environment interactions typically require that the performance of individuals from each family can be measured in different environments, where the number and size of families are important design parameters (Sae-Lim et al., 2010). Studies investigating genetic correlations between traits measured under different diets (mainly growth and other traits such as body shape, feed intake, lipid content, body protein etc.), have reported moderate to low re-ranking due to genotype-by-diet interactions, and the genetic correlations were found to range from 0.48 to 1 for rainbow trout (*Oncorhynchus mykiss*), and 0.51 to 0.99 for European seabass (Sae-Lim et al., 2016).

References

Alfiko, Y., Xie, D., Tri Astuti,R., Wong, J. & Wang, L. (2022) Insects as a feed ingredient for fish culture: Status and trends. *Aquaculture and Fisheries*, 7(2), 166–178, https://doi.org/10.1016/j.aaf.2021.10.004.

Alltech (2019) 8th annual Alltech global feed survey. www.alltechfeedsurvey.com.

Arru, B., Furesi, R. Gasco, L. Madau, F.A. & Pulina, P. (2019) The introduction of insect meal into fish diet: the first economic analysis on European sea bass farming. *Sustainability*, 11(6), 1697. https://doi.org/10.3390/su11061697.

Ask, B., Van Der Waaij, E.H., Glass, E.J. & Bishop, S.C. (2007) Modeling immunocompetence development and immunoresponsiveness to challenge in chicks. *Poultry Science*, 86, 1336–1350. https://doi.org/10.1093/ps/86.7.1336.

Auchterlonie, N. (2019) Fish meal and fish oil as essential components in aquafeed. Presented at the Aquafarm Conference, 13 February, Pordenone, Italy. www.iffo.net/iffo-presentations.

Azaza, M.S., Khiari, N., Dhraief, M.N., Aloui, N., Kraïem, M.M. & Elfeki, A. (2015) Growth performance, oxidative stress indices and hepatic carbohydrate metabolic enzymes activities of juvenile Nile tilapia, Oreochromis niloticus L., in response to dietary starch to protein ratios. *Aquaculture Research*, 46, 14–27. https://doi.org/10.1111/are.12153.

Bangera, R., Thorland, I., Hillestad, B., Avitua, S.V., Johansen, H. et al., (2018) Accuracy of genomic predictions for yield traits, flesh colour and fillet fat in Atlantic salmon (Salmo salar). In: *Proceedings of the World Congress on Genetics Applied to Livestock Production.*

Barría, A., Christensen, K.A., Yoshida, G.M., Correa, K., Jedlicki, A., Lhorente, J.P., Davidson, W.S. & Yáñez, J.M. (2018) Genomic predictions and genome-wide association study of resistance against Piscirickettsia salmonis in coho salmon (Oncorhynchus kisutch) using ddRAD sequencing. *G3*, 8, 1183–1194. https://doi.org/10.1534/g3.118.200053.

Baskerville-Bridges, B. & Kling, L.J. (2000a) Early weaning of Atlantic cod (Gadus morhua) larvae onto a microparticulate diet. *Aquaculture*, 189, 109–117. https://doi.org/10.1016/S0044-8486(00)00356-2.

Baskerville-Bridges, B. & Kling, L.K. (2000b) Development and evaluation of microparticulate diets for early weaning of Atlantic cod Gadus morhua larvae. *Aquaculture Nutrition*, 6, 171–182. https://doi.org/10.1046/j.1365-2095.2000.00149.x.

Becker, E.W. (2007) Micro-algae as a source of protein. *Biotechnology Advances*, 25, 207–210. https://doi.org/10.1016/j.biotechadv.2006.11.002.

Becker, E.W. (2013) Microalgae for human and animal nutrition. In: *Handbook of Microalgal Culture: Applied Phycology and Biotechnology*, 2nd edn (edited by A. Richmond and Q. Hu). Wiley-Blackwell, pp. 461–503. https://doi.org/10.1002/9781118567166.ch25.

Belghit, I., Liland, N.S., Gjesdal, P., Biancarosa, I., Menchetti, E., Li, Y., Waagbø, R., Krogdahl, Å & Lock, E.J. (2019) Black soldier fly larvae meal can replace fish meal in diets of sea-water phase Atlantic salmon (Salmo salar). *Aquaculture*, 503, 609–619. https://doi.org/10.1016/j.aquaculture.2018.12.032.

Bell, J.G. & Waagbø, R. (2008) Safe and nutritious aquaculture produce: Benefits and risks of alternative sustainable aquafeeds. In: *Aquaculture in the Ecosystem* (edited by M. Holmer, K. Black, C. M. Duarte, N. Marbà & I. Karakassis). Springer: Dordrecht, The Netherlands, pp. 185–225.

Béné, C., Arthur, R., Norbury, H., Allison, E.H., Beveridge, M., Bush, S., Campling, L., Leschen, W., Little, D., Squires, D., Thilsted, S.H., Troell, M. & Williams, M. (2016) Contribution of fisheries and aquaculture to food security and poverty reduction: Assessing the current evidence. *World Development*, 79, 177–196. https://doi.org/10.1016/j.worlddev.2015.11.007.

Benedito-Palos, L., Navarro, J.C., Kaushik, S. & Pèrez-Sánchez, J. (2010) Tissue-specific robustness of fatty acid signatures in cultured gilthead sea bream (Sparus aurata L.) fed practical diets with a combined high replacement of fish meal and fish oil. *Journal of Animal Science*, 88, 1759–1770. https://doi.org/10.2527/jas.2009-2564.

Bishop, S., Axford, R., Nicholas, F. & Owen, J. (2010) *Breeding for Disease Resistance in Farm Animals*, 3rd edn. CABI Publishing: Wallingford.

Biswas, G., Thirunavukkarasu, A.R., Sundaray, J.K. & Kailasam, M. (2010) Optimization of feeding frequency of Asian seabass (Lates calcarifer) fry reared in net cages under brackishwater environment. *Aquaculture*, 305, 26–31. https://doi.org/10.1016/j.aquaculture.2010.04.002.

Booth, M.A., Tucker, B.J., Allan, G.L. & Fielder, D.S. (2008) Effect of feeding regime and fish size on weight gain, feed intake and gastric evacuation in juvenile Australian snapper Pagrus auratus. *Aquaculture*, 282, 104–110. https://doi.org/10.1016/j.aquaculture.2008.06.027.

Boyd, C.E. (1982). *Water Quality Management for Pond Fish Culture*. Elsevier: Amsterdam.

Brown, M.R., Jeffrey, S.W., Volkman, J.K. & Dunstan, G.A. (1997) Nutritional properties of microalgae for mariculture. *Aquaculture*, 151, 315–331. https://doi.org/10.1016/S0044-8486(96)01501-3.

Byelashov, O.A. & Griffin, M.E. (2014) Fish in, fish out: Perception of sustainability and contribution to public health. *Fisheries*, 39, 531–535. https://doi.org/10.1080/03632415.2014.967765.

Cahu, C.L. & Zambonino-Infante, J.L. (2001) Substitution of live food by formulated diets in marine fish larvae. *Aquaculture*, 200, 161–180. https://doi.org/10.1016/S0044-8486(01)00699-8.

Cai, J., Zhou, X., Yan, X., Lucente, D. & Lagana, C. (2017) Top 10 species groups in global aquaculture 2017. Available: www.fao.org/3/ca5224en/CA5224EN.pdf.

Campos, I., Matos, E., Marques, A. & Valente, L.M.P. (2017) Hydrolyzed feather meal as a partial fishmeal replacement in diets for European seabass (Dicentrarchus labrax) juveniles. *Aquaculture*, 476, 152–159. https://doi.org/10.1016/j.aquaculture.2017.04.024.

Castillo, S. & Gatlin, D.M. (2015) Dietary supplementation of exogenous carbohydrase enzymes in fish nutrition: A review. *Aquaculture*, 435, 286–292. https://doi.org/10.1016/j.aquaculture.2014.10.011.

Castillo-Juárez, H., Casares, J.C.Q., Campos-Montes, G., Villela, C.C., Ortega, A.M. & Montaldo, H.H. (2007) Heritability for body weight at harvest size in the Pacific white shrimp, Penaeus (litopenaeus) vannamei, from a multi-environment experiment using univariate and multivariate animal models. *Aquaculture*, 273, 42–49. https://doi.org/10.1016/j.aquaculture.2007.09.023.

CFS (2014) Sustainable fisheries and aquaculture for food security and nutrition. https://www.fao.org/3/av032e/av032e.pdf.

Chi, Z., Pyle, D., Wen, Z., Frear, C. & Chen, S. (2007) A laboratory study of producing docosahexaenoic acid from biodiesel-waste glycerol by microalgal fermentation. *Process Biochemistry*, 42, 1537–1545. https://doi.org/10.1016/j.procbio.2007.08.008.

Chiu, Y.N., Sumagaysay, N.S. & Sastrillo, S.M.A. (1987) Effect of feeding frequency and feeding rate on the growth and feed efficiency of milkfish, Chanos Chanos Forsskal, juveniles. *Asian Fisheries Science*, 1, 27–31.

Chung, Y.-S.S., Lee, J.-W.W. & Chung, C.-H.H. (2017) Molecular challenges in microalgae towards cost-effective production of quality biodiesel. *Renewable and Sustainable Energy Reviews*, 74, 139–144. https://doi.org/10.1016/j.rser.2017.02.048.

Combs, G.F., Jr (1992). *The Vitamins – Fundamental Aspects in Nutrition and Health*. Academic Press: San Diego.

Conceição, L.E.C., Yúfera, M., Makridis, P., Morais, S. & Dinis, M.T. (2010) Live feeds for early stages of fish rearing. *Aquaculture Research*, 41, 613–640. https://doi.org/10.1111/j.1365-2109.2009.02242.x.

Correa, K., Bangera, R., Figueroa, R., Lhorente, J.P. & Yáñez, J.M. (2017a) The use of genomic information increases the accuracy of breeding value predictions for sea louse (Caligus rogercresseyi) resistance in Atlantic salmon (Salmo salar). *Genetics, Selection, Evolution*, 49, 15. https://doi.org/10.1186/s12711-017-0291-8.

Correa, K., Lhorente, J.P., Bassini, L., López, M.E., Di Genova, A., Maass, A., Davidson, W.S. & Yáñez, J.M. (2017b) Genome wide association study for resistance to Caligus rogercresseyi in Atlantic salmon (Salmo salar L.) using a 50K SNP genotyping array. *Aquaculture*, 472, 61–65. https://doi.org/10.1016/j.aquaculture.2016.04.008.

Cowey, C.B. (1988) The nutrition of fish: The developing scene. *Nutrition Research Reviews*, 1, 255–280. https://doi.org/10.1079/NRR19880017.

Dabrowski, K. (1984) The feeding of fish larvae: Present 'state of the art ' and perspectives. *Reproduction Nutrition Developpement*, 24, 807–833. https://doi.org/10.1051/rnd:19840701.

Darias, M.J., Mazurais, D., Koumoundouros, G., Cahu, C.L. & Zambonino-Infante, J.L. (2011) Overview of vitamin D and C requirements in fish and their influence on the skeletal system. *Aquaculture*, 315, 49–60. https://doi.org/10.1016/j.aquaculture.2010.12.030.

De Silva, S.S. (1995) Supplementary feeding in semi-intensive aquaculture systems. In: *Farm-Made Aquafeeds*. FAO fisheries technical Paper no. 343, Rome (edited by M. B. New, A. G. J. Tacon & I. Csavas). Food and Agriculture Organization: Rome, pp. 24–60.

Dekkers, J.C. & Hospital, F. (2002) The use of molecular genetics in the improvement of agricultural populations. *Nature Reviews. Genetics*, 3, 22–32. https://doi.org/10.1038/nrg701.

den Hartog, L.A. & Sijtsma, S.R. (2013) Sustainable feed ingredients. Paper presented at the 12th International Symposium of Australian Renderers Association 'Rendering for Sustainability', Victoria, Australia, 23–26 July 2013.

Dhont, J. & Van Stappen, G. (2003) Biology, tank production and nutritional value of artemia. In: *Live Feeds in Marine Aquaculture* (edited by J. G. Støttrup & L. A. McEvoy). Blackwell Publishing Science Ltd: Oxford, pp. 65–121.

Dufflocq, P., Pérez-Enciso, M., Lhorente, J.P. & Yáñez, J.M. (2019) Accuracy of genomic predictions using different imputation error rates in aquaculture breeding programs: A simulation study. *Aquaculture*, 503, 225–230. https://doi.org/10.1016/j.aquaculture.2018.12.061.

Dwyer, K.S., Brown, J.A., Parrish, C. & Lall, S.P. (2002) Feeding frequency affects food consumption, feeding pattern and growth of juvenile yellowtail flounder (Limanda ferruginea). *Aquaculture*, 213, 279–292. https://doi.org/10.1016/S0044-8486(02)00224-7.

Edwards, S.M., Sørensen, I.F., Sarup, P., Mackay, T.F. & Sørensen, P. (2016) Genomic prediction for quantitative traits is improved by mapping variants to gene ontology categories in Drosophila melanogaster. *Genetics*, 203, 1871–1883. https://doi.org/10.1534/genetics.116.187161.

Elshire, R.J., Glaubitz, J.C., Sun, Q., Poland, J.A., Kawamoto, K., Buckler, E.S. & Mitchell, S.E. (2011) A robust, simple genotyping-by-sequencing (GBS) approach for high diversity species. *PLOS ONE*, 6, e19379. https://doi.org/10.1371/journal.pone.0019379.

Falconer, D.S. & Mackay, T.F. (1996) *Introduction to Quantitative Genetics*. Longman: London.

FAO (Food and Agriculture Organization of the United Nations) (2014) *The State of World Fisheries and Aquaculture 2014*. Food and Agriculture Organization: Rome.

FAO (Food and Agriculture Organization of the United Nations) (2018) *The State of World Fisheries and Aquaculture 2018*. Food and Agriculture Organization: Rome.

FAO (Food and Agriculture Organization of the United Nations) (2019) Aquaculture feed and fertilizer resources information system. www.fao.org/fishery/affris/species-profiles/en/.

FAO (2022) In Brief to The State of World Fisheries and Aquaculture 2022. Towards Blue Transformation. Rome, FAO. https://doi.org/10.4060/cc0463en.

Folkvord, A. & Otterå, H. (1993) Effects of initial size distribution, day length, and feeding frequency on growth, survival, and cannibalism in juvenile Atlantic cod (Gadus morhua L.). *Aquaculture*, 114, 243–260. https://doi.org/10.1016/0044-8486(93)90300-N.

Fry, J.P., Love, D.C., MacDonald, G.K., West, P.C., Engstrom, P.M., Nachman, K.E. & Lawrence, R.S. (2016) Environmental health impacts of feeding crops to farmed fish. *Environment International*, 91, 201–214. https://doi.org/10.1016/j.envint.2016.02.022.

Fry, J.P., Mailloux, N.A., Love, D.C., Milli, M.C. & Cao, L. (2018) Corrigendum: Feed conversion efficiency in aquaculture: Do we measure it correctly? *Environmental Research Letters*, 13, 024017. https://doi.org/10.1088/1748-9326/aaa273.

Garaway, C.J. (2005) Fish, fishing and the rural poor: A case study of the household importance of small scale fisheries in the Lao PDR. *Aquatic Resources, Culture and Development*, 1(2), 131–144. https://doi.org/10.1079/ARC20059.

Gasco, L., Józefiak, A. & Henry, M. (2021). Beyond the protein concept: health aspects of using edible insects on animals. *Journal of Insects as Food and Feed*, 7, 715–741. https://doi.org/ 10.3920/JIFF2020.0077.

Gilmour, A., Gogel, B., Cullis, B. & Thompson, R. (2009) *ASReml User Guide*, Release 3.0. VSN International Ltd: Hemel Hempstead, UK, Vol. HP1, p. 1ES. UK.

Gjedrem, T. & Rye, M. (2018) Selection response in fish and shellfish: A review. *Reviews in Aquaculture*, 10, 168–179. https://doi.org/10.1111/raq.12154.

Gjedrem, T., Gjøen, H.M. & Gjerde, B. (1991a) Genetic origin of Norwegian farmed Atlantic salmon. *Aquaculture*, 98, 41–50. https://doi.org/10.1016/0044-8486(91)90369-I.

Gjedrem, T., Salte, R. & Gjøen, H.M. (1991b) Genetic variation in susceptibility of Atlantic salmon to furunculosis. *Aquaculture*, 97, 1–6. https://doi.org/10.1016/0044-8486(91)90274-B.

Gjedrem, T., Robinson, N. & Rye, M. (2012) The importance of selective breeding in aquaculture to meet future demands for animal protein: A review. *Aquaculture*, 350–353, 117–129. https://doi.org/10.1016/j.aquaculture.2012.04.008.

Gjerde, B., Nielsen, H.M. & Sae-Lim, P. (2014) Compensation of loss in genetic gain due to genotype by environment interaction by increasing the size of the breeding nucleus in an aquaculture population. In: *Proceedings of the World Congress on Genetics Applied to Livestock Production.*

Gjerde, B., Ødegård, J. & Thorland, I. (2011) Estimates of genetic variation in the susceptibility of Atlantic salmon (Salmo salar) to the salmon louse Lepeophtheirus salmonis. *Aquaculture*, 314, 66–72. https://doi.org/10.1016/j.aquaculture.2011.01.026.

Gjøen, H.M. & Bentsen, H.B. (1997) Past, present, and future of genetic improvement in salmon aquaculture. *ICES Journal of Marine Science*, 54, 1009–1014. https://doi.org/10.1006/jmsc.1997.0299.

Glover, K.A., Solberg, M.F., Besnier, F. & Skaala, Ø. (2018) Cryptic introgression: Evidence that selection and plasticity mask the full phenotypic potential of domesticated Atlantic salmon in the wild. *Scientific Reports*, 8, 13966. https://doi.org/10.1038/s41598-018-32467-2.

Goddard, S. (1996). *Feed Management in Intensive Aquaculture.* Chapman & Hall: New York.

Guillaume, J., Kaushik, S., Bergot, P. & Metailler, R. (2001). *Nutrition and Feeding of Fish and Crustaceans.* Springer-Verlag London: London.

Guimarães, A.M., Dias Schleder, D., Nagata, M., Nóbrega, R.O., Fracalossi, D.M., Quadros Seiffert, W. & do Nascimento Vieira, F. (2019) Aurantiochytrium sp. meal can replace fish oil in practical diets for the juvenile Pacific white shrimp. *Aquaculture Nutrition*, 25, 798–807. https://doi.org/10.1111/anu.12897.

Gunathilaka, G.L.B.E., Kim, M.-G., Lee, C., Shin, J., Lee, B.-J. & Lee, K.-J. (2019) Effects of taurine supplementation in low fish meal diets for red seabream (Pagrus major) in low water temperature season. *Fisheries and Aquatic Sciences*, 22, 23. https://doi.org/10.1186/s41240-019-0138-z.

Gutierrez, A.P., Matika, O., Bean, T.P. & Houston, R.D. (2018) Genomic selection for growth traits in Pacific oyster (Crassostrea gigas): Potential of low-density marker panels for breeding value prediction. *Frontiers in Genetics*, 9, 391–391. https://doi.org/10.3389/fgene.2018.00391] [PubMed: 30283494.

Gutierrez, A.P., Turner, F., Gharbi, K., Talbot, R., Lowe, N.R., Peñaloza, C., McCullough, M., Prodöhl, P.A., Bean, T.P. & Houston, R.D. (2017) Development of a medium density combined-species SNP array for Pacific and European oysters (Crassostrea gigas and Ostrea edulis). *G3*, 7, 2209–2218. https://doi.org/10.1534/g3.117.041780] [PubMed: 28533337.

Halver, J. & Hardy, R. (2002). *Fish Nutrition*, 3rd edn. Academic Press: San Diego, USA, p. 824.

Hamre, K., Yúfera, M., Rønnestad, I., Boglione, C., Conceição, L.E.C. & Izquierdo, M. (2013) Fish larval nutrition and feed formulation: Knowledge gaps and bottlenecks for advances in larval rearing. *Reviews in Aquaculture*, 5, S26–S58. https://doi.org/10.1111/j.1753-5131.2012.01086.x.

Hamzah, A., Ponzoni, R.W., Nguyen, N.H., Khaw, H.L., Yee, H.Y. et al., (2014) Performance of the genetically improved tilapia (GIFT) strain over ten generations of selection in Malaysia. *Pertanika Journal of Tropical Agricultural Science*, 37, 411–429.

Hardy, R.W. (2010) Utilization of plant proteins in fish diets: effects of global demand and supplies of fishmeal. *Aquaculture Research*, 41, 770–776. http://dx.doi.org/10.1111/j.1365-2109.2009.02349.x.

Hemaiswarya, S., Raja, R., Ravi Kumar, R.R., Ganesan, V. & Anbazhagan, C. (2011) Microalgae: A sustainable feed source for aquaculture. *World Journal of Microbiology and Biotechnology*, 27, 1737–1746. https://doi.org/10.1007/s11274-010-0632-z.

Hendrix Genetics (n.d.) Salmon genetics and research. https://www.landcatch.co.uk/landcatch---salmon-breeding/.

Henry, M., Gasco, L., Piccolo, G. & Fountoulaki, E. (2015) Review on the use of insects in the diet of farmed fish: Past and future. *Animal Feed Science and Technology*, 203, 1–22. https://doi.org/10.1016/j.anifeedsci.2015.03.001.

Holt, G.J. (1993) Feeding larval red drum on microparticulate diets in a closed recirculating water system. *Journal of the World Aquaculture Society*, 24, 225–230. https://doi.org/10.1111/j.1749-7345.1993.tb00011.x.

Hossain, M.A.R., Haylor, G.S. & Beveridge, M.C.M. (2001) Effect of feeding time and frequency on the growth and feed utilization of African catfish Clarias gariepinus (Burchell 1822) fingerlings. *Aquaculture Research*, 32, 999–1004. https://doi.org/10.1046/j.1365-2109.2001.00635.x.

Hou, Y., Yin, Y. & Wu, G. (2015) Dietary essentiality of 'nutritionally non-essential amino acids' for animals and humans. *Experimental Biology and Medicine*, 240, 997–1007. https://doi.org/10.1177/1535370215587913.

Houston, R.D., Haley, C.S., Hamilton, A., Guy, D.R., Mota-Velasco, J.C., Gheyas, A.A., Tinch, A.E., Taggart, J.B., Bron, J.E., Starkey, W.G., McAndrew, B.J., Verner-Jeffreys, D.W., Paley, R.K., Rimmer, G.S., Tew, I.J. & Bishop, S.C. (2010) The susceptibility of Atlantic salmon fry to freshwater infectious pancreatic necrosis is largely explained by a major QTL. *Heredity*, 105, 318–327. https://doi.org/10.1038/hdy.2009.171.

Houston, R.D., Taggart, J.B., Cézard, T., Bekaert, M., Lowe, N.R., Downing, A., Talbot, R., Bishop, S.C., Archibald, A.L., Bron, J.E., Penman, D.J., Davassi, A., Brew, F., Tinch, A.E., Gharbi, K. & Hamilton, A. (2014b) Development and validation of a high density SNP genotyping array for Atlantic salmon (Salmo salar). BMC Genomics, 15, 90. https://doi.org/10.1186/1471-2164-15-90.

Houston, R., Bishop, S.C., Woolliams, J. & Haley, C. (2014a) Marker-assisted selection to breed for resistance to infectious pancreatic necrosis in salmon. Research Excellence Framework (REF), impact case studies. impact.ref.ac.uk/CaseStudies/CaseStudy.aspx?Id=23913.

Hua, K., Cobcroft, J.M., Cole, A., Condon, K., Jerry, D.R., Mangott, A., Praeger, C., Vucko, M.J., Zeng, C., Zenger, K. & Strugnell, J.M. (2019) The future of aquatic protein: Implications for protein sources in aquaculture diets. One Earth, 1, 316–329. https://doi.org/10.1016/j.oneear.2019.10.018.

Ibeas, C., Izquierdo, M.S. & Lorenzo, A. (1994) Effect of different levels of n – 3 highly unsaturated fatty acids on growth and fatty acid composition of juvenile gilthead sea bream (Sparus aurata). Aquaculture, 127, 177–188. https://doi.org/10.1016/0044-8486(94)90424-3.

IFFO (2020) The marine ingredients organisation. A more sophisticated way of looking at fish in: fish out ratios. Available at: https://www.iffo.com/more-sophisticated-way-looking-fish-infish-out (accessed 9 May 2023).

IPIFF (2018). International Platform of Insects for Food and Feed (IPIFF). The European Insect Sector Today: Challenges, Opportunities and Regulatory Landscape. IPIFF: Brussels.

Izquierdo, M. (2005) Essential fatty acid requirements in Mediterranean fish species. In: Mediterranean Fish Nutirion, Vol. 2005 (edited by D. Montero, B. Basurco, I. Nengas, M. Alexis & M. Izquierdo). CIHEAM: Zaragoza, pp. 91–102.

Izquierdo, M.S., Domínguez, D., Jiménez, J.I., Saleh, R., Hernández-Cruz, C.M., Zamorano, M.J. & Hamre, K. (2019) Interaction between taurine, vitamin E and vitamin C in microdiets for gilthead seabream (Sparus aurata) larvae. *Aquaculture*, 498, 246–253. https://doi.org/10.1016/j.aquaculture.2018.07.010.

Jackson, A. (2012) Fish meal and fish oil and its role in sustainable aquaculture. *International Aquafeed*, September–October. Parendale Publishers Ltd: Cheltenham, pp. 18–21.

Jauncey, K. (1998). *Tilapia: Feeds and Feeding*. Pisces Press Ltd: Stirling, p. 240.

Jones, D.B., Jerry, D.R., Khatkar, M.S., Raadsma, H.W., Steen, H.V.D., Prochaska, J., Forêt, S. & Zenger, K.R. (2017) A comparative integrated gene-based linkage and locus ordering by linkage disequilibrium map for the Pacific white shrimp, litopenaeus vannamei. *Scientific Reports*, 7, 10360. https://doi.org/10.1038/s41598-017-10515-7.

Joshi, R., Arnyasi, M., Lien, S., Gjøen, H.M., Alvarez, A.T. et al., (2018) A high density linkage map for Nile tilapia. In: *Proceedings of the World Congress on Genetics Applied to Livestock Production*.

Kamlangdee, N. & Fan, K. (2003) Polyunsaturated fatty acids production by Schizochytrium sp. isolated from mangrove. *Songklanakarin Journal of Science and Technology*, 25, 643–650.

Karapanagiotidis, I.T., Psofakis, P., Mente, E., Malandrakis, E. & Golomazou, E. (2019) Effect of fishmeal replacement by poultry by-product meal on growth performance, proximate composition, digestive enzyme activity, haematological parameters and gene expression of gilthead seabream (Sparus aurata). *Aquaculture Nutrition*, 25, 3–14. https://doi.org/10.1111/anu.12824.

Kolkovski, S. (2001) Digestive enzymes in fish larvae and juveniles – implications and applications to formulated diets. *Aquaculture*, 200, 181–201. https://doi.org/10.1016/S0044-8486(01)00700-1.

Krossøy, C., Waagbø, R. & Ørnsrud, R. (2011) Vitamin K in fish nutrition. *Aquaculture Nutrition*, 17, 585–594. https://doi.org/10.1111/j.1365-2095.2011.00904.x.

Lall, S. & Milley, J. (2008) Trace mineral requirements of fish and crustaceans. In: *Trace Elements in Animal Production Systems* (edited by P. Schlegel, S. Durosoy & A. W. Jongbloed). Wageningen Academic: Wageningen.

Lall, S.P. (2002) The minerals. In: *Fish Nutrition*, 3rd edn (edited by E. H. John & W. H. Ronald). Academic Press: San Diego, pp. 259–308.

Lavens, P. & Sorgeloos, P. (1996). *Manual on the Production and Use of Live Food for Aquaculture*. FAO Technical Papers 361. Food and Agriculture Organization: Rome.

Lee, C.S. (2003) Biotechnological advances in finfish hatchery production: A review. *Aquaculture*, 227, 439–458. https://doi.org/10.1016/S0044-8486(03)00522-2.

Lee, S., Hwang, U. & Cho, S.H. (2000) Effects of feeding frequency and dietary moisture content on growth, body composition and gastric evacuation of juvenile Korean rockfish (Sebastes schlegeli). *Aquaculture*, 187, 399–409. https://doi.org/10.1016/S0044-8486(00)00318-5.

Lekang, O. (2022) *Salmon Farming*. 5m Books: Great Easton, UK.

Li, Y., Willer, C.J., Ding, J., Scheet, P. & Abecasis, G.R. (2010) Mach: Using sequence and genotype data to estimate haplotypes and unobserved genotypes. *Genetic Epidemiology*, 34, 816–834. https://doi.org/10.1002/gepi.20533.

Lien, S., Gidskehaug, L., Moen, T., Hayes, B.J., Berg, P.R., Davidson, W.S., Omholt, S.W. & Kent, M.P. (2011) A dense SNP-based linkage map for Atlantic salmon (Salmo salar) reveals extended chromosome homeologies and striking differences in sex-specific recombination patterns. *BMC Genomics*, 12, 615. https://doi.org/10.1186/1471-2164-12-615.

Lipschutz-Powell, D., Woolliams, J.A. & Doeschl-Wilson, A.B. (2014) A unifying theory for genetic epidemiological analysis of binary disease data. *Genetics, Selection, Evolution*, 46, 15. https://doi.org/10.1186/1297-9686-46-15.

Lock, E.-J., Waagbã˝, R., Wendelaar Bonga, S. & Flik, G. (2010) The significance of vitamin D for fish: A review. *Aquaculture Nutrition*, 16, 100–116. https://doi.org/10.1111/j.1365-2095.2009.00722.x.

Lough, G., Kyriazakis, I., Bergmann, S., Lengeling, A. & Doeschl-Wilson Andrea, B. (2015) Health trajectories reveal the dynamic contributions of host genetic resistance and tolerance to infection outcome. *Proceedings of the Royal Society of London. Series B*, 282, 20152151. https://doi.org/10.1098/rspb.2015.2151.

Lubzens, E. & Zmora, O. (2003) Production and nutritional value of rotifers. In: *Live Feeds in Marine Aquaculture* (edited by J. G. Støttrup & L. A. McEvoy). Blackwell Publishing Science Ltd: Oxford, pp. 17–64.

Luthada-Raswiswi R, Mukaratirwa S, O'Brien G (2021) Animal Protein Sources as a Substitute for Fishmeal in Aquaculture Diets: A Systematic Review and Meta-Analysis. *Applied Sciences*, 11(9), 3854. https://doi.org/10.3390/app11093854.

Ma, Z., Qin, J.G., Hutchinson, W., Chen, B.N. & Song, L. (2014) Responses of digestive enzymes and body lipids to weaning times in yellowtail kingfish Seriola lalandi (Valenciennes, 1833) larvae. *Aquaculture Research*, 45, 973–982. https://doi.org/10.1111/are.12039.

Macleod, I.M., Bowman, P.J., Vander Jagt, C.J., Haile-Mariam, M., Kemper, K.E., Chamberlain, A.J., Schrooten, C., Hayes, B.J. & Goddard, M.E. (2016) Exploiting biological priors and sequence variants enhances QTL discovery and genomic prediction of complex traits. *BMC Genomics*, 17, 144. https://doi.org/10.1186/s12864-016-2443-6.

Madsen, P., Su, G., Labouriau, R. & Christensen, O. (2006). *DMU – A Package for Analysing Multivariate Mixed Models. The Proceedings of the 8th World Congress on Genetics Applied to Livestock Production; Brazil.*

Makkar, H.P.S., Tran, G., Heuzé, V. & Ankers, P. (2014) State of the art on use of insects as animal feed. *Animal Feed Science and Technology*, 197, 1–33. https://doi.org/10.1016/j.anifeedsci.2014.07.008.

Mata-Sotres, J.A., Lazo, J.P. & Baron-Sevilla, B. (2015) Effect of age on weaning success in Totoaba (Totoaba macdonaldi) larval culture. *Aquaculture*, 437, 292–296. https://doi.org/10.1016/j.aquaculture.2014.11.037.

Matos, E., Dias, J., Dinis, M.T. & Silva, T.S. (2017) Sustainability vs. quality in gilthead seabream (Sparus aurata L.) farming: Are trade-offs inevitable? *Reviews in Aquaculture*, 9, 388–409. https://doi.org/10.1111/raq.12144.

Mcleod, R., Buschman, E., Arbuckle, L.D. & Skamene, E. (1995) Immunogenetics in the analysis of resistance to intracellular pathogens. *Current Opinion in Immunology*, 7, 539–552. https://doi.org/10.1016/0952-7915(95)80100-6.

Merino, G., Barange, M., Blanchard, J.L., Harle, J., Holmes, R., Allen, I., Allison, E.H., Badjeck, M.C., Dulvy, N.K., Holt, J., Jennings, S., Mullon, C. & Rodwell, L.D. (2012) Can marine fisheries and aquaculture meet fish demand from a growing human population in a changing climate? *Global Environmental Change*, 22, 795–806. https://doi.org/10.1016/j.gloenvcha.2012.03.003.

Metsoviti, M.N., Gkalogianni, E.Z., Katouni, A.M., Rougkas, G., Savvaki, E., Psofakis, P., Katsoulas, N., Papapolymerou, G. & Karapanagiotidis, I.T. (2018) Effects of replacing fish meal by Chlorella vulgaris and fish oil by Nannochloropsis gaditana and Schizochytrium sp. blend on growth and feed efficiency of gilthead sea bream (Sparus aurata). In: *Proceedings of the European Aquaculture Society and World Aquaculture Society (Aqua 2018)*, 25–29 August, Montpellier, France.

Meuwissen, T.H., Hayes, B.J. & Goddard, M.E. (2001) Prediction of total genetic value using genome-wide dense marker maps. *Genetics*, 157, 1819–1829. https://doi.org/10.1093/genetics/157.4.1819.

Monge-Ortiz, R., Martínez-Llorens, S., Márquez, L., Moyano, F.J., Jover-Cerdá, M. & Tomás-Vidal, A. (2016) Potential use of high levels of vegetal proteins in diets for market-sized gilthead sea bream (Sparus aurata). *Archives of Animal Nutrition*, 70, 155–172. https://doi.org/10.1080/1745039X.2016.1141743.

Moutinho, S., Martínez-Llorens, S., Tomás-Vidal, A., Jover-Cerdá, M., Oliva-Teles, A. & Peres, H. (2017) Meat and bone meal as partial replacement for fish meal in diets for gilthead seabream (Sparus aurata) juveniles: Growth, feed efficiency, amino acid utilization, and economic efficiency. *Aquaculture*, 468, 271–277. https://doi.org/10.1016/j.aquaculture.2016.10.024.

Muller-Feuga, A. (2000) The role of microalgae in aquaculture: Situation and trends. *Journal of Applied Phycology*, 12, 527–534. https://doi.org/10.1023/A:1008106304417.

Muller-Feuga, A., Moal, J. & Kaas, R. (2003) The microalgae of aquaculture. In: *Live Feeds in Marine Aquaculture* (edited by J. G. Støttrup & L. A. McEvoy). Blackwell Publishing Science Ltd: Oxford, pp. 206–252.

Müller, M. & Kersten, S. (2003) Nutrigenomics: Goals and strategies. *Nature Reviews. Genetics*, 4, 315–322. https://doi.org/10.1038/nrg1047.

Nagappan, S., Das, P., Quadir, M.A., Thaher, M., Khan, S, Mahata, C., Al-Jabri, H, Vatland, A.K. & Kumar, G. (2021) Potential of microalgae as a sustainable feed ingredient for aquaculture, *Journal of Biotechnology*, 341, 1-20, https://doi.org/10.1016/j.jbiotec.2021.09.003.

Navarrete, P. & Tovar-Ramírez, D. (2014) Use of yeasts as probiotics in fish aquaculture. *Sustainable Aquaculture Techniques, Hernandez-Vergara MP and Perez-Rostro CI.* https://www.intechopen.com/books/sustainable-aquaculture-techniques/use-of-yeasts-as-probiotics-in-fish-aquaculture. https://doi.org/10.5772/57196.

Naylor, R.L., Goldburg, R.J., Primavera, J.H., Kautsky, N., Beveridge, M.C.M., Clay, J., Folke, C., Lubchenco, J., Mooney, H. & Troell, M. (2000) Effect of aquaculture on world fish supplies. *Nature*, 405, 1017–1024. https://doi.org/10.1038/35016500.

Neely, K.G., Myers, J.M., Hard, J.J. & Shearer, K.D. (2008) Comparison of growth, feed intake, and nutrient efficiency in a selected strain of coho salmon (Oncorhynchus kisutch) and its source stock. *Aquaculture*, 283, 134–140. https://doi.org/10.1016/j.aquaculture.2008.06.038.

Nevejan, N., De Schryver, P., Wille, M., Dierckens, K., Baruah, K. & Van Stappen, G. (2018) Bacteria as food in aquaculture: Do they make a difference? *Reviews in Aquaculture*, 10, 180–212. https://doi.org/10.1111/raq.12155.

Nogales-Mérida, S., Gobbi, P., Józefiak, D., Mazurkiewicz, J., Dudek, K., Rawski, M., Kierończyk, B. & Józefiak, A. (2019) Insect meals in fish nutrition. *Reviews in Aquaculture*, 11, 1080–1103. https://doi.org/10.1111/raq.12281.

NRC (National Research Council) (1993). *Nutrient Requirements of Fish*. The National Academies Press: Washington, DC, USA. https://doi.org/10.17226/2115.

NRC (National Research Council) (2011). *Nutrient Requirements of Fish and Shrimp*. The National Academies Press: Washington, DC, USA. https://doi.org/10.17226/13039.

Ødegård, J., Olesen, I., Gjerde, B. & Klemetsdal, G. (2007a) Evaluation of statistical models for genetic analysis of challenge-test data on ISA resistance in Atlantic salmon (Salmo salar): Prediction of progeny survival. *Aquaculture*, 266, 70–76. https://doi.org/10.1016/j.aquaculture.2007.02.012.

Ødegård, J., Olesen, I., Gjerde, B. & Klemetsdal, G. (2007b) Positive genetic correlation between resistance to bacterial (furunculosis) and viral (infectious salmon anaemia) diseases in farmed Atlantic salmon (Salmo salar). *Aquaculture*, 271, 173–177. https://doi.org/10.1016/j.aquaculture.2007.06.006.

Ødegård, J., Moen, T., Santi, N., Korsvoll, S.A., Kjøglum, S. & Meuwissen, T.H. (2014) Genomic prediction in an admixed population of Atlantic salmon (Salmo salar). *Frontiers in Genetics*, 5, 402. https://doi.org/10.3389/fgene.2014.00402.

Oh, S.Y. & Maran, B.A.V. (2015) Feeding frequency influences growth, feed consumption and body composition of juvenile rock bream (Oplegnathus fasciatus). *Aquaculture International*, 23, 175–184. https://doi.org/10.1007/s10499-014-9806-2.

Oh, S.Y., Maran, B.A.V. & Park, J.W. (2019) Optimum feeding frequency for juvenile short barbeled velvetchin Hapalogenys nigripinnis reared in floating sea cages. *Fisheries Science*, 85, 379–385. https://doi.org/10.1007/s12562-019-01288-1.

Palti, Y., Gao, G., Liu, S., Kent, M.P., Lien, S., Miller, M.R., Rexroad, C.E. & Moen, T. (2015) The development and characterization of a 57K single nucleotide polymorphism array for rainbow trout. *Molecular Ecology Resources*, 15, 662–672. https://doi.org/10.1111/1755-0998.12337.

Panserat, S. & Kaushik, S.J. (2010) Regulation of gene expression by nutritional factors in fish: Review article. *Aquaculture Research*, 41, 751762. https://doi.org/10.1111/j.1365-2109.2009.02173.x.

Peres, H. & Oliva-Teles, A. (2006) Effect of the dietary essential to non-essential amino acid ratio on growth, feed utilization and nitrogen metabolism of European sea bass (Dicentrarchus labrax). *Aquaculture*, 256, 395–402. https://doi.org/10.1016/j.aquaculture.2006.02.010.

Prchal, M., Kause, A., Vandeputte, M., Gela, D., Allamellou, J.M., Kumar, G., Bestin, A., Bugeon, J., Zhao, J. & Kocour, M. (2018) The genetics of overwintering performance in two-year old common carp and its relation to performance until market size. *PLOS ONE*, 13, e0191624. https://doi.org/10.1371/journal.pone.0191624.

Price, C., Black, K.D., Hargrave, B.T. & Morris, Jr, J.A. (2015) Marine cage culture and the environment: Effects on water quality and primary production. *Aquaculture Environment Interactions*, 6, 151–174. https://doi.org/10.3354/aei00122.

Psofakis, P., Karapanagiotidis, I.T., Malandrakis, E.E., Golomazou, E., Exadactylos, A. & Mente, E. (2020) Effect of fishmeal replacement by hydrolyzed feather meal on growth performance, proximate composition, digestive enzyme activity, haematological parameters and growth-related gene expression of gilthead seabream (*Sparus aurata*). *Aquaculture*, 521, 735006. https://doi.org/10.1016/j.aquaculture.2020.735006.

Rahman, M.M., Varga, I. & Chowdhury, S.N. (1992) Manual on polyculture and integrated fish farming in Bangladesh. *Field*, document BGD/87/045/91/11, FAO of the UN & United Nations Development Programme Department of fisheries. Government of Bangladesh. www.fao.org/3/AC375E/AC375E00.htm#TOC.

Rainuzzo, J.R., Reitan, K.I. & Olsen, Y. (1997) The significance of lipids at early stages of marine fish: A review. *Aquaculture*, 155, 103–115. https://doi.org/10.1016/S0044-8486(97)00121-X.

Rasdi, N.W. & Qin, J.G. (2016) Improvement of copepod nutritional quality as live food for aquaculture: A review. *Aquaculture Research*, 47, 1–20. https://doi.org/10.1111/are.12471.

Robledo, D., Matika, O., Hamilton, A. & Houston, R.D. (2018a) Genome-wide association and genomic selection for resistance to amoebic gill disease in Atlantic salmon. *G3*, 8, 1195–1203. https://doi.org/10.1534/g3.118.200075.

Robledo, D., Palaiokostas, C., Bargelloni, L., Martínez, P. & Houston, R. (2018b) Applications of genotyping by sequencing in aquaculture breeding and genetics. *Reviews in Aquaculture*, 10, 670–682. https://doi.org/10.1111/raq.12193.

Rosenlund, G., Stoss, J. & Talbot, C. (1997) Co-feeding marine fish larvae with inert and live diets. *Aquaculture*, 155, 183–191. https://doi.org/10.1016/S0044-8486(97)00116-6.

Ruiz, M.A., Hernández-Cruz, C.M., Caballero, M.J., Fernández-Palacios, H., Saleh, R., Izquierdo, M.S. & Betancor, M.B. (2019) Appearance of systemic granulomatosis is modulated by the dietary supplementation of vitamin E and C in meagre (Argyrosomus regius) larvae fed inert microdiets. *Aquaculture*, 506, 139–147. https://doi.org/10.1016/j.aquaculture.2019.03.032.

Sae-Lim, P., Komen, H. & Kause, A. (2010) Bias and precision of estimates of genotype-by-environment interaction: A simulation study. *Aquaculture*, 310, 66–73. https://doi.org/10.1016/j.aquaculture.2010.10.020.

Sae-Lim, P., Gjerde, B., Nielsen, H.M., Mulder, H. & Kause, A. (2016) A review of genotype-by-environment interaction and micro-environmental sensitivity in aquaculture species. *Reviews in Aquaculture*, 8, 369–393. https://doi.org/10.1111/raq.12098.

Sae-Lim, P., Kause, A., Mulder, H.A. & Olesen, I. (2017) Breeding and genetics symposium: Climate change and selective breeding in aquaculture. *Journal of Animal Science*, 95, 1801–1812. https://doi.org/10.2527/jas.2016.1066.

Sánchez-Muros, M.-J., Barroso, F.G. & Manzano-Agugliaro, F. (2014) Insect meal as renewable source of food for animal feeding: A review. *Journal of Cleaner Production*, 65, 16–27. https://doi.org/10.1016/j.jclepro.2013.11.068.

Sanna, S., Jackson, A.U., Nagaraja, R., Willer, C.J., Chen, W.M., Bonnycastle, L.L., Shen, H., Timpson, N., Lettre, G., Usala, G., Chines, P.S., Stringham, H.M., Scott, L.J., Dei, M., Lai, S., Albai, G., Crisponi, L., Naitza, S., Doheny, K.F., Pugh, E.W., Ben-Shlomo, Y., Ebrahim, S., Lawlor, D.A., Bergman, R.N., Watanabe, R.M., Uda, M., Tuomilehto, J., Coresh, J., Hirschhorn, J.N., Shuldiner, A.R., Schlessinger, D., Collins, F.S., Davey Smith, G., Boerwinkle, E., Cao, A., Boehnke, M., Abecasis, G.R. & Mohlke, K.L. (2008) Common variants in the GDF5-UQCC region are associated with variation in human height. *Nature Genetics*, 40, 198–203. https://doi.org/10.1038/ng.74.

Sargent, J.R., Henderson, R.J. & Tocher, D.R. (1989) The lipids. In: *Fish Nutrition* (edited by J. Halver & R. Hardy). Interscience Publishers: New York, USA, 153–218.

Sargent, J.R., Bell, J.G., Bell, M.V., Henderson, R.J. & Tocher, D.R. (1995) Requirement criteria for essential fatty acids. *Journal of Applied Ichthyology*, 11, 183–198. https://doi.org/10.1111/j.1439-0426.1995.tb00018.x.

Sargent, J.R., McEvoy, L.A. & Bell, J.G. (1997) Requirements, presentation and sources of polyunsaturated fatty acids in marine fish larval feeds. *Aquaculture*, 155, 117–127. https://doi.org/10.1016/S0044-8486(97)00122-1.

Sargent, J.R., Bell, G., McEvoy, L.A., Tocher, D.R. & Estèvez, A. (1999a) Recent developments in the essential fatty acid nutrition of fish. *Aquaculture*, 177, 191–199. https://doi.org/10.1016/S0044-8486(99)00083-6.

Sargent, J., McEvoy, L., Estevez, A., Bell, G., Bell, M., Henderson, J. & Tocher, D. (1999b) Lipid nutrition of marine fish during early development: Current status and future directions. *Aquaculture*, 179, 217–229. https://doi.org/10.1016/S0044-8486(99)00191-X.

Sargent, J.R., Tocher, D.R. & Bell, J.G. (2002) The lipids. In: *Fish Nutrition* (edited by J. Halver & R. Hardy). Elsevier: San Diego, USA, pp. 181–257.

Sarker, P.K., Kapuscinski, A.R., Bae, A.Y., Donaldson, E., Sitek, A.J., Fitzgerald, D.S. & Edelson, O.F. (2018) Towards sustainable aquafeeds: Evaluating substitution of fishmeal with lipid-extracted microalgal co-product (Nannochloropsis oculata) in diets of juvenile nile tilapia (Oreochromis niloticus). *PLOS ONE*, 13, e0201315. https://doi.org/10.1371/journal.pone.0201315.

Sevatdal, S. & Horsberg, T.E. (2003) Determination of reduced sensitivity in sea lice (Lepeophtheirus salmonis KrØyer) against the pyrethroid deltamethrin using bioassays and probit modelling. *Aquaculture*, 218, 21–31. https://doi.org/10.1016/S0044-8486(02)00339-3.

Shalem, O., Sanjana, N.E., Hartenian, E., Shi, X., Scott, D.A., Mikkelson, T., Heckl, D., Ebert, B.L., Root, D.E., Doench, J.G. & Zhang, F. (2014) Genome-scale CRISPR-Cas9 knockout screening in human cells. *Science*, 343, 84–87. https://doi.org/10.1126/science.1247005.

Shields, R.J. (2001) Larviculture of marine finfish in Europe. *Aquaculture*, 200, 55–88. https://doi.org/10.1016/S0044-8486(01)00694-9.

Shields, R.J. & Lupatsch, I. (2012) Algae for aquaculture and animal feeds. Technol. Assess. *Theory into Practice*, 21, 23–37.

Shimeno, S., Kheyyali, D. & Shikata, T. (1995) Metabolic response to dietary carbohydrate to protein ratios in carp. *Fisheries Science*, 61, 277–281. https://doi.org/10.2331/fishsci.61.277.

Shumway, S. (2021) *Molluscan Shellfish Aquaculture: A Practical Guide*. 5m Books Ltd., Nottingham, UK.

Silva, C.R., Gomes, L.C. & Brandão, F.R. (2007) Effect of feeding rate and frequency on tambaqui (Colossoma macropomum) growth, production and feeding costs during the first growth phase in cages. *Aquaculture*, 264, 135–139. https://doi.org/10.1016/j.aquaculture.2006.12.007.

Sorgeloos, P., Dhert, P. & Candreva, P. (2001) Use of the brine shrimp, artemia spp., in marine fish larviculture. *Aquaculture*, 200, 147–159. https://doi.org/10.1016/S0044-8486(01)00698-6.

Steffens, W. (1989) *Principles of Fish Nutrition*. Ellis Horwood Limited: Chichester.

Tacon, A.G.J. (1988) *The Nutrition and Feeding of Farmed Fish and Shrimp- A Training Manual. 3. Feeding Methods*. FAO Field Document No. 7, Brazil.

Tacon, A.G.J. & Metian, M. (2008) Global overview on the use of fish meal and fish oil in industrially compounded aquafeeds: Trends and future prospects. *Aquaculture*, 285, 146–158. https://doi.org/10.1016/j.aquaculture.2008.08.015.

Tacon, A.G.J. & Metian, M. (2015) Feed matters: Satisfying the feed demand of aquaculture. *Reviews in Fisheries Science and Aquaculture*, 23, 1–10. https://doi.org/10.1080/23308249.2014.987209.

Tacon, A. and Tran, L. (2022) *Nutritional Fish and Shrimp Pathology: A Handbook,* 5m Books Ltd., Nottingham, UK.

Talbot, C. & Hole, R. (1994) Fish diets and the control of eutrophication resulting from aquaculture. *Journal of Applied Ichthyology*, 10, 258–270. https://doi.org/10.1111/j.1439-0426.1994.tb00165.x.

Teshima, S., Ishikawa, M. & Koshio, S. (2000) Nutritional assessment and feed intake of microparticulate diets in crustaceans and fish. *Aquaculture Research*, 31, 691–702. https://doi.org/10.1046/j.1365-2109.2000.318490.x.

Thodesen, J., Grisdale-Helland, B., Helland, S.J. & Gjerde, B. (1999) Feed intake, growth and feed utilization of offspring from wild and selected Atlantic salmon (Salmo salar). *Aquaculture*, 180, 237–246. https://doi.org/10.1016/S0044-8486(99)00204-5.

Tiezzi, F. & Maltecca, C. (2015) Accounting for trait architecture in genomic predictions of US Holstein cattle using a weighted realized relationship matrix. *Genetics, Selection, Evolution*, 47, 24. https://doi.org/10.1186/s12711-015-0100-1.

Tocher, D.R. (2003) Metabolism and functions of lipids and fatty acids in teleost fish. *Reviews in Fisheries Science*, 11, 107–184. https://doi.org/10.1080/713610925.

Tran, G., Heuzè, V. & Makkas, H. (2015) Insect in fish diets. *Animal Frontiers*, 5, 37–44.

Treasurer, J., Bricknell, I. & Bron, J. (2022) *Sealice Biology and Control*. 5m Books: Great Easton, UK.

Troell, M., Naylor, R.L., Metian, M., Beveridge, M., Tyedmers, P.H., Folke, C., Arrow, K.J., Barrett, S., Crépin, A.S., Ehrlich, P.R., Gren, A., Kautsky, N., Levin, S.A., Nyborg, K., Österblom, H., Polasky, S., Scheffer, M., Walker, B.H., Xepapadeas, T. & de Zeeuw, A. (2014) Does aquaculture add resilience to the global food system? *Proceedings of the National Academy of Sciences of the United States of America*, 111, 13257–13263. https://doi.org/10.1073/pnas.1404067111.

Tsai, H.Y., Hamilton, A., Tinch, A.E., Guy, D.R., Gharbi, K., Stear, M.J., Matika, O., Bishop, S.C. & Houston, R.D. (2015) Genome wide association and genomic prediction for growth traits in juvenile farmed Atlantic salmon using a high density SNP array. *BMC Genomics*, 16, 969. https://doi.org/10.1186/s12864-015-2117-9.

Tsai, H.Y., Hamilton, A., Tinch, A.E., Guy, D.R., Bron, J.E., Taggart, J.B., Gharbi, K., Stear, M., Matika, O., Pong-Wong, R., Bishop, S.C. & Houston, R.D. (2016) Genomic prediction of host resistance to sea lice in farmed Atlantic salmon populations. *Genetics, Selection, Evolution*, 48, 47. https://doi.org/10.1186/s12711-016-0226-9.

Tsai, H.-Y., Matika, O., Edwards, S.M., Antolín-Sánchez, R., Hamilton, A. et al., (2017) Genotype imputation to improve the cost-efficiency of genomic selection in farmed Atlantic salmon. *G3: Genes|Genomes|Genetics*, 7, 1377–1383.

Tsairidou, S., Anacleto, O., Woolliams, J.A. & Doeschl-Wilson, A. (2019a) Enhancing genetic disease control by selecting for lower host infectivity and susceptibility. *Heredity*, 122, 742–758. https://doi.org/10.1038/s41437-018-0176-9.

Tsairidou, S., Hamilton, A., Robledo, D., Bron, J.E. & Houston, R.D. (2019b) Optimizing low-cost genotyping and imputation strategies for genomic selection in Atlantic salmon. *G3: Genes|Genomes|Genetics*, 10(1), 581–590. https://doi.org/10.1534/g3.119.400800.

Tucker, B.J., Booth, M.A., Allan, G.L., Booth, D. & Fielder, D.S. (2006) Effects of photoperiod and feeding frequency on performance of newly weaned Australian snapper Pagrus auratus. *Aquaculture*, 258, 514–520. https://doi.org/10.1016/j.aquaculture.2006.03.033.

Tully, O. & Mcfadden, Y. (2000) Variation in sensitivity of sea lice [Lepeophtheirus salmonis (Krøyer)] to dichlorvos on Irish salmon farms in 1991–92. *Aquaculture Research*, 31, 849–854. https://doi.org/10.1046/j.1365-2109.2000.00518.x.

Turchini, G.M., Torstensen, B.E. & Ng, W.-K. (2009) Fish oil replacement in finfish nutrition. *Reviews in Aquaculture*, 1, 10–57. https://doi.org/10.1111/j.1753-5131.2008.01001.x.

Van Huis, A. (2020) Insects as food and feed, a new emerging agricultural sector: A review. *Journal of Insects as Food and Feed*, 6, 27–44. https://doi.org/10.3920/JIFF2019.0017.

Vanraden, P.M. (2008) Efficient methods to compute genomic predictions. *Journal of Dairy Science*, 91, 4414–4423. https://doi.org/10.3168/jds.2007-0980.

Vizcaíno, A.J., López, G., Sáez, M.I., Jiménez, J.A., Barros, A., Hidalgo, L., Camacho-Rodríguez, J., Martínez, T.F., Cerón-García, M.C. & Alarcón, F.J. (2014) Effects of the microalga Scenedesmus almeriensis as fishmeal alternative in diets for gilthead sea bream, Sparus aurata, juveniles. *Aquaculture*, 431, 34–43. https://doi.org/10.1016/j.aquaculture.2014.05.010.

Vuori, K., Strandén, I., Lidauer, M. & Mäntysaari, E. (2006) MiX99-effective solver for large and complex linear models. In: *Proceedings of the 8th World Congr. Genet. Appl. Livest.* Prod. (WCGALP), 27–33.

Wargelius, A., Leininger, S., Skaftnesmo, K.O., Kleppe, L., Andersson, E., Taranger, G.L., Schulz, R.W. & Edvardsen, R.B. (2016) Dnd knockout ablates germ cells and demonstrates germ cell independent sex differentiation in Atlantic salmon. *Scientific Reports*, 6, 21284. https://doi.org/10.1038/srep21284.

WHO (2019) *World Health Organization: Global and Regional Food Consumption Patterns and Trends; Availability and Consumption of Fish*. https://www.who.int/nutrition/topics/3_foodconsumption/en/index5.html.

Wiener, G., Lee, C.J. & Woolliams, J.A. (1994) Consequences of inbreeding for financial returns from sheep. *Animal Science*, 59, 245–249. https://doi.org/10.1017/S000335610000773X.

Wiener, G., Lee, G.J. & Woolliams, J.A. (1992) Effects of rapid inbreeding and of crossing of inbred lines on conception rate, prolificacy and ewe survival in sheep. *Animal Science*, 55, 115–121. https://doi.org/10.1017/S0003356100037338.

Willer, C.J., Sanna, S., Jackson, A.U., Scuteri, A., Bonnycastle, L.L., Clarke, R., Heath, S.C., Timpson, N.J., Najjar, S.S., Stringham, H.M., Strait, J., Duren, W.L., Maschio, A., Busonero, F., Mulas, A., Albai, G., Swift, A.J., Morken, M.A., Narisu, N., Bennett, D., Parish, S., Shen, H., Galan, P., Meneton, P., Hercberg, S., Zelenika, D., Chen, W.M., Li, Y., Scott, L.J., Scheet, P.A., Sundvall, J., Watanabe, R.M., Nagaraja, R., Ebrahim, S., Lawlor, D.A., Ben-Shlomo, Y., Davey-Smith, G., Shuldiner, A.R., Collins, R., Bergman, R.N., Uda, M., Tuomilehto, J., Cao, A., Collins, F.S., Lakatta, E., Lathrop, G.M., Boehnke, M., Schlessinger, D., Mohlke, K.L. & Abecasis, G.R. (2008) Newly identified loci that influence lipid concentrations and risk of coronary artery disease. *Nature Genetics*, 40, 161–169. https://doi.org/10.1038/ng.76.

Woolliams, J.A. (2007) Genetic contributions and inbreeding. In: *Utilization and Conservation of Farm Animal Genetic Resources* (edited by K. Oldenbroek). Wageningen Academic: Wageningen.

Woolliams, J.A., Berg, P., Dagnachew, B.S. & Meuwissen, T.H.E. (2015) Genetic contributions and their optimization. *Journal of Animal Breeding and Genetics*, 132, 89–99. https://doi.org/10.1111/jbg.12148.

Xie, F., Ai, Q., Mai, K., Xu, W. & Ma, H. (2011) The optimal feeding frequency of large yellow croaker (Pseudosciaena crocea, Richardson) larvae. *Aquaculture*, 311, 162–167. https://doi.org/10.1016/j.aquaculture.2010.12.005.

Xu, J., Zhao, Z., Zhang, X., Zheng, X., Li, J., Jiang, Y., Kuang, Y., Zhang, Y., Feng, J., Li, C., Yu, J., Li, Q., Zhu, Y., Liu, Y., Xu, P. & Sun, X. (2014) Development and evaluation of the first high-throughput SNP array for common carp (Cyprinus carpio). *BMC Genomics*, 15, 307. https://doi.org/10.1186/1471-2164-15-307.

Yáñez, J.M., Houston, R.D. & Newman, S. (2014) Genetics and genomics of disease resistance in salmonid species. *Frontiers in Genetics*, 5, 415. https://doi.org/10.3389/fgene.2014.00415.

Yáñez, J.M., Naswa, S., López, M.E., Bassini, L., Correa, K., Gilbey, J., Bernatchez, L., Norris, A., Neira, R., Lhorente, J.P., Schnable, P.S., Newman, S., Mileham, A., Deeb, N., Di Genova, A. & Maass, A. (2016) Genomewide single nucleotide polymorphism discovery in Atlantic salmon (Salmo salar): Validation in wild and farmed American and European populations. *Molecular Ecology Resources*, 16, 1002–1011. https://doi.org/10.1111/1755-0998.12503.

Yoshida, G.M., Carvalheiro, R., Lhorente, J.P., Correa, K., Figueroa, R., Houston, R.D. & Yáñez, J.M. (2018) Accuracy of genotype imputation and genomic predictions in a two-generation farmed Atlantic salmon population using high-density and low-density SNP panels. *Aquaculture*, 491, 147–154. https://doi.org/10.1016/j.aquaculture.2018.03.004.

Yoshida, G.M., Lhorente, J.P., Correa, K., Soto, J., Salas, D. & Yáñez, J.M. (2019) Genome-wide association study and cost-efficient genomic predictions for growth and fillet yield in nile tilapia (Oreochromis niloticus). *G3*, 9, 2597–2607. https://doi.org/10.1534/g3.119.400116.

Zenger, K.R., Khatkar, M.S., Jones, D.B., Khalilisamani, N., Jerry, D.R. & Raadsma, H.W. (2018) Genomic selection in aquaculture: Application, limitations and opportunities with special reference to marine shrimp and pearl oysters. *Frontiers in Genetics*, 9, 693. https://doi.org/10.3389/fgene.2018.00693.

Zhang, Y., Øverland, M., Shearer, K.D., Sørensen, M., Mydland, L.T. & Storebakken, T. (2012) Optimizing plant protein combinations in fish meal-free diets for rainbow trout (Oncorhynchus mykiss) by a mixture model. *Aquaculture*, 360–361, 25–36. https://doi.org/10.1016/j.aquaculture.2012.07.003.

Zhao, W., Lu, L. & Tang, Y. (2010) Research and application progress of insect antimicrobial peptides on food industry. *International Journal of Food Engineering*, 6, article 10. https://doi.org/10.2202/1556-3758.1943.

Zheng, X., Kuang, Y., Lv, W., Cao, D., Sun, Z. & Sun, X. (2016) Genome-wide association study for muscle fat content and abdominal fat traits in common carp (Cyprinus carpio). *PLOS ONE*, 11, e0169127. https://doi.org/10.1371/journal.pone.0169127.

4

Aquaculture production now and in the future – an ecosystem perspective

Trevor Telfer, Lynne Falconer and Malcolm Beveridge

Introduction

Aquaculture can be considered as an ecologically open system; meaning it is heavily reliant on the ecosystem providing the resources needed to maintain it (environmental goods) and the functions performed by the ecosystem to offset its effects (environmental services). Aquatic farming systems adapt and exploit these goods and services to provide food for humans, but to ensure long-term sustainability this ecosystem-based management approach should be achieved within the capacity of the environment that provides and sustains them.

The Food and Agriculture Organization (FAO) of the United Nations employs a Code of Conduct of Responsible Fisheries (CCRF), which promotes an ecosystem-based management approach to sustainable development of seafood production. As part of this the following should be considered.

1. Aquaculture should be developed in the context of ecosystem functions and services ensuring there is no degradation of these beyond their resilience capacity.
2. Aquaculture should improve human well-being and equity for all relevant stakeholders.
3. Aquaculture should be developed in the context of (and integrated to) other relevant sectors. These can be implemented at three scales; farm, water body and watershed or aquaculture zone. This approach takes into consideration the ecosystem perspective and sustainability of goods and services for aquaculture systems in the changing environment. (Brugère et al., 2018)

Figure 4.1 presents a simplified way of thinking about the ecosystem perspective of aquaculture and its utilization of environmental good and services. As can be seen, and with all food sectors, aquaculture utilizes resources provided by the environment, while at the same time relies on the environment to assimilate nutrient, chemical and biological wastes. Use of such aquaculture systems in a changing environment means that there may be issues affecting provision of environmental goods and services. At the same time the industry is evolving, and new aquaculture technology is being developed, allowing previously unavailable locations to be used for aquaculture while maximizing resource use in existing locations.

Margaret Crumlish and Rachel Norman (eds) **Aquatic Food Security**
DOI: 10.52517/9781789183368.004, © 5m Books Ltd, 2023

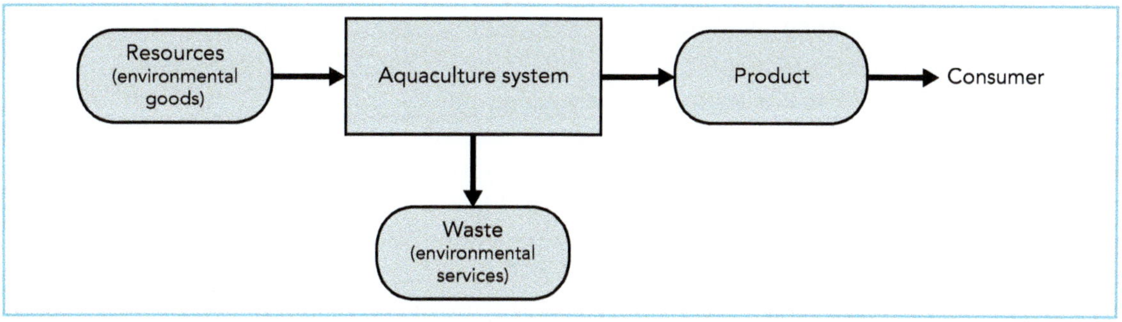

Figure 4.1 The resources required for aquaculture from an ecosystem perspective.

This chapter will investigate and discuss the concepts of environmental goods and services related to sustainable aquaculture development in the context of a changing environment and technologies for the future.

Resource use and the aquaculture process (environmental goods)

Aquaculture exploits three key environmental goods: space, water and feed. Efficient use of these resources helps enhance its sustainability.

Space

All enterprises require space, but aquaculture production requires space in both land and water environments. For example, coastal fish farms require water-based space for the net-pens and land-based space to locate piers, slipways, storage sheds and offices. For pond culture land is required on which to build fish or shrimp ponds. Aquaculture is often competing with other users for space and this can lead to conflict, especially in locations where space and/or natural resources are limited.

Inland pond-based aquaculture often requires a lot of space. Extensive aquaculture uses much more land per unit production than intensive aquaculture and often more than agriculture per kilogram of production. This is an interesting comparison as most inland extensive or semi-intensive aquaculture is associated or in conflict with agriculture, meaning that though they can complement each other, there may be considerable reduction in area available for crops. There are advantages and disadvantages to relationships between aquaculture and agriculture. A benefit is that aquaculture/agriculture systems can increase habitat and landscape biodiversity, creating a new wetland or enhancing an existing wetland to provide multiple habitats. However, use of space for aquaculture, particularly in coastal fringe regions has also caused ecological problems in some countries by supplanting mangroves as the key coastal habitat.

Mangroves are small trees or shrubs growing in coastal saline or brackish waters in tropical or sub-tropical regions. These are highly productive areas, second only to coral reefs, which are of major ecosystem significance (Hamilton, 2013). Mangrove environments have been exploited for many purposes over a long period of time, aquaculture being one the most significant. For example, it is estimated that shrimp farming is responsible for around one-quarter of the mangrove destruction that has taken place since 1980, about one-fifth of the total mangrove areas in the world have been lost in that time (Hamilton, 2013).

More recently, mangrove destruction has decreased due to conservation measures and our greater understanding of their ecological and ecosystem importance. An often-asked question when considering the destruction of mangroves is, why and at what cost? Investment in aquaculture and other coastal activities, such as tourism, has meant coastal development has taken place very quickly, providing rapid profits, with estimates of the direct value of the supplanting of mangroves lying between US$730 and US$16,870 per hectare (Hamilton, 2013). However, the true indirect ecological value of mangroves in supporting fisheries, as source of fuel, in coastal protection and as nursery areas for wild fish is often not considered. Mangroves have been found to be essential to the success of local fisheries for species such as crab, fish, shellfish and shrimp (Carrasquilla-Henao and Juanes, 2017). More specifically, in Indonesia local shrimp catches have been shown to be directly associated with the area of mangroves near to the fishing grounds (Dwipongoo, 1986) (Figure 4.2).

Use of space on land for pond culture can result in excess levels of nutrients entering the soil leading to anaerobic conditions and so to acidification of the soil. This causes lowered productivity, which degrades soils and affects its response to pond fertilizers in the future. The acid conditions can also release toxic heavy metals (e.g. zinc, aluminium and manganese), which are normally complexed with organic particulate in sediments and therefore not bioavailable, causing potential toxicity to fish, shrimp and other organisms. It also may make surrounding soil unusable for agriculture in the future.

Salinization is another form of land degradation caused by pond farming that uses brackish or sea waters. This can make the adjacent land 'salty' making it unsuitable for agriculture and having the potential to contaminate drinking and irrigation water (Brattan and Flaherty, 2001). Recent studies though have suggested that well-managed shrimp ponds only lead to highly localized and limited salinization effects (Ayers et al., 2017).

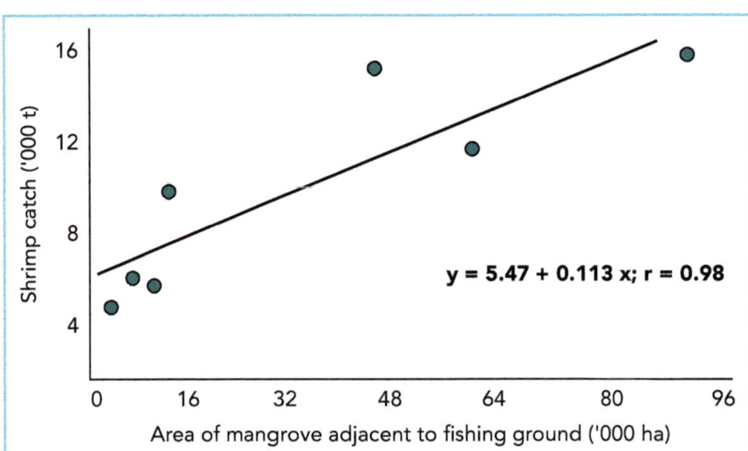

Figure 4.2 Relationship between mangrove area and commercial shrimp capture in Indonesia (redrawn after Dwipongoo, 1986).

Water

Water has economic value, which is a function of the amount needed against the value of the commodity produced (Muir and Beveridge, 1987). The economic value of water used for aquaculture production can therefore be compared against other industries that require water. As an example, Table 4.1 shows differences in amount of water used by different industries and its comparative value in 1992.

Apart from recirculating aquaculture systems (RAS), aquaculture uses large amounts of water to physically support the animals or plants grown, to supply dissolved oxygen for respiration and to remove or dilute any wastes that may be developed from the production process. The volume of water used is determined by the species being grown, the life cycle stage of the species and the intensity of the aquaculture operation. In aquaculture, demand for water can also change over time, either annually, seasonally or even daily, leading to supply problems in areas where water is limited or where seasonal requirements vary.

Aquaculture sometimes uses rather than 'borrows' water, rendering it unusable for other purposes in the immediate locality. For example, a sub-tropical shrimp pond can lose about one-fifth of its water per day through ground seepage and evaporation (Funge-Smith and Briggs, 1998). This means a pond of 1 ha area and 1 m depth could lose 1.8 million litres of water per day. Such losses and usage of water can lead to conflicts and competing demands, most often with agriculture. It can also drive aquaculture intensification in which higher stocking densities are grown in less water and ponds are redesigned for water conservation. While this leads to greater efficiency of water use, there could be environmental resource implications related to amount of waste discharged and the potential for disease transmission (Phuoc et al., 2020).

Food

The provision of feed in aquaculture production is dependent on the system used: extensive, semi-intensive, or intensive aquaculture. Extensive aquaculture normally uses natural sources of food directly from the environment (e.g. mussel production). If carefully managed this can be a highly sustainable form of culture, however, over production can lead to overexploitation

Table 4.1 Water requirements and value for industry and aquaculture. Compiled from Muir and Beveridge (1987) and Phillips et al. (1991).

Product	Water use (m^3 t^{-1} production)	Water value (US\$ m^{-3} water)
Alcohol	125–170	12–16
Cotton	90–450	2.2–11
Paper	9–450	0.7–33
Steel	8–250	0.8–250
Beef	42	48
Petrol	21.6–810	0.6–23
Shrimp ponds	11–15,000	0.1–1.1
Salmonid in cages	252,000	0.006–0.018
Channel catfish	6470	0.25
Clarias catfish	50–200	5–20

of the food resource. For example, in the Ria de Arosa in Spain, one of Europe's largest mussel production areas, grazing of phytoplankton and suspended particulates has led to not only depletion of food for the mussels but to changes in food web character (Outeiro et al., 2018). This has implications for the ecosystem, nutrient transfer through different trophic levels in the food web and aquatic productivity. Semi-intensive aquaculture systems, which include most carp and tilapia culture in ponds, rely on a mixture of natural food and supplementary foods given by the farmer. The latter may simply be fertilizer to stimulate food production within the ponds or artificial diets. These systems are not normally considered an issue for environmental sustainability when operated at low production levels. However, development funding has often encouraged higher levels of production and intensification of these systems, which then require more food and fertilizer. In intensive aquaculture, organisms are cultured in large quantities and all food comes from largely artificial diets. This is often practised for higher value products in highly managed systems, for example, salmon farming in coastal net-pens. In these systems, higher stocking densities may be used, and culture practice is completely reliant on farmer supplied feed.

One of the main sources of protein and lipid for these artificial diets, fish meal and oil, can come from marine capture fisheries leading to a situation where a number of tonnes of lower trophic fish may be used to make food for a fewer number of tonnes of higher trophic level (piscivorous) fish (Naylor et al., 2000). Opinion is divided as to the relative amounts and there is often debate, even between scientists, on the true implications of this for the sustainability of marine resources. Substitution of protein and lipid from terrestrial plant sources has increased recently and now this provides a significant amount of these ingredients, for example, in salmonid feeds (see Aas et al., 2019), making it less reliant on capture fisheries (see Chapter 3).

Aquaculture waste and their effects

The definition of waste is 'that which is left over after use or thrown away because it is not wanted'. In aquaculture this means what is left of the inputs to production that are not removed (i.e. at harvest) as all eventually find their way into the wider environment (Beveridge, 2004). Here we address nutrient-based wastes, which can enter the environment. These can come from uneaten feed, faecal material and excretory wastes (Figure 4.3). All contain three primary nutrients (carbon, nitrogen and phosphorus), which can affect the environment and wider ecosystem.

Wastes can be both particulate (solids) or soluble (liquids). The former tends to fall through the water column and be incorporated into sediments, whereas the latter tend to remain in the water column and contribute to the pelagic environment.

There are several reasons that food is not eaten by the cultured animal. This type of waste is most relevant to artificial foods, which are normally in the form of dry or wet pellets. Between 1% and 4% of the total feed used is 'wasted' through transport and handling causing part of the feed to go to dust, which in turn is then not suitable to be eaten by fish. Once the feed is put into the culture system there is potential further waste as animals often cannot detect it or find it unattractive to eat. For pond culture this waste accounts for 1–5% of the dry food fed (Beveridge, 2004). For fish in net-pens, such as salmon or sea bass, as the environment is more dynamic it can be between 3% and 10% for dry pelleted feed (Cromey et al., 2002). For trash fish used in net-pens (e.g. yellow croaker or sea perch, or some tuna farms) the uneaten waste could be between 15% to

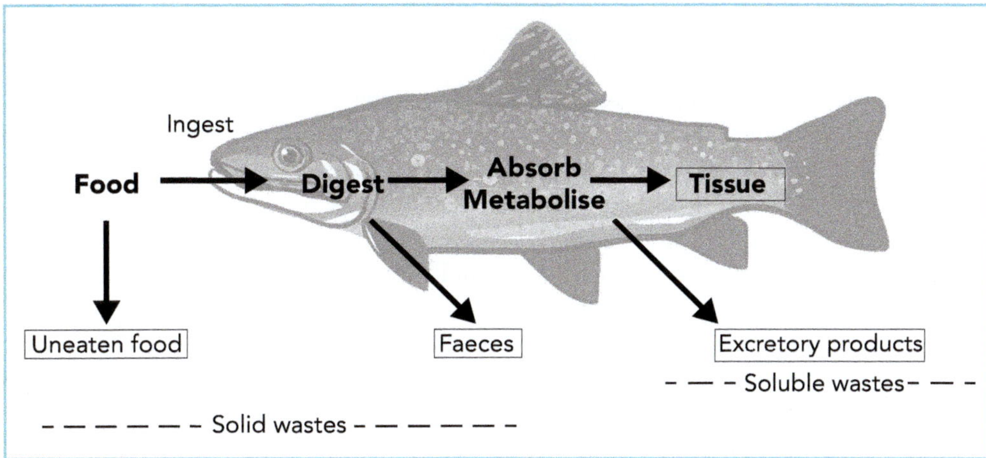

Figure 4.3 Mechanisms of waste production from fish feed entering the environment.

20% of food fed by wet weight (Cai et al., 2016). In coastal systems, wild fish may also consume some of the uneaten feed.

Another solid waste entering the environment is faecal material, formed from water, undigested food, dead cells from the lining of the intestines and dead bacteria. In aquatic organisms the amount of faeces produced can be estimated from digestibility studies and the dietary constituents of the food. For example, percentage faecal production for salmonids, carp and shrimp is very similar when related to unit body weight (~27 g/100 g by dry weight). Catfish, however, produce much higher levels of faeces (~41 g/100 g by dry weight) (Beveridge, 2004), however, this can change with specific diets. There are two key nutrients within solid particulate wastes that contribute to environmental productivity: organic carbon and inorganic nitrogen.

The excretory products are largely soluble and dilute within the water column after release. There is great variability between species in amount per unit weight, but the most important nutrient within these products is inorganic nitrogen usually excreted as ammonia, which is toxic to fish and other organisms when in high quantities in its unionized form (NH_3). Bacterial action within the water column readily breaks this down to firstly nitrite, also toxic, and then rapidly to nitrate, which is more stable within the environment. This can contribute to productivity of the water column, if in excess, through enhanced production of phytoplankton.

How do we quantify these nutrients in the environment? We can use computer-based environmental models or direct measurement. The most common form of modelling of the nutrient output is to use a mass balance model. This uses the fact that the total nutrient entering the system, through input of food, is the same as the nutrient exiting the system, albeit in a different form. In its simplest form mass balance modelling for aquaculture requires information on the food fed or available, the nutrient content of this food, the biomass of fish or shellfish harvested and the nutrient content in their flesh. There are also normally several assumptions used in such models which rely on published information and as they can be difficult to measure, for example, estimating how much of the food available is ingested, the presumed digestibility of the individual nutrients and how these vary with age and behaviour of the fish. The accuracy of the data and the assumptions used can impact on the accuracy of the model. The fate of the nutrients once in the environment are often more difficult to model and require detailed data on local environmental conditions and degradation characteristics of the wastes to assess their behaviour (Bannister et al., 2016).

The second method is to use direct measurement of samples taken from the environment. This gives very accurate results and reflects the actual situation within the water column or sediments at the time of sampling. This is used to monitor the environmental condition and for management by comparing the measurements with set environmental quality standards. However, a major disadvantage of direct measurement is that an impact is not detected until it is happening. Conversely, predictive models of nutrient inputs, such as mass balance and environmental fate models, can be used to estimate potential impacts before they happen.

The most successful approach for environmental assessment for aquaculture sustainability is to use a combined approach. The environmental measurements can be used to quantify the mass balance for different parts of the system for unit mass/production for a pond. If the nutrient input quantity (e.g. through feed) is known this can be used to quantify the nutrient loading in different environmental partitions. Once developed, models can also be validated for a location or set of environmental conditions using measured data. Once validated the same model can be used for other locations and environmental conditions as appropriate (providing that sufficient local environmental and production data is available).

What are the environmental consequences of the wastes? A good example is open net-pen farming of fish as it is one of the best studied aquaculture systems. All wastes go directly to the adjacent environment. In such systems, soluble wastes within the water column can have highly variable effects. Freshwaters containing net-pens are situated within enclosed systems, such as lakes, which can be slow to flush any additional nutrient released from aquaculture, resulting in an increase in algal biomass and consequent changes in the lake food web. Recovery can be slow from this type of change, illustrating the need to carefully manage aquaculture to ensure sustainability in these systems. Freshwater systems can be particularly sensitive to change, especially if the nutrient inputs change its trophic nature (e.g. from oligotrophic (low nutrient) to mesomorphic (moderate nutrient) (OECD, 1982)). Change in trophic levels are often used as an environmental threshold or quality standard to manage waste inputs and assess the capacity of such systems to support aquaculture production.

In marine systems there is little evidence of direct hypernutrification especially in relation to fish cage farming. This may be due to the very high dilution factor of the oceans and coastal seas into which the waste from cage aquaculture is discharged. However, this waste is certainly a contributory factor to total coastal nutrient levels among other inputs from land run-off, tidal oceanic inputs and sewage. Solid wastes, such as uneaten food and faecal materials, are discharged from the fish cages directly into the water column. Being heavier than water they settle quickly to the seabed and are distributed by the tidal current (though faeces spread over a wider area than uneaten feed). Deposition increases the nutritional content of sediments mostly beneath the fish cages, with a gradient of decreasing deposition moving away from the pens. A gradient of sediment effects and nutrient enhancement generally occurs away from the fish cages. Nearer the cages the sediment can become oxygen depleted due to high bacterial productivity, with increased levels of sulphide, as shown in Figure 4.4.

The different environmental tolerances of the animals living in the sediments mean that the benthic (sediment dwelling) communities change with the sediment condition. Beneath or near the cages, only the few animals that can withstand the lowered sediment oxygen conditions (e.g. the opportunist polychaetes *Capitella capitata* and *Malacoceros fuliginosa*),

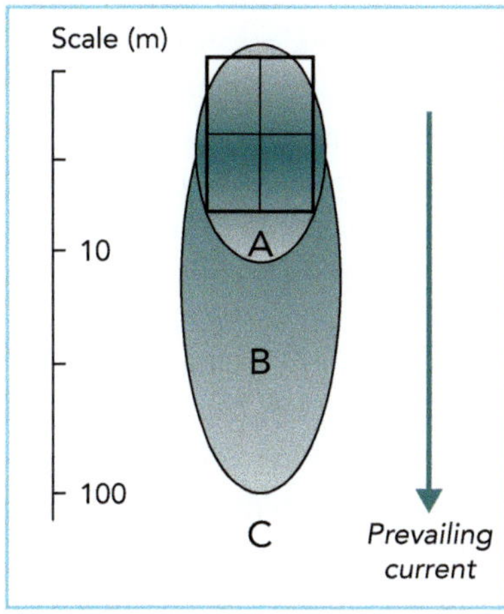

Figure 4.4 Zones of sediment nutrient distribution and effect near coastal Atlantic salmon cages: A. oxygen depleted zone with opportunist animal species; B. nutrient enhanced zone, transitional animal species; C. background zone.

which can then exploit the enormous food source from waste materials and proliferate (up to 10,000 individuals/m^2). Further away from the cages where conditions ease there is more competition for the food resource and other species dominate, until eventually the 'background' or the usual community is found.

There are many ways of quantifying this change in community, for example, looking at the diversity of the community using a recognized index of diversity such as the Shannon Function (Hs), or by measuring the biomass, abundance or species number with a distance of the cages.

The impacts of solid wastes also varies with the type of aquaculture and the nature of the culture method. For example, much tropical cage aquaculture still is reliant on trash fish (from by-catch) as a food source, rather than pelleted feed, such as that used in salmon farming. There is more waste associated with trash fish and the neutral density of the fish flesh tends to allow dispersal of this waste over a wider area than uneaten pellets.

As discussed previously, pond culture has fundamentally different environmental issues to cages, as the discharge to the environment is often a point source entering rivers or coastal environments. These are less diffuse and often more controllable. There are three types of pond farming systems which have different effects on the environment: intensive, semi-intensive and extensive.

Animals in intensive pond systems, such as for freshwater trout production in Denmark, are completely fed using artificial feeds, and often have a low water exchange resulting in high levels of nutrients in the discharge. Though water exchange is highly controllable and the effluent can be readily treated, a disadvantage of this type of system is that the sediment in the bottom of the pond can trap phosphorus, the limiting nutrient in freshwaters, which is retained for considerable periods of time.

In semi-intensive aquaculture the natural food supply is supplemented through either fertilization or addition of artificial feed (pellets, trash fish). This is a compromise between the very low stocking density of extensive systems and the high biomass of intensive systems. Most of the world's farmed fish production is achieved in semi-intensive aquaculture systems, particularly in China and South-East Asia. One advantage of this system for environmental sustainability is there is comparatively little waste to treat or release as most of the waste remains trapped in the sediments at the bottom of the ponds and is assimilated. A pulse of highly enriched effluent may be released at harvest, but this is normally assimilated well by the environment (Briggs and Funge-Smith, 1994).

In extensive aquaculture the whole culture relies on natural resource for provision of feed, and in some cases, seed. There is usually no supplementing of nutrients so the naturally limited food leads to lower stocking densities than other types of aquaculture. Extensive aquaculture can be carried out in large ponds (fish or shrimp) or coastal environments.

Along the coasts, mollusc and seaweed culture obtains 'food' directly from the water column as particulate or soluble nutrients, respectively. Occasionally, the density of farmed organisms grown in these extensive systems can exceed the capacity of the environment to produce natural feed, that is a high density of mussels can extract most of the phytoplankton from an area causing changes in the plankton community and food web structure.

Interactions between culture and wild organisms

Cultured aquatic organisms have the potential to interact with indigenous wild stocks in a number of ways. They can escape and interbreed, they can introduce new or alien species and they have the potential to spread disease.

Shellfish are often collected naturally from the local area or transplanted from other areas of recruitment. Though there is less consideration of inter-species interactions, there can be issues with transfer of diseases if this is not managed effectively. Transplanting and bivalve culture have increased the distribution of a number of species, often increasing their geographical range (McKindsey et al., 2007).

Escapees and farmed fish have received considerable attention, especially salmon. There are a variety of ways that farmed fish can escape. The most obvious way is from large releases due to cage failure in storms and equipment breakage (Jensen et al., 2010). Here potentially hundreds of thousands of fish can escape. However, these can remain near the escape point for a period of time and potentially many can be recovered, through implementation of the industry codes of practice in place in many nations. These often outline methods of action on catastrophic escape. Another potential mechanism of escape is a slow trickle of fish escaping as a result of routine operations, such as smolt delivery, bath treatment or grading. This is becoming less of an issue though as technology improves and methods, such as well-boat transfer, become industry standard. It is suspected that about 65% of the species introduced through aquaculture become established in the wild, though omnivores are more likely to become established than carnivores or herbivores. Cultured fish strains are selected for their high fecundity, rapid development, wide tolerances and quick growth. In reality though, when in the wild they can be less successful as they are often unable to feed effectively or hunt for food or compete for mates (Deverill, 2000).

Studies have shown that there are high numbers of salmon caught by fishermen which are of farmed origin, showing they are becoming established, and affecting wild salmon stocks. However, the inter-relationships are complex and may not be straightforward and may result from a number of interactive factors such as over-fishing, habitat regeneration (this is especially true in areas where there is a lot of forestry, which destroys or affects many of the head streams) and global warming (diverting the migration/return routes of salmon that use temperature clues for navigation in the open ocean), not just aquaculture escapees.

Environmental management of aquaculture

There are a range of measures available to effectively manage the impact of aquaculture wastes. There are two basic approaches, which are driven by legislative and economic factors: prevention and treatment/remediation. More sustainable use of resources is occurring through better use of space and water, and through reuse and conservation measures. In addition, much of the food sourced from the environment for intensive aquaculture, such as cage fish farming, are being substituted by alternatives (Egerton et al., 2020).

Wastes from aquaculture entering the environment are still considered an issue and affect both environmental quality and the ability of that same environment to support aquaculture. There has been considerable research and implementation of management and regulation to mediate these issues over the recent decades, which encompass a number of strategies.

If wastes are not reduced, then the aquatic resources used for aquaculture will be affected and thus reduce the ability of businesses to maximize profitability in the future. There is a delicately balanced relationship between environmental sustainability and effective aquatic resource management.

Computer-based predictive modelling is becoming ever more important in the assessment of the environmental impacts of aquaculture. Most models investigate the potential for change in the productivity of the local or regional water bodies or associated sediments. They focus on the direction of change and the magnitude of change. Environmental models have mostly been used in intensive aquaculture, which, by its nature, tends to have most effect on local environments through input of wastes. Nutritional requirements for such culture techniques are often met through external feed supply. In other words, nutrients are put into the environment that were not there before relying on the assimilative capacity of the environment to remove or use them. However, each environment has a finite capacity to assimilate these additional nutrients; it has a finite capacity for aquaculture. This is often called the carrying capacity of the environment to sustain aquaculture, and it is what most models attempt to predict; the environmental capacity of a local and regional area and at what point this will be exceeded through aquaculture production (Ross et al., 2013).

Before such modelling can effectively take place there are a number of requirements to be met.

- Knowing what determines the productivity of the environment, as this may change depending on the particular environmental partition (water, sediments, etc.).
- The food consumption and waste production characteristics of the fish species to be modelled.
- The nature of the wastes released and how the environment responds to them.
- How much change is permissible or acceptable, usually as a function of assimilation potential of the local environment, plus a 'safety' factor. This is usually defined within regulation and governance practice.

Aquaculture development in a region involves socioeconomics and society acceptance, and these are also important in defining what is acceptable. However, there is a cut-off point where the environment cannot easily recover and thus the environmental resources available are lessened.

The national or regional authority has a duty to promote development and aquatic resource use, including aquaculture, if it is sustainable and the responsibility to protect and conserve the natural environment. In addition, it has an obligation to protect the consumer. In doing all of this, it must balance costs and benefits of exploiting the aquatic environment in a sustainable way. Many countries use legislation and regulation to monitor and manage the effective use of aquatic resources, including aquaculture. Though not used in all countries for all forms of aquaculture, an environmental impact assessment can be used to aid decision making. The environmental statement, a report or document produced from the environmental impact assessment process, is a public document and contains all information relevant to the development. Normally for aquaculture this contains (Telfer et al., 2009):

- a description and design of the site and production processes
- details of the emissions
- what the existing environment is like – a baseline survey
- an analysis of the likely impact on this environment from the development using predictive models of effect
- mitigation measures that can be used to minimize impact, including details of where alternative sites for this development have been investigated.

The statement has to be written so as to be understood by three types of stakeholders: planners and decision makers, technical specialists and the public. One of the issues with the environmental impact assessment process is that it is often done locally and on a site by site basis so rarely takes cumulative effects into account. As a consequence, the authority should also implement integrated coastal management or catchment management in its regulation and control of aquaculture, in which coastal or freshwater catchment areas are managed holistically rather than each development or resource user considered in isolation. Integrated management also has considerable stakeholder participation and takes a more strategic view of developments and their consequences, with both regulation and incentives used to encourage implementation.

The enterprise or industry also has an obligation for sustainability, within which effective farm management is imperative. Feeds and feeding strategies are of huge importance in terms of environmental waste, in particular, the way fish are fed, and how the feed is managed. This can use advanced feeding technology for control, but often employs the simple principles such as feeding efficiency using food conversion ratio (FCR). Contrast these two farms A and B. Each are producing 1000 t of fish per year.

Farm A is well managed. It uses low 'pollution' feeds (low phosphorus levels = 0.8%). Feeding is done very carefully so the FCR (1.1 : 1) is low as little food is wasted. The phosphorus content of fish tissue is 4.8%. A simple mass balance calculation gives an annual output of 4.0 t of phosphorus going directly into the water.

$$(1000 \times 0.008 \times 1.1) - (1000 \times 0.48) = 4.0$$

Farm B is poorly managed. It uses 'standard' feed (phosphorus content = 1.3%), which is fed carelessly using more food and resulting a higher FCR (1.5 : 1). The same mass balance calculation gives an annual output of 14.7 t of phosphorus to the environment.

$$(1000 \times 0.013 \times 1.5) - (1000 \times 0.48) = 14.7$$

Though simplified this example shows the influence of effective management on waste inputs into the environment.

Figure 4.5 (Left) A marine farm feeding system, which recovers uneaten food and recycles it back into the fish pens. (Right) A freshwater drum filter system, which collects particulates for processing and recycling.

Another method of minimizing waste inputs involves its removal using technological solutions. An example of this is the use of feeding systems that detect and collect uneaten food falling from a marine net pen (Figure 4.5(a)) and recycles this back into the pen. Other examples, such as a rotary drum filter used for inland-based aquaculture, have cylindrical mesh filters that remove particles from the waste effluents (Figure 4.5(b)). Here the waste is removed from the mesh by a 'scraper' arm and the resulting sludge transported for treatment and recycling.

Tarpaulins or fine mesh nets in fish pens can also be used to collect particulate materials or dead fish at the bottom of marine cages, where it is pumped to the surface for processing and disposal. Though in such systems often the volume of waste and cost of processing is prohibitive.

Aquaculture in a changing environment

Freshwater, marine and terrestrial ecosystems across the globe are transforming at an unprecedented rate due to climate change and ocean acidification (IPCC, 2014, 2019). As aquaculture is fundamentally linked with the natural environment, such changes will bring opportunities and threats to production. However, impacts will not be uniform throughout the world. Differences occur due to variations in the rate and magnitude of environmental change (IPCC, 2014, 2019), as well as in the biological tolerances of the farmed species and the ability of stakeholders to develop and implement adaptation measures. There has been increased focus on adaptation in recent years as the inevitability of climate change has become apparent (Biesbroek et al., 2010), and adaptation will be a necessary process if the growing demands for aquatic food is to be met. Adaptation can be implemented across different spatial, temporal and socio-political scales (Adger et al., 2005), but relevant information is required to assess potential risks and then identify the most appropriate response.

Although the general trends of climate change are understood (IPCC, 2014, 2019), assessing the direct and indirect implications for aquaculture production is complex. Climate projections simulate potential futures but there are many uncertainties associated with this, which pose challenges to short- and long-term planning. Furthermore, climate models do not capture local conditions influencing production, which can make it difficult to assess potential risks (Falconer et al., 2020).

In the real world, aquaculture is affected by multiple stressors at the same time and there are many uncertainties about how these combine to affect production, and the health and welfare of the farmed species (Sarà et al., 2018). Multiple stressors are difficult to include in assessments, but they are challenges that aquaculture producers will face.

It is the local environmental conditions within and surrounding a farm that influence production, and a range of impacts will be experienced across the sector. In low-lying deltas, sea level rise and associated coastal flooding could lead to a loss of suitable production sites (Trieu and Phong, 2015). Some areas may be vulnerable to increased extreme events, such as cyclones, that can damage farm infrastructure, risk the health and welfare of the farmed species and endanger farm operators (Ahmed et al., 2013). Ocean acidification could reduce the growth and shell integrity of shellfish, affecting marketability (Fitzer et al., 2018). While increased temperatures can lead to health and welfare issues, or even mortalities (Hvas et al., 2017). These few examples highlight some of the challenges that aquaculture will face in the future as environmental changes affect almost all aspects of production, with impacts on growth, health and welfare, economics and environmental interactions.

Stakeholders across the sector can develop and implement adaptation measures to minimize potential risks (Reid et al., 2019). These can range from relatively simple farm-level measures to more complex, industry-scale strategies. However, there is also a danger that adaptation measures used to minimize the risks to aquaculture lead to increased emissions. For example, moving from open-pen culture in the sea to land-based RAS would minimize exposure to climate stressors, but would increase energy use (Badiola et al., 2018). Therefore, there is a need to develop effective adaptation strategies that limit the negative effects on aquaculture but at the same time do not adversely affect the wider ecosystem. It is also important to recognize that not all aquaculture producers will have the necessary resources available to adapt to the magnitude of the impact they will face. Governments and international development programmes may need to provide assistance to some of the more vulnerable aquaculture producing areas in order to minimize disruption to global food supply and continue contributions to food and nutrition security (Handisyde et al., 2017).

In addition to direct impacts, the changing environment will also affect the wider supply chain. For example, climate change may affect the availability of feed ingredients as the changing environment will also impact capture fisheries, which could have implications for fishmeal and fish oil (Merino et al., 2012), while terrestrial ingredients may also be affected by increased temperatures or changing rainfall patterns (Rosenzweig et al., 2014). It is difficult to assess the potential impact of climate change on feed ingredients as they are part of a complex global supply chain that has many socio-political considerations as well as ecosystem influences. However, climate change will introduce new challenges to the feed ingredient sector causing supply constraints due to increased need from a growing aquaculture industry as well as higher demand from other food production sectors.

Aquaculture has a role to play in climate mitigation and efforts to reduce emissions of greenhouse gases (GHG). At farm and industry level, this may involve the development and use of technology that is powered by renewable sources or less energy intensive machinery. Alternative methods of production should also be explored. For example, studies have shown that some integrated systems such as rice-fish co-culture can lower emissions (Li et al., 2019). As RAS systems develop, it may be possible to grow non-native species closer to their market, reducing transport emissions. Some forms of aquaculture are also suggested as potential climate mitigation strategies. Seaweed can act as a carbon dioxide sink so seaweed farms could be developed and used for carbon sequestration (Duarte et al., 2017). However, new production technologies or alternative farming methods are not always simple to implement and there may be technical, economic, social and legislative barriers that prevent establishment at a commercial scale.

Clearly, aquaculture production will face challenges from climate change and ocean acidification that will affect contributions to global food and nutrition security. Changes to the global and local environmental systems will have implications for aquatic food production and will require adaptation measures to maximize the opportunities and minimize the risks. There is a need for more research into the effects of climate change on aquaculture, particularly at the farm and regional scale, that can support short- and long-term planning and management decisions.

New technologies for aquaculture in the future

A number of newer technologies potentially offer farm- and industry-level solutions to the aquaculture sector's demands on resources and ecosystem services.

Offshore aquaculture

It is claimed that offshore aquaculture (or open ocean aquaculture), if properly sited and managed, minimizes waste nutrient impacts, avoids conflict in the congested coastal zone, reduces the incidence of disease and provides environmental conditions that support better growth and production (Buck and Langan, 2017).

There is little consensus on the definition of offshore aquaculture (Froehlich et al., 2017). While a framework of consistent metrics (distance from shore, depths, currents, wave climate, winds, sea bottom characteristics) would help guide research and development, a legal framework that aligns with the United Nations Convention on the Law of the Sea (UNCLOS) and supports zoning is essential if the sub-sector is to prosper.

Despite a lack of study, offshore aquaculture appears to have potential to mitigate pollution and disease across species and taxonomic groups (Froehlich et al., 2017). The sub-sector remains small, although is growing. Capital and operating costs can be daunting, and economies of scale may be important (i.e. fewer, larger operations), as might co-location with other sectors, such as wind power, to mitigate development and operating costs (Buck and Langan, 2017). Some promote the advantages of offshore integrated multi-trophic aquaculture (see below) but highlight the need to improve knowledge of resource flows between integrated species in the hydrodynamically challenging conditions that characterize offshore waters (Buck et al., 2018).

Integrated multi-trophic aquaculture

Integrated multi-trophic aquaculture (IMTA) refers to the co-cultivation of fed (e.g. finfish, shrimp) with extractive aquatic species, such as suspension-feeding mussels and oysters, and deposit-feeding sea cucumbers and sea urchins, and macroalgae that consume or absorb the organic and inorganic effluents generated by the fed species. While the sustainability of aquaculture may thus be increased, each of the components must be marketable and profitable and/or add value through the ecosystem services they provide in order for IMTA to be economically viable (Buck et al., 2018; Neori et al., 2004).

Commercial IMTA production remains small; satisfying the above conditions is difficult. Future success may lie in the adoption by policymakers of the FAO Ecosystem Approach to Aquaculture, which advocates zoning and licensing systems that set targets for different types of aquaculture – fed and extractive – within each zone (Brugère et al., 2018).

RAS and aquaponics

RAS create highly controlled and pathogen free salt- and freshwater environments for rearing farmed aquatic organisms, use limited water and discharge few wastes (Badiola et al., 2012). Filters and biofilters remove waste solids and reduce nitrite and ammonia toxicity and UV is used to control pathogens. Given their minimal land and water use, RAS can be located almost anywhere, especially in urban and peri-urban areas, facilitating access to inputs and to higher value fresh and live market products. Economic viability depends upon high stocking densities, intensive production methods and securing the limited premium market segments, including production of Atlantic salmon smolts.

RAS suffer from two disadvantages, however; technical complexity and high energy use (and thus carbon footprint). Because they are complex, RAS are expensive to build and require skilled operating personnel. Issues of reliability remain too but have been steadily improving.

The credentials of RAS as an environmentally sound production method are undermined by their high energy consumption and reliance on complete feed, which can account for up to 90% of total carbon dioxide emissions (Song et al., 2019). Impacts may be reduced through energy sourcing, reformulation of feeds and by utilizing wastes.

Aquaponics combines RAS with hydroponics, the soil-less culture of plants, in a single system. Wastes from aquatic food production are filtered and the dissolved waste nutrients used to produce a crop, usually of salad vegetables or herbs (Somerville et al., 2014). The concept of aquaponics is very attractive: two crops, not one; water conservation; waste recycling (circular economy); the use of non-agricultural land; high yields. However, aquaponic systems are capital intensive, complex, require detailed knowledge of both farmed plants and animals, and require daily maintenance. While the technology is evolving there is as yet little commercial production.

Closed containment systems

Closed containment systems (CCS) describes a variety of floating, rigid or flexible wall aquaculture enclosures. CCS have primarily been developed in Norway in response to the government's challenge of developing technologies that address long-standing environment- and welfare-related issues associated with conventional cage aquaculture: sea lice, farmed fish welfare and their control; escapees and their impacts on wild Atlantic salmon; aquaculture waste and their impacts on the marine environment (Hersoug et al., 2019).

A number of commercial CCS, varying in size and design, exist. Most pump unfiltered, lice-free seawater from depth (~25 m), which is circulated throughout the CCS by pumps, often with oxygen added via diffusers. Waste particles are separated at the outlet and stored prior to disposal. Thus, fish remain lice-free, escapes are minimal and few wastes are released.

CCS design remains a work in progress seeking to optimize the production environment (Nilsen et al., 2019) and reduce capital and operational costs.

Lab-grown seafood

Lab-grown seafood (also known as cell-based seafood, cellular seafood and clean seafood) refers to the potentially game-changing production of seafood through cell culture. To date, shrimp, lobster, crab, Atlantic and coho salmon and tuna have been piloted (Rubio et al., 2019; Marwaha et al., 2022). Lab-based seafood is based on the isolation and multiplication of muscle cells from aquatic animals in closed loop bioreactors where they are provided with the ideal temperatures, nutrients, dissolved oxygen and pH levels. A cellular slurry is produced that can be used in such products as seafood burgers, fish fingers or, using 3D printer technology, to produce commodities that emulate the appearance, texture and taste of conventional seafood items.

Little land is needed; industrial sites in urban and peri-urban areas can be used – indeed, may be preferable, because of access to skilled labour, energy and markets. Little water is required compared to conventional aquaculture. The biggest outstanding issues are those of energy use, and thus GHG emissions, and production costs. Few figures for energy use are available but almost as important in GHG emission terms as energy quantity is energy source (i.e. whether the energy is carbonized or not). The high costs per unit production can be expected to reduce markedly as better, cheaper growth media are developed and with economies of scale.

Lab-grown meat and seafood are largely unlinked to other global food supply chains and thus add resilience to the global food system. There are also fewer biosafety or biosecurity issues – there are no contaminants, such as antimicrobial residues, mercury (tuna) or PCBs (Atlantic salmon) – and no farmed animal welfare issues to consider. There are no wastes – bones, scales or skin – either. There remain, however, a number of technical barriers to producing lab-grown seafood products readily recognizable to consumers. Rubio et al. (2019) and Marwaha et al. (2022) summarize these as a greater understanding of fish muscle cell and tissue cultivation, less costly, serum-free media formulations optimized for fish cell culture and bioreactor designs that better meet the needs of fish cells for large-scale production. To these one must add the attraction of significant investment and winning of consumer acceptance.

Nevertheless, many believe that lab-grown seafood has distinct advantages over lab-grown meat: production is less energy intensive (aquatic animal cells are grown at lower temperatures), fish cells are more tolerant of limited dissolved oxygen levels and fluctuating pH conditions. Moreover, seafood is perceived by consumers as a 'healthier' choice. Marketeers also anticipate strong environmental product attributes relating to protection of the marine environment and that there may be less of a 'yuck' factor associated with lab-grown seafood than meat.

Summary

To ensure the environmental sustainability of aquaculture development it is important to consider a wider ecosystem perspective, and consider the goods and services provided. Environmental goods provide resources for aquaculture and other users but risk being overexploited unless managed. This is to the benefit of the wider environment, the multiple users of the resources and aquaculture. The key environmental resources that aquaculture uses are feed to provide nutrients for production, seed to provide new stocks for culture, land to provide space for ponds and other associated structures, and water to support the cultured organisms. Aquaculture is also reliant on services provided by the environment to sustain it by assimilating its waste products, whether from nutrients or chemical discharge or release of non-indigenous organisms.

Maintenance of these goods and services are essential to ensure a resilient environment both now and in the future. Measures to ensure this should include environmental modelling and sampling and monitoring to confirm adherence to the carrying capacity for aquaculture and to aid in its management. Such measures are often included in governance and regulation of aquaculture and define the amount of aquaculture licensed within a location or area. Waste can also be minimized and managed using treatment and collection technology.

Aquaculture practice and the environment are changing, so future systems need to be resilient to future changes brought by climate change and responsive to exploit new locations and environmental conditions. Modern systems technology, such as RAS and offshore production units, can be used to exploit new environments for aquaculture. This can include alternative aquaculture methods and sources of aquatic proteins. There is huge potential to increase sustainable production of aquatic food in the future to aid global food security while maintaining the ecosystem's goods and services.

References

Aas, T.S., Ytrestøyl, T. & Åsgård, T. (2019) Utilization of feed resources in the production of Atlantic salmon (Salmo salar) in Norway: An update for 2016. *Aquaculture Reports*, 15, 100216. https://doi.org/10.1016/j.aqrep.2019.100216.

Adger, W.N., Arnell, N.W. & Tompkins, E.L. (2005) Successful adaptation to climate change across scales. *Global Environmental Change*, 15, 77–86. https://doi.org/10.1016/j.gloenvcha.2004.12.005.

Ahmed, N., Occhipinti-Ambrogi, A. & Muir, J.F. (2013) The impact of climate change on prawn postlarvae fishing in coastal Bangladesh: Socioeconomic and ecological perspectives. *Marine Policy*, 39, 224–233. https://doi.org/10.1016/j.marpol.2012.10.008.

Ayers, J.C., George, G., Fry, D., Benneyworth, L., Wilson, C., Auerbach, L., Roy, K., Karim, M., R., Akter, F. & Goodbred, S. (2017) Salinization and arsenic contamination of surface water in southwest Bangladesh. Geochemical Transactions, 18, 4. https://doi.org/10.1186/s12932–017–0042–3.

Badiola, M., Basurko, O.C., Piedrahita, R., Hundley, P. & Mendiola, D. (2018) Energy use in Recirculating Aquaculture Systems (RAS): A review. *Aquacultural Engineering*, 81, 57–70. https://doi.org/10.1016/j.aquaeng.2018.03.003.

Badiola, M., Mendiola, D. & Bostock, J. (2012) Recirculating Aquaculture Systems (RAS) analysis: Main issues on management and future challenges. *Aquacultural Engineering*, 51, 26–35. https://doi.org/10.1016/j.aquaeng.2012.07.004.

Bannister, R.J., Johnsen, I.A., Hansen, P.K., Kutti, T. & Asplin, L. (2016) Near- and far-field dispersal modelling of organic waste from Atlantic salmon aquaculture in fjord systems. *ICES Journal of Marine Science*, 73, 2408–2419. https://doi.org/10.1093/icesjms/fsw027.

Beveridge, M.C.M. (2004) *Cage Aquaculture*, 3rd edn. Blackwell Publishing Ltd: Oxford.

Biesbroek, G.R., Swart, R.J., Carter, T.R., Cowan, C., Henrichs, T., Mela, H., Morecroft, M.D. & Rey, D. (2010) Europe adapts to climate change: Comparing national adaptation strategies. *Global Environmental Change*, 20, 440–450. https://doi.org/10.1016/j.gloenvcha.2010.03.005.

Braaten, R.O. & Flaherty, M. (2001) Salt balances of inland shrimp ponds in Thailand: Implications for land and water salinization. *Environmental Conservation*, 28, 357–367. https://doi.org/10.1017/S0376892901000388.

Briggs, M.R.P. & Funge-smith, S.J. (1994) A nutrient budget of some intensive marine shrimp ponds in Thailand. *Aquaculture Research*, 25, 789–811. https://doi.org/10.1111/j.1365-2109.1994.tb00744.x.

Brugère, C., Aguilar-Manjarrez, J., Beveridge, M.C.M. & Soto, D. (2018) The ecosystem approach to aquaculture 10 years on – A critical review and consideration of its future role in blue growth. *Reviews in Aquaculture*, 9. https://doi.org/10.1111/raq.12242].

Buck, B. & Langan, R., editors (2017) *Aquaculture Perspective of Multi-use Sites in the Open Ocean: The Untapped Potential for Marine Resources in the Anthropocene*. Springer: Heidelberg. https://doi.org/10.1007/978-3-319-51159-7.

Buck, B.H., Troell, M.F., Krause, G., Angel, D.L., Grote, B. & Chopin, T. (2018) State of the art and challenges for offshore integrated multi-trophic aquaculture (IMTA). *Frontiers in Marine Science*, 5, 165. https://doi.org/10.3389/fmars.2018.00165.

Cai, H., Ross, L.G., Telfer, T.C., Wu, C., Zhu, A., Zhao, S. & Xu, M. (2016) Modelling the nitrogen loadings from large yellow croaker (Larimichthys crocea) cage aquaculture. *Environmental Science and Pollution Research International*, 23, 7529–7542. https://doi.org/10.1007/s11356-015-6015-0.

Carrasquilla-Henao, M. & Juanes, F. (2017) Mangroves enhance local fisheries catches: A global meta-analysis. *Fish and Fisheries*, 18, 79–93. https://doi.org/10.1111/faf.12168.

Cromey, C.J., Nickell, T.D. & Black, K.D. (2002) DEPOMOD – Modelling the deposition and biological effects of waste solids from marine cage farms. *Aquaculture*, 214, 211–239. https://doi.org/10.1016/S0044-8486(02)00368-X.

Deverill J.I. (2000) The post-stocking behaviour of hatchery-reared brown trout (Salmo trutta L.). PhD thesis. University of Stirling: Stirling.

Duarte, C.M., Wu, J., Xiao, X., Bruhn, A. & Krause-Jensen, D. (2017) Can seaweed farming play a role in climate change mitigation and adaptation? *Frontiers in Marine Science*, 4. https://doi.org/10.3389/fmars.2017.00100.

Dwipongoo, A. (1986). *Growth, Mortality and Recruitment of Commercially Important Fishes and Penaeid Shrimp in Indonesian Waters*. ICLARM Technical Report 17.

Egerton, S., Wan, A., Murphy, K., Collins, F., Ahern, G., Sugrue, I., Busca, K., Egan, F., Muller, N., Whooley, J., McGinnity, P., Culloty, S., Ross, R.P. & Stanton, C. (2020) Replacing fishmeal with plant protein in Atlantic salmon (Salmo salar) diets by supplementation with fish protein hydrolysate. *Scientific Reports*, 10, 4194. https://doi.org/10.1038/s41598-020-60325-7.

Falconer, L., Hjøllo, S.S., Telfer, T.C., McAdam, B.J., Hermansen, Ø. & Ytteborg, E. (2020) The importance of calibrating climate change projections to local conditions at aquaculture sites. *Aquaculture*, 514, 734487. https://doi.org/10.1016/j.aquaculture.2019.734487.

Fitzer, S.C., Torres Gabarda, S., Daly, L., Hughes, B., Dove, M., O'Connor, W., Potts, J., Scanes, P. & Byrne, M. (2018) Coastal acidification impacts on shell mineral structure of bivalve mollusks. *Ecology and Evolution*, 8, 8973–8984. https://doi.org/10.1002/ece3.4416.

Froehlich, H.E., Smith, A., Gentry, R.R. & Halpern, B.S. (2017) Offshore aquaculture: I know it when I see it. *Frontiers in Marine Science*, 4, 156.

Funge-Smith, S.J. & Briggs, M.R.P. (1998) Nutrient budgets in intensive shrimp ponds: Implication for sustainability. *Aquaculture*, 164, 117–133. https://doi.org/10.1016/S0044-8486(98)00181-1.

Hamilton, S. (2013) Assessing the role of commercial aquaculture in displacing mangrove forest. *Bulletin of Marine Science*, 89, 585–601. https://doi.org/10.5343/bms.2012.1069.

Handisyde, N., Telfer, T.C. & Ross, L.G. (2017) Vulnerability of aquaculture-related livelihoods to changing climate at the global scale. *Fish and Fisheries*, 18, 466–488. https://doi.org/10.1111/faf.12186.

Hersoug, B., Mikkelsen, E. & Karlsen, K.M. (2019) 'Great expectations' – Allocating licenses with special requirements in Norwegian salmon farming. *Marine Policy*, 100, 152–162. https://doi.org/10.1016/j.marpol.2018.11.019.

Hvas, M., Folkedal, O., Imsland, A. & Oppedal, F. (2017) The effect of thermal acclimation on aerobic scope and critical swimming speed in Atlantic salmon, Salmo salar. *Journal of Experimental Biology*, 220, 2757–2764. https://doi.org/10.1242/jeb.154021.

IPCC (2014) 2014: Synthesis report. Contribution of working groups I, II and III to the fifth assessment report of the Intergovernmental Panel on Climate Change. *Climate Change* (edited by Core Writing Team, R. K. Pachauri & L. A. Meyer). IPCC: Geneva.

IPCC (2019) *IPCC Special Report on the Ocean and Cryosphere in a Changing Climate* (edited by H.-O. Pörtner, D. C. Roberts, V. Masson-Delmotte, P. Zhai, M. Tignor, E. Poloczanska, K. Mintenbeck, A. Alegría, M. Nicolai, A. Okem, J. Petzold, B. Rama & N. M. Weyer). IPCC: Geneva.

Jensen, Ø, Dempster, T., Thorstad, E.B., Uglem, I. & Fredheim, A. (2010) Escapes of fishes from Norwegian Sea-cage aquaculture: Causes, consequences and prevention. *Aquaculture Environment Interactions*, 1, 71–83. https://doi.org/10.3354/aei00008.

Li, F., Feng, J., Zhou, X., Xu, C., Haissam Jijakli, M., Zhang, W. & Fang, F. (2019) Impact of rice-fish/shrimp co-culture on the N2O emission and NH3 volatilization in intensive aquaculture ponds. *Science of the Total Environment*, 655, 284–291. https://doi.org/10.1016/j.scitotenv.2018.10.440.

Marwaha, N., Beveridge, M. C. M., & Phillips, M. J. (2022). Food, fad or feed: Alternative seafood and its contribution to food systems. *Frontiers in Sustainable Food Systems*, 6. https://doi.org/10.3389/fsufs.2022.750253.

McKindsey, C.W., Landry, T., O'Beirn, F.X. & Davies, I.M. (2007) Bivalve aquaculture and exotic species: A review of ecological considerations and management issues. *Journal of Shellfish Research*, 26, 281–294. https://doi.org/10.2983/0730-8000(2007)26[281:BAAESA]2.0.CO;2.

Merino, G., Barange, M., Blanchard, J.L., Harle, J., Holmes, R., Allen, I., Allison, E.H., Badjeck, M.C., Dulvy, N.K., Holt, J., Jennings, S., Mullon, C. & Rodwell, L.D. (2012) Can marine fisheries and aquaculture meet fish demand from a growing human population in a changing climate? *Global Environmental Change*, 22, 795–806. https://doi.org/10.1016/j.gloenvcha.2012.03.003.

Muir, J.F. & Beveridge, M.C.M. (1987) Water resources and aquaculture development. Arch. Hydrobiol. Beih. Ergebn Limnol, 28, 321–324.

Naylor, R.L., Goldburg, R.J., Primavera, J.H., Kautsky, N., Beveridge, M.C.M., Clay, J., Folke, C., Lubchenco, J., Mooney, H. & Troell, M. (2000) Effect of aquaculture on world fish supplies. *Nature*, 405, 1017–1024. https://doi.org/10.1038/35016500.

Neori, A., Chopin, T., Troell, M., Buschmann, A.H., Kraemer, G.P., Halling, C., Shpigel, M. & Yarish, C. (2004) Integrated aquaculture: Rationale, evolution and state of the art emphasizing seaweed biofiltration in modern mariculture. *Aquaculture*, 231, 361–391. https://doi.org/10.1016/j.aquaculture.2003.11.015.

Nilsen, A., Hagen, Ø, Johnsen, C.A., Prytz, H., Zhou, B., Nielsen, K.V. & Bjørnevik, M. (2019) The importance of exercise: Increased water velocity improves growth of Atlantic salmon in closed cages. *Aquaculture*, 501, 537–546. https://doi.org/10.1016/j.aquaculture.2018.09.057.

OECD (1982). *Eutrophication of Waters: Monitoring, Assessment and Control.* OECD: Paris.

Outeiro, L., Byron, C. & Angelini, R. (2018) Ecosystem maturity as a proxy of mussel aquaculture carrying capacity in Ria de Arousa (NW Spain): A food web modeling perspective. *Aquaculture*, 496, 270–284. https://doi.org/10.1016/j.aquaculture.2018.06.043.

Phillips, M. J., Beveridge, M. C. M., & Clarke, R. (1991). *Impact of aquaculture on water resources. Advances inworld aquaculture.* D. E. Bruno & J. R. Tomasso (Eds.), 3 (pp. 568–591).

Phuoc, N.N., Richards, R. & Crumlish, M. (2020) Environmental conditions influence susceptibility of striped catfish Pangasianodon hypophthalmus (Sauvage) to Edwardsiella ictaluri. *Aquaculture*, 523, 735226. https://doi.org/10.1016/j.aquaculture.2020.735226.

Reid, G.K., Gurney-Smith, H.J., Flaherty, M., Garber, A.F., Forster, I., Brewer-Dalton, K., Knowler, D., Marcogliese, D.J., Chopin, T., Moccia, R.D., Smith, C.T. & De Silva, S. (2019) Climate change and aquaculture: Considering adaptation potential. *Aquaculture Environment Interactions*, 11, 603–624. https://doi.org/10.3354/aei00333.

Rosenzweig, C., Elliott, J., Deryng, D., Ruane, A.C., Müller, C., Arneth, A., Boote, K.J., Folberth, C., Glotter, M., Khabarov, N., Neumann, K., Piontek, F., Pugh, T.A.M., Schmid, E., Stehfest, E., Yang, H. &

Jones, J.W. (2014) Assessing agricultural risks of climate change in the 21st century in a global gridded crop model intercomparison. *Proceedings of the National Academy of Sciences of the United States of America*, 111, 3268–3273. https://doi.org/10.1073/pnas.1222463110.

Ross, L.G., Telfer, T.C., Falconer, L., Soto, D. & Aguilar-Manjarrez, J., editors (2013). *Site Selection and Carrying Capacities for Inland and Coastal Aquaculture. FAO/Institute of Aquaculture, University of Stirling, Expert Workshop, 6–8 December 2010.* Stirling, UK Fisheries and Aquaculture Proceedings No. 21. Food and Agriculture Organization: Rome.

Rubio, N., Datar, I., Stachura, D., Kaplan, D. & Krueger, K. (2019) Cell-based fish: A novel approach to seafood production and an opportunity for cellular agriculture. *Frontiers in Sustainable Food Systems*, 3, 43–57. https://doi.org/10.3389/fsufs.2019.00043.

Sarà, G., Mangano, M.C., Johnson, M. & Mazzola, A. (2018) Integrating multiple stressors in aquaculture to build the blue growth in a changing sea. *Hydrobiologia*, 809, 5–17. https://doi.org/10.1007/s10750-017-3469-8.

Somerville, C., Cohen, M., Pantanella, E., Stankus, A. & Lovatelli, A. (2014) Small-scale aquaponic food production. Integrated fish and plant farming. FAO Fisheries and Aquaculture Technical Paper 589. Food and Agriculture Organization: Rome. www.fao.org/3/a-i4021e.pdf.

Song, X.Q., Liu, Y., Pettersen, J.B., Brandão, M., Ma, X., Røberg, S. & Frostell, B. (2019) Life cycle assessment of recirculating aquaculture systems: A case of Atlantic salmon farming in China. *Journal of Industrial Ecology*, 23, 1077–1086. https://doi.org/10.1111/jiec.12845.

Telfer, T.C., Atkin, H. & Corner, R.A. (2009) Review of environmental impact assessment and monitoring in aquaculture in Europe and North America. In: *Environmental Impact Assessment and Monitoring in Aquaculture*. FAO Fisheries and Aquaculture Technical Paper, Vol. 527. Food and Agriculture Organization: Rome.

Trieu, T.T.N. & Phong, N.T. (2015) The impact of climate change on salinity intrusion and Pangasius (Pangasianodon hypophthalmus) farming in the Mekong Delta, Vietnam. *Aquaculture International*, 23, 523–534. https://doi.org/10.1007/s10499-014-9833-z.

5

Production-level diseases and public health considerations in aquaculture

Margaret Crumlish and Brian Austin

Introduction

There has been a dramatic increase in aquaculture in the years since the end of the Second World War, both in terms of diversification in the range of aquatic species produced and in overall quantities. The monoculture situation, which is a common feature of intensive aquaculture production, is also unfortunately ideal for the development and spread of infectious diseases. Compared with terrestrial agriculture, farmed aquatic species are often in close proximity with wild animals and their microflora among which are causal agents of disease. Therefore, there is potential for the exchange of pathogens between wild and farmed species, and between them and the environment. With a large number of tightly packed individuals in confined spaces, this is ideal for the transmission and spread of disease. The situation is exacerbated when species, which are exotic to the geographical area, are introduced and farmed, and acquire pathogens from native animals. Here, the farmed species may be unlikely to have resistance to the pathogen, and the resulting disease may be more serious leading to higher mortalities than recorded in native animals. Where aquaculture sites are in close proximity on the same water source then the rapid spread of disease is to be anticipated. Examples are given below, and may explain the occurrence and spread of diseases, such as coho salmon syndrome in Chile (Cvitanich et al., 1991). The outcome may be heavy losses in production, issues with the safe and sanitary disposal of large numbers of corpses, and adverse effects on consumer confidence. Losses in production may have disastrous consequences for communities in less affluent countries, such as Bangladesh, where there is greater reliance on aquaculture to provide high-quality protein to the population. Clearly, rapid and accurate diagnoses are essential to determine exactly what is infecting farmed aquatic species so that meaningful control measures may be implemented as quickly as possible. If there is any delay such that the stock is showing signs of clinical disease with increasing mortalities then the effectiveness of any control measure is reduced. The presence of disease may have numerous effects on sustainability.

Clearly, there are the economic issues related to mortalities, and thus the loss of production, a source of protein to the human consumers and income to the aquaculturists. Moribund animals with clearly defined clinical manifestations of disease, for example, ulcerations and haemorrhaging, will

Margaret Crumlish and Rachel Norman (eds) **Aquatic Food Security**
DOI: 10.52517/9781789183368.005, © 5m Books Ltd, 2023

be unsaleable. Also, some diseases may result in poor feed conversion maybe leading to reduced/stunted growth, which will adversely affect production costs. Some pathogens, for example, *Vibrio vulnificus*, which occurs in bivalves, may have implications for human health, including fatal septicaemias, leading to embargoes on the sale of the infected animals for (human) consumption (Lipp and Rose, 1997; Raszl et al., 2016). *V. cholerae*, which is the cause of cholera in humans, has been associated with some diseases of fish and shellfish. *V. cholerae* serogroup O139, which has significance for human health, has been associated with mass mortalities in *Penaeus monodon* (tiger shrimp) in India (Joseph et al., 2015). Also, *V. cholerae* has been regarded as a possible cause of white faeces syndrome in *Litopenaeus vannamei* (whiteleg shrimp) in China (Cao et al., 2015). The presence of the pathogen has been used as a reason to ban imports of shrimp, which may have major impacts on exporting countries (WHO, 2006). Then, there is the category of disease, such as the Office International des Epizooties (OIE) listed and notifiable viral haemorrhagic septicaemia (VHS) of salmonids notably trout, that is regarded as so serious that its presence inevitably leads to eradication and sanitary disposal of the infected stock (Hoferer et al., 2019). However, aquaculture has been resilient, and recovered from the losses of production associated with disease. Certainly, in Asia, shrimp production is recovering from the problems of acute hepatopancreatic necrosis syndrome that has blighted the industry since its first recognition in 1990.

Overall, there is a diverse range of pathogens and parasites that may cause losses to farmed stock (Table 5.1). The significance of some of these organisms in terms of food security will be discussed below.

Bacterial fish pathogens

A diverse range of bacterial taxa have been associated with vertebrates and invertebrates of relevance to aquaculture worldwide (Table 5.1). Some pathogens have been restricted to only a few species, for example, *Pasteurella skyensis* has been reported only in Atlantic salmon (*Salmo salar*) farmed in marine sites in Scotland (Birkbeck et al., 2002), whereas others occur in many species across wide geographical areas, for example, *Aeromonas hydrophila* has been found extensively in freshwater fish species in most countries involved with aquaculture (Austin and Austin, 2016). New pathogens appear with alarming regularity. For example, in 1989, a new disease of salmon emerged, and was named as coho salmon syndrome, Huito disease (Schaefer et al., 1991) and salmonid rickettsial septicaemia (Cvitanich et al., 1991), with the pathogen named as *Piscirickettsia salmonis* (Fryer et al., 1992). From Chile, the disease spread widely to North America and Europe (Birkbeck et al., 2004). Currently, the disease has become well established in salmon farming areas. Losses were in the region of 3–7% each week, with cumulative mortalities of ~90%.

Overall, the impact of bacterial pathogens on production may occur in all stages from egg to adult. Mortalities among juvenile stages may be offset by increased production whereas problems in market-ready individuals will impact greatly on sales, the availability of high-quality protein for consumers (this has great significance in underdeveloped countries), and thus the finances of the aquacultural operation.

Eggs

Egg disease has been attributed to a range of microbial species, including *Hahella chejuensis* (red egg disease/syndrome) in tilapia (*Oreochromis niloticus*) from Thailand (Senapin et al., 2016),

Table 5.1 Important pathogens and parasite of farmed fish and shellfish.

Pathogen	Disease	Host	Geographical range
Bacteria			
Aeromonas hydrophila	Motile aeromonas septicaemia	Most freshwater fish	Worldwide
A. salmonicida	Furunculosis, carp erythrodermatitis	Salmonids, carp	Europe, North and South America, Japan
Allivibrio salmonicida	Coldwater vibriosis	Atlantic salmon	Norway, North America
Candidatus Hepatobacter penaei	Necrotizing hepatopancreatitis	Shrimp	South, Central and North America
Edwardsiella ictaluri	Enteric septicaemia of catfish, Bacillary necrosis of *pangasius*	Catfish	USA, Vietnam
E. tarda	Edwardsiellosis	Various	Japan, USA
Flavobacterium columnare	Columnaris	Many freshwater fish species	Extensive
Flavobacterium psychrophilum	Rainbow trout fry syndrome, coldwater disease	Rainbow trout	Europe, USA
Francisella noatunensis	Francisellosis	Atlantic salmon, tilapia	North and South America, Europe
Lactococcus garvieae	Lactococcosis	Many fish species	Extensive
Moritella viscosa	Winter ulcer disease	Atlantic salmon	Iceland, Norway
Photobacterium damselae subsp. *damselae*	Photobacteriosis	Many marine fish species	Mediterranean countries, Asia
Photobacterium damselae subsp. *piscicida*	Pasteurellosis, pseudotuberculosis	Many marine fish species	Mediterranean countries
Piscirickettsia salmonis	Coho salmon syndrome	*Salmo* spp.	Chile, North America, Europe
Renibacterium salmoninarum	Bacterial kidney disease	Trout	Europe, North and South America, Japan
Streptococcus iniae	Streptococcosis	Many fish species	Extensive
Tenacibaculum maritimum	Gill disease		
Vibrio anguillarum	Vibriosis	Most marine fish species	Worldwide
V. harveyi	Luminous vibriosis	Shrimp	Asia, South America
V. parahaemolyticus	Acute hepatopancreatic necrosis, shrimp vibriosis	Shrimp	Asia, North America
V. vulnificus	Septicaemia	Eels, bivalves	Europe, Japan, North America
Yersinia ruckeri	Enteric redmouth	Rainbow trout	
Fungi			
Aphanomyces invadans	Epizootic ulcerative syndrome	Many freshwater fish species	Africa, Asia, Australia and North America
Saprolegnia spp.	Saprolegniasis	Most freshwater fish species	Worldwide

Table 5.1 (continued)

Pathogen	Disease	Host	Geographical range
Viruses			
Infectious hypodermal and haematopoietic necrosis virus	Infectious hypodermal and haematopoietic necrosis	Shrimp	Asia, Central and South America, Pacific islands, Middle East
Infectious haematopoietic necrosis virus	Infectious haematopoietic necrosis	Salmon and trout	Asia, Europe, North America
Infectious pancreatic necrosis virus	Infectious pancreatic necrosis	Trout	Europe, North America
Infectious salmon anaemia virus	Infectious salmon anaemia	Atlantic salmon	Chile, Europe, North America
Koi herpes virus	Koi herpes virus disease	Carp	Asia, Europe
Red sea bream iridovirus	Red sea bream iridovirus disease	Marine fish species	Asia
Spring viraemia of carp virus	Spring viraemia of carp		Asia, Europe, North and South America
Taura syndrome virus	Taura syndrome	Shrimp	Asia, North and South America
Viral haemorrhagic septicaemia virus	Viral haemorrhagic septicaemia	Many fish species	Asia, Europe, North America
White spot syndrome virus	White spot syndrome	Shrimp	America, Asia, Middle East
White tail disease virus	White tail disease	*Macrobrachium rosenbergii*	Asia
Yellowhead disease virus	Yellowhead disease	Shrimp	East Africa, Asia, Australia and Mexico
Parasites			
Lepeophtheirus salmonis	Sea lice	Atlantic salmon	Europe
Neoparamoeba perurans	Amoebic gill disease	Atlantic salmon	Australia, Scotland

Pseudoalteromonas piscicida (Nelson and Ghiorse, 1999) and *P. undina* (Pujalte et al., 2007) egg disease in damselfish, sea bass and sea bream in Spain and the USA and *Tenacibaculum ovolyticum* in egg mortalities in halibut from Norway (Hansen et al., 1992). Red egg disease/syndrome was recognized initially in 2000 when during incubation eggs became yellow to red in colour, and failed to hatch. Half the eggs were lost over a 4 month period to March 2014, leading to a 10% reduction in fry production, although 50% losses were recorded during cold weather (Senapin et al., 2016). *T. ovolyticum* was attributed with mortalities in eggs and larvae. Thus, the chorion essentially dissolved, and the zona radiata became damaged by the action of exotoxins leading to punctures of the egg, leaking of cell constituents and mortalities (Hansen et al., 1992). To counteract the threat to egg production aquaculture relies on effective disinfecting techniques, manually removal of damaged and/or dead eggs and, where feasible, overproduction of eggs to ensure a sufficient number for hatching and subsequent growth.

Larvae/fingerlings

Larval/fingerling stages may be infected by a wider range of organisms. A topical example is rainbow trout fry syndrome (RTFS) and coldwater disease, both conditions of which are attributed to *Flavobacterium psychrophilum.* The pathogen affects juveniles to adults, with heavy losses of over half of the stock. For example, 40–60% mortalities were recorded in farmed perch (*Perca fluviatilis*) in Finland (Lönnström et al., 2008). However, substantially higher mortalities (i.e. ~90%) occurred in rainbow trout fry in Norway (Nilsen et al., 2011a). Here disease signs centred on anorexia, abdominal distension and melanosis particularly around the caudal peduncle. Larger fish became lethargic and revealed the presence of ulceration (Nilsen et al., 2011a). In addition, 7% of Atlantic salmon smolts in the size range of 60–10 g from a cool freshwater site in Norway died with clinical signs of septicaemia and myositis (Nilsen et al., 2011b). In mitigation, aquaculturists have used broodstock screening (Long et al., 2014) and egg disinfection techniques to successfully control RTFS (Madsen and Dalsgaard, 2008). Vaccines have shown signs of success, although research has emphasized uptake by intraperitoneal injection (Dumetz et al., 2006; Fredriksen et al., 2013), which is not practical for large numbers of small fish. However, there is potential for live attenuated vaccines, which could be applied by immersion or orally. One example conferred protection from challenge (Ghosh et al., 2015). In addition, a polyvalent, formalin-inactivated whole cell vaccine was applied by immersion for 30 seconds followed by a booster dose, and led to commendable protection (relative percent survival [RPS] = 84%) (Hoare et al., 2017).

Adults

Any losses that occur when fish are approaching market size may have significant economic effects on the owners. Examples of important bacterial pathogens that may affect larger fish [the disease names are given in brackets] include *Aeromonas salmonicida* [furunculosis; carp erythrodermatitis [CE]; ulcer disease], *Aliivibrio salmonicida* [coldwater vibriosis/Hitra disease], *Edwardsiella ictaluri* [enteric septicaemia of catfish; ESC], *Edwardsiella tarda* [edwardsiellosis, emphysematous putrefactive disease of catfish; EPD], *Francisella noatunensis* [francisellosis], *Lactococcus garvieae* [lactococcosis], *Moritella viscosa* [winter ulcer disease], *Photobacterium damselae* subsp. *piscicida* [pasteurellosis, pseudotuberculosis], *Piscirickettsia salmonis* [coho salmon syndrome, salmonid rickettsial septicaemia; SRS], *Renibacterium salmoninarum* [bacterial kidney disease; BKD], *Streptococcus iniae* [streptococcosis], *Tenacibaculum maritimum* [gill disease, black patch necrosis; BPN], *Vibrio anguillarum* [vibriosis] and *Yersinia ruckeri* [enteric redmouth; ERM] (see Austin and Austin, 2016).

Bacterial diseases of the dominant categories of farmed fish

Carp

Atypical strains of *Aeromonas salmonicida* have been linked to an ulcerative condition in carp, known as carp erythrodermatitis (CE). Characterized as a contagious skin disease, there are varying levels of mortality, but survivors are physically damaged with scar tissue leading to serious deformities, and are virtually unsaleable or of greatly reduced commercial value. Secondary invasion by other bacteria or fungi is a likely scenario (Austin and Austin, 2016). Vaccines have been researched, and in one example liposome-entrapped antigens from the pathogen were fed to carp leading to stimulation of an antibody response.

Catfish

Enteric septicaemia of catfish (ESC) was recognized initially among catfish farmed in the USA (Hawke, 1979), and subsequently in Asia (Hassan et al., 2012). In Vietnam, the disease is referred to as 'bacillary necrosis of *Pangasius*' (Crumlish et al., 2002). Losses of up to 50% may ensue (Plumb and Sanchez, 1983). Vaccines have offered excellent protection against the disease (Aarattuthodiyil et al., 2020; Kordon et al., 2020). A second condition, edwardsiellosis/EPD, leads to the development of abscesses, which develop and extend into gas-filled hollow areas. If damaged/punctured, a very unpleasant odour is released. The disease does not cause large-scale losses, but the severe economic impact results from the odours, which lead to cessation of activity in the processing plants while disinfection and deodoration are carried out. The result is that substantial financial losses to the processors may ensue from the presence of a comparatively few diseased fish (Meyer and Bullock, 1973). Fortunately, vaccination is successful for the control of the condition (Takano et al., 2011).

Salmonids

Numerous bacterial diseases have been reported in salmonids. Some pathogens infect principally salmonids in freshwater (e.g. *Yersinia ruckeri*), others are marine pathogens, notably *Aliivibrio salmonicida,* and some occur in both fresh- and seawater, such as *Aeromonas salmonicida.* Financial losses ensue as a result of mortalities, such as the case of bacterial kidney disease (BKD), and if the stock become physically damaged as with winter ulcer disease. All sizes of salmonids in freshwater, specifically rainbow trout (*Oncorhynchus mykiss*), have suffered heavy mortalities resulting from enteric redmouth (ERM), which is caused by *Yersinia ruckeri,* although disease is less severe in larger fish (Altinok and Grizzle, 2001; Pajdak-Czaus et al., 2019). However, effective vaccines have been commercialized and are readily available (Chettri et al., 2013; Villumsen et al., 2014). Trout and salmon in both fresh- and seawater have experienced infections and progressive low-level mortalities with the nutritionally fastidious, intracellular *Renibacterium salmoninarum,* the cause of BKD (Yoshimizu, 2016). Nevertheless, large-scale mortalities attributable to BKD of up to 56% have been documented in Japanese chum salmon (*Oncorhynchus keta*) farms 10–15 months after rearing eggs and asymptomatically infected fry (Suzuki et al., 2018). Unfortunately, BKD has been very difficult to control; long-term antibiotic treatment with penicillin or erythromycin has not been particularly successful and raises environmental and public health concerns (Austin and Austin, 2016). Slaughter of infected stock with site disinfection where possible (Austin and Austin, 2016) has been used successfully as an extreme control measure (Gudmundsdóttir et al., 2000). Vaccines have been developed and evaluated; a live product is commercially available (Salonius et al., 2005). *Aeromonas salmonicida,* the causal agent of furunculosis, has on occasions led to severe levels of mortalities in trout and salmon in fresh- and seawater. Vaccines have been developed, and a whole cell formalin-inactivated product based on the immunogenic iron regulated outer membrane proteins (IROMPs) commercialized (Santos et al., 2005; Austin and Austin, 2016; Marana et al., 2017).

Winter ulcer disease, which was characterized by the presence of large ulcers on the flanks of infected Atlantic salmon during winter, was recognized initially in Iceland and Norway (Benediktsdóttir et al., 1998; MacKinnon et al., 2019; Salte et al., 1994). Two organisms were considered as responsible, namely *Vibrio wodanis* and *Moritella viscosa* (Benediktsdóttir et al., 2000; MacKinnon et al., 2019). Apart from mortalities, survivors were unsaleable because of the gross physical damage. Vaccines have been researched; a formalized, adjuvanted whole cell product of *M. viscosa* administered by intraperitoneal injection led to an RPS of 97% (Greger and Goodrich, 1999). Coldwater vibriosis/Hitra disease is another disease of Atlantic salmon that is problematic in colder temperatures during late autumn to early spring. The disease was initially recognized as a generalized haemorrhagic septicaemia in Atlantic salmon farms located near the island of Hitra in Norway (Egidius et al., 1981) before spreading throughout Norway. Coldwater vibriosis is now widespread in salmon farming areas in Europe and North America (Austin and Austin, 2016). Substantial mortalities have occurred, but vaccines have achieved considerable success (Karlsen et al., 2011).

Tilapia

Francisellas, which are intracellular bacterial pathogens associated with systemic granulomatous disease, may cause very high levels of mortalities and severe economic losses in susceptible fish species (Pulpipat et al., 2020). For example, up to 60% mortalities were reported in Costa Rican tilapia farms (Soto et al., 2009). The pathogen has been linked with granulomas in Nile tilapia from Brazil (Leal et al., 2014) and hybrid Nile tilapia in Hawaii (Soto et al., 2013). Whereas effective treatment has been achieved with florfenicol (Soto et al., 2010), the need for a successful vaccine has spurred researchers. An attenuated mutant showed promise. Specifically, the attenuated cells were administered to tilapia by immersion, and resulted in commendable protection (RPS of up to 87.5%) following waterborne challenge with a virulent isolate (Soto et al., 2011). A formalin-inactivated whole cell preparation of *Francisella noatunensis* subsp. *orientalis* adjuvanted with Montanide was injected into tilapia followed by a booster dose 2 weeks later. After challenge, the RPS was >70% (Pulpipat et al., 2020). A similar approach using a formalized autogenous whole cell vaccine adjuvanted with Montanide led to an even higher RPS of 100% when challenged with the homologous isolate (Ramirez-Paredes et al., 2019). However, lower RPS values of 69.8% and 65.9% resulted after challenge with heterologous cultures (Shahin et al., 2019).

Marine fish species

A diverse range of marine fish species have been reported to have been diseased with *Photobacterium damselae* subsp. *piscicida* and *Vibrio anguillarum*. Pasteurellosis has developed into a serious disease with heavy losses of up to half the stock among a range of marine fish species (Egusa, 1983; Kanai, 2017). Since the initial occurrence in yellowtail, pasteurellosis has been recognized in sea bream (Balebona et al., 1998), sole *(Solea senegalensis)* (Zorrilla et al., 1999) and golden pompano (*Trachinotus ovatus*) (Wang et al., 2013). There has been a spread to farmed and wild fish in the Mediterranean, notably France, Italy and Spain (Magariños et al., 1992; Mladineo et al., 2006). Again, mitigation is possible by means of vaccination. For example, formalin-inactivated cells when administered to sea bream by intraperitoneal injection led to an RPS of 96% (Hanif et al., 2005). Furthermore, a bivalent formalized whole cell vaccine couple with bacterial extracellular products in combination with *V. harveyi* was applied to sole by immersion with a subsequent booster dose leading to an RPS of ~82% at 4 months (Arijo et al., 2005). Vibriosis attributed to *V. anguillarum* is a common pathogen in mariculture and has become a major constraint in the successful culture of numerous species resulting in high levels of mortalities (Austin and Austin, 2016). Fortunately, the disease may be controlled by vaccines (Angelidis et al., 2006) of which several examples of formalin-inactivated whole cell preparations are available commercially.

Bacterial shrimp pathogens

It has been estimated that approximately 20% of shrimp production is lost annually as a result of bacterial pathogens of which *Vibrio* spp. are the most important (Table 5.1; Flegel, 2102). *Vibrio harveyi* has developed into a significant problem of mariculture insofar as the pathogen infects fish and invertebrates, particularly shrimp. The pathogen has been associated with acute hepatopancreatic necrosis disease (along with *V. parahaemolyticus*, Muthukrishnan et al., 2019), mass mortalities in bacterial white tail disease in Chinese cultured whiteleg shrimp (Zhou et al., 2012), black shell disease in Indian sites rearing tiger shrimp (Selvin et al., 2005), luminous vibriosis (Prayitno and Latchford, 1995) and *Bolitas nigricans* (Robertson et al., 1998), particularly in Asia and South America. With luminous vibriosis, the shrimp appear to glow in the dark when the density of bacterial populations has increased. This is the prelude to the expression of clinical disease before the onset of mortalities (Prayitno and Latchford, 1995). Significant mortalities occurred in the economically important shrimp industry in Ecuador. Furthermore, a second manifestation of *V. harveyi* infection occurred in Ecuador when shrimp succumbed to a condition known as *Bolitas nigricans* (Spanish for small balls). Here the digestive tract became blocked with balls of sloughed-off epidermal tissue that impacted greatly on nutrition, that is, the shrimp were unable to feed (Robertson et al., 1998). Because of the economic importance of shrimp culture, much effort has been expended in developing effective disease control measures. From the earlier dominance on chemotherapy, current approaches have focused on the use of bacteriophages, probiotics, biofloc and vaccines. Research identified bacteriophages that controlled *V. harveyi* diseases, including luminous vibriosis (Choudhury et al. 2019; Lal et al., 2017; Stalin and Srinivasan, 2017). Probiotics, notably *V. alginolyticus,* were used successfully on Ecuadorian sites, and were allegedly responsible for reducing the need for antibiotics by as much as 95% (Robertson et al., 1998). In addition, biofloc has been considered to reduce lesions in the hepatopancreas of Pacific white shrimp caused by *V. harveyi* (Aguilera-Rivera et al., 2017).

V. parahaemolyticus has been recognized as a serious pathogen of shrimp vibriosis before the emergence of a new disease (acute hepatopancreatic necrosis syndrome/disease) as discussed below. The pathogen has been recovered from mass mortalities among shrimp farmed especially in the Americas, notably Mexico (Lopez-Tellez et al., 2019) and Asia, including India (Kumar et al., 2014; Nelapati et al., 2012). A complication is that the pathogen may also be a leading cause of human food-borne infections (i.e. acute gastroenteritis), associated with the consumption of contaminated/improperly prepared shrimp (Nelapati et al., 2012).

In 1990, a new disease with high (40–100%) mortality levels, which was coined as acute hepatopancreatic necrosis syndrome (AHPNS) and early mortality syndrome (EMS) was recognized in early stages of shrimp (with in the first 40 days) (Hong et al., 2016), involving whiteleg shrimp and tiger shrimp in China, Thailand and Vietnam (Flegel, 2012) with spread to India, Indonesia and Malaysia (Aranguren et al., 2017), and beyond Asia to Mexico in 2013 (Nunan et al., 2014; Tang et al., 2020). A characteristic pathology has been reported in acute stages of the disease, namely the profound sloughing off of epithelial cells in the hepatopancreas in the absence of any signs of the pathogen. Other signs of the disease included lethargy, slow growth, empty digestive tract and a pale atrophied hepatopancreas (Hong et al., 2016). The link with EMS was later removed to avoid confusion because of the association with other pathogens and environmental conditions (Hong et al., 2016). The economic loss in terms of reduced shrimp production in just one country, Thailand in 2010–2017, was estimated to be a staggering amount of US$7.38 billion with an additional US$4.2 billion of export losses (Shinn et al., 2018). For the Mekong Delta region of Vietnam in 2015, losses were thought to be >US$26 million (Shinn et al., 2018), and occurred in ~59,000 ha of farms (Dang et al., 2018). Indeed, AHPND was regarded as the most serious disease affecting shrimp aquaculture in Vietnam (Dang et al., 2018) with the microsporidian *Enterocytozoon hepatopenaei* (EHP) regarded as a risk factor (Aranguren et al., 2017). However, there is now evidence that the Vietnamese shrimp industry is recovering from the disease (Dang et al., 2018). Overall, AHPND may have depressed worldwide shrimp production by ~20% (Hong et al., 2016).

The causal agent was linked initially to *V. parahaemolyticus* (Hong et al., 2016; Lopez-Leon et al., 2016; Soto-Rodriguez et al., 2015) when the condition was referred to as a 'disease' (AHPND rather than the initial recognition as a syndrome, viz. AHPNS). Pathogenic cultures contain the pVA plasmid, which codes for genes specifying the PirA(vp) and PirB(vp) toxins that cause rapid death in infected shrimp (Flegel and Sritunyaluucksana, 2018). Isolates were identified by molecular methods, e.g. repetitive extragenic palindromic element-PCR (rep-PCR) and immersion challenge led to the development of disease reminiscent of the signs witnessed on shrimp farms (Soto-Rodriguez et al., 2015). However, other vibrios have since been linked with AHPND (Kumar et al., 2020), and include *V. campbellii* (Dong et al., 2017; Han et al., 2017), *V. harveyi* (Muthukrishnan et al., 2019), *V. owensii* (Xiao et al., 2017) and *V. punensis* (Restrepo et al., 2018). Therefore, the initial description of syndrome seems appropriate after all.

Mention will be made about necrotizing hepatopancreatitis of juvenile, adult and broodstock shrimp which is caused by an as yet uncultured obligately intracellular organism, coined *Candidatus* Hepatobacter penaei, with infection centring in the heptopancreas (Nunan et al., 2013). Incidences have been recorded throughout South, Central and North America (Morales-Covarrubias et al., 2011). Mortalities have been reported to approach 100% of the stock. Antibiotic therapy with florphenicol and oxytetracycline has been successful at controlling the disease (Morales-Covarrubias et al., 2012). However, there are concerns about the use of antibiotics in any nonmedical situation.

Fungal pathogens

A comparatively low range of fungal pathogens has been linked with serious diseases of aquaculture leading to large-scale losses and issues with sustainability (Table 5.1). Saprolegniasis, which is attributed to the oomycete fungi *Saprolegnia diclina* and *S. parasitica*, has been regarded as a serious disease of freshwater fish, notably trout, causing large-scale losses particularly to eggs and fry (Songe et al., 2016; Willoughby, 1970). It has been estimated that >10% of salmonid eggs become infected in hatcheries with saprolegniasis annually (Bruno et al., 2011). Fish and eggs may be observed to develop white-grey cotton-wool like growth that spreads progressively covering most (80%) of the surface area leading to destruction of the (fish) epidermidis (Hussein et al., 2001; Richards and Pickering, 1979; Willoughby, 1989). Disease signs include lethargy and loss of equilibrium. Malachite green was a very effective control agent (Kitancharoen et al., 1997), and since its ban as a result of its toxicity and carcinogenicity, research has been directed at finding a suitable replacement. Disinfectants, such as hydrogen peroxide, have shown potential (Kitancharoen et al., 1997). Another approach has been to use plant products, of which lawsone (Shin et al., 2017) and the aqueous methanolic extracts of garlic skin (0.4–1.6 g/l) and stems (0.8 and 1.6 g/l) reduced saprolegniasis significantly (Ozcelik et al., 2020).

Epizootic ulcerative syndrome (EUS) was first reported in farmed ayu (*Plecoglossus altivelis*) in Japanese farms during 1971, and has since spread widely in freshwater and estuarine fish in Africa, Asia, Australia and North America resulting in substantial losses, even up to 100% (Iberahim et al., 2018; Kamilya and Baruah, 2014; Pradham et al., 2014; Sumithra et al., 2020). EUS has developed into a sufficiently serious condition to be included in the list of notifiable diseases by the World Organization for Animal Health (OIE). There has been some disagreement about the nature of the causal agent: bacteria (*Aeromonas hydrophila* and *Pseudomonas fluorescens*) have been implicated (Mastan and Qureshi, 2001) although these organisms may well have been secondary invaders/opportunists of already damaged tissues, but the overwhelming evidence pointed to a fungal cause. Thus, *Aphanomyces invaderis* was proposed as the causal agent of EUS (Willoughby et al., 1995), with a later name change to *A. invadans* (Iberahim et al., 2018). The organism has been regarded as highly virulent and invasive leading to the development of necrotic dermal ulcers, with penetration to the underlying tissues. Granulomas have been noted in the visceral organs (Kumar et al., 2020; Vishwanath et al., 1998). Unfortunately, control measures have not been particularly successful to date. However, the ability of infected fish to develop immune responses suggest that vaccines may offer hope for the future (Kumar et al., 2020).

Viral pathogens of fish

The most important viral diseases of salmonids, cyprinids and sea bream are included in the list of OIE notifiable diseases, and include infectious haematopoietic necrosis (IHN), infectious pancreatic necrosis (IPN), infectious salmon anaemia (ISA), red sea bream iridovirus disease (RSBID), spring viraemia of carp (SVC) and VHS (Table 5.1). The detection of any of these viruses in aquaculture would inevitably trigger a rapid response involving strict movement controls, particularly the culling of infected stock, hygienic disposal of the corpses, thorough disinfection of the facilities and surveillance at neighbouring sites. If compensation is available, the impact on the fish farmer would be reduced otherwise livelihoods would be threatened. Undoubtedly, the viruses are transferred from wild populations of fish in the vicinity of fish farms (e.g. Karreman, 2006).

IHN was first recognized in the early 1950s in farmed sockeye salmon on the west coast of North America (Rucker et al., 1953). The disease is caused by a rhabdovirus, which causes an acute, systemic disease that affects freshwater – particularly fry – and seawater stages of Atlantic salmon and rainbow trout causing economic problems for the infected aquaculture sites (Dixon et al., 2016). Often, infection leads to mortality because of impaired osmotic function, oedema and widespread haemorrhaging especially at a water temperature of 3–18°C (Bootland and Leong, 1999). With acute outbreaks of IHN, several percentages of the stock will be lost each day, with the cumulative mortalities in the region of 90–95% (Bootland and Leong, 1999). With chronic disease, there will be progressive low-level mortalities. The disease was first recorded in the USA, and spread to Europe and Asia (China, Japan and Korea) with the movement of infected fish and eggs. Once in a fish farm, the virus is likely to become established in wild fish populations in the direct vicinity. Mayflies (Shors and Winston, 1989) and sea lice (Jakob et al., 2011) may be potential vectors in the transmission of the disease. Evidence points to the ability of the virus to survive outside of the fish host in fresh water containing organic material for >1 month at cool temperatures. The virus is likely to enter the host via the gills and at the base of the fins, and is to be found in highest quantities in the internal organics, notably the kidney and spleen (Wolf, 1988; Bootland and Leong, 1999). Disease may be prevented by the disinfection of eggs (Bovo et al., 2005), the use of virus-free water, and the implementation of strict biosecurity procedures (Winton, 1991). The disease progression may be halted by controlling the movement of infected stock and by eradication with appropriate disposal. Fortunately, vaccination is effective at controlling IHN (Garver and Wade, 2017).

IPN, which is caused by an aquabirnavirus, was recognized initially in trout from freshwater sites in North America during the 1950s (Wood et al., 1955) and then in Europe from the early 1970s (Ball et at., 1971). The disease is regarded as a highly contagious disease of young fish, notably rainbow trout fry and fingerlings with mortalities reaching 80–90% of the stock (Dopazo 2020; Roberts and Pearson, 2005; Ruane et al., 2007). With the expansion of the salmon farming industry during the 1970s in Norway and Scotland, the incidence of IPN increased such that it is now widespread in salmon juvenile stages in freshwater and in post-smolts in seawater (Ruane et al., 2007). The first signs of an outbreak in fry is a sudden and progressive increase in daily mortalities. Disease signs include melanosis, distended abdomen and characteristic spiral/corkscrew swimming. Survivors in seawater may exhibit weak if any growth and appear chronically emaciated, and prone to sea lice infestation (Roberts and Pearson, 2005).

ISA, which is caused by an Orthomyxoviridae isavirus, was first recognized as a serious anaemia in farmed Atlantic salmon in Norway during 1984, and became regarded as a deadly disease comprising a serious threat to seawater aquaculture having spread to Canada, Chile, Faroe Islands, Ireland, Scotland and the USA (Aamelfot et al., 2014; Rimstad and Markussen, 2020; Shchelkanov et al., 2017). Total mortalities may reach 100% (Kilbenge et al., 2004). Disease signs include lethargy, pale gills, exophthalmia and haemorrhaging in the anterior eye chamber. An added complication is that ISA may be carried by sea lice, and therefore contribute to its spread to other salmon farms (Shchelkanov et al., 2017). The most effective disease control strategy involves culling the infected stock (McClure et al., 2005).

RSBID has caused severe economic damage to mariculture since its initial recognition in farmed red sea bream during 1990, and has been detected in at least 20 farmed marine fish species of which juveniles were more susceptible than adults (Matsuoka et al., 1996), and subsequently spread to devil stinger (*Inimicus japonicus*) hatcheries in Japan (Kawato et al., 2017) and in perch (*Perca fluviatilis*), stone flounder (*Kareius bicoloratus*) and turbot (*Scophthalmus maximus*) in Shandong Province, China (Zhao et al., 2011). The disease has become widely distributed across Asia. Disease signs include lethargy, petechia on the gills, severe anaemia, enlarged spleen and enlarged cells, which stain by Giemsa's methods, in the gills, intestine, heart, kidney and spleen (Inoue et al., 1992; Jung et al., 1997; Nakajima and Maeno, 1998). There is a correlation between mortalities and water temperature, with heavy losses at 18°C, but none at all at 13°C (Lyu et al., 2009). An effective formalin-inactivated vaccine is available commercially in Japan (Kawato et al., 2017).

SVC, which is caused by a representative of the Rhabdoviridae, Sprivivirus, has developed into a devastating disease of farmed carp (up to 70% mortalities were reported in goldfish (*Carassius auratus*) in Sao Paulo, Brazil (Alexandrino et al., 1998) particularly during spring when the water temperatures increase after winter. Specifically, disease outbreaks in carp tend to occur in water temperatures of 11–17°C, and rarely <10°C (Fijan, 1988). In Europe, losses of up to 70% (usually 1–40%) have been reported in young carp (Ahne et al., 2002). SVC is characterized by acute haemorrhaging on the surface and in the internal organs, deterioration of the gill lamellae, abdominal distension and enteritis, exophthalmia, swollen spleen, pericarditis and liver necrosis (Ashraf et al., 2016), leading to severe economic losses in numerous countries in Asia and Europe, including China (Shao and Zhao, 2017; Zheng et al., 2018) and Poland (Maj-Paluch et al., 2019), and also in North and South America, namely Brazil (Alexandrino et al., 1998), Mexico (Ortega et al., 2019), Canada and the USA (Ahne et al., 2002; Ashraf et al., 2016; Garver et al., 2007; Goodwin, 2002). The disease caused >35% mortalities in the cyprinid *Percocypris pingi* farmed in Sichuan Province, China during April 2016 (Zheng et al., 2018). SVC was detected initially in the UK (England and Wales, but not Scotland where cyprinids are less common) during 1977 leading to a programme of eradication, which resulted in the country being free of the disease in 2010 (Taylor et al., 2013). The virus is capable of surviving outside the host for 5 weeks in 10°C river water, and >6 weeks in 4°C (4 days at 10°C) pond sediment (Ahne, 1976). Once SVC becomes established, it is very difficult to eradicate without completely destroying all living organisms on the site.

VHS, which is caused by a virus of the Rhabdoviridae, is regarded as an extremely devastating disease that affects >140 species of freshwater and marine fish, worldwide, and constitutes an important threat to aquatic food security particularly of farmed Japanese flounder (*Paralichthys olivaceus*), rainbow trout and turbot (Skall et al., 2005; Escobar et al., 2018). There may be rapid onset of mortalities especially in smaller fish. Thus, fry may experience 100% mortalities. The highest mortalities have been documented in spring among fish in cool/cold water (i.e. 9–12°C) (Skall et al., 2005). Disease signs include anaemic gills, distended abdomen, exophthalmia, lethargy, melanosis and extensive haemorrhages on the body surface. Chronic infections may not result in external disease signs. The virus may survive outside the fish hosts in freshwater but less so in seawater (Hawley and Garver, 2008).

Koi herpesvirus (KHV) is a serious disease of juvenile to adult carp with widespread distribution affecting particularly the gills, kidney and spleen (Sano et al., 2004). Mortalities are typically in the range of 70–90% (Bergmann et al., 2010).

Viral pathogens of shrimp

It has been estimated that 60% of shrimp production is lost as a result of viral diseases of which white spot syndrome virus (WSSV) is among the most serious (Table 5.1; Flegel, 2012).

The European Commission Council Directive 2006/88/EC adopted during 2008 has listed white spot syndrome (WSS) as a serious crustacean disease. WSS, which is characterized by the presence of white spots up to 3 mm in diameter on the exoskeleton, was reported initially as a cause of very high or 100% mortalities with rapid onset, that is, within 3–10 days of outbreaks first occurring, in farmed shrimp in China, Taiwan and Thailand during 1991–1992. Since then, WSS has spread throughout shrimp producing countries, including India, Japan, Korea, the Middle East and the Americas, resulting in considerable losses to production particularly of *Penaeus monodon, L. vannamei* and *L. stylirostris* and thus profound economic effects (Dey et al., 2020; Lo and Kou, 1998; Sanchez-Martinez et al., 2007). As an example, WSS was estimated to have caused >US$11 million of losses to shrimp production in the Mekong Delta, Vietnam in 2015 (Shinn et al., 2018). In the presence of external stressors, the virus is capable of rapid replication with resulting mortalities (Lo and Kou, 1998). The virus, which is a member of the Nimaviridae, affects a diverse range of aquatic crustaceans resulting in varying levels of mortality. The rapid spread of the disease may result from the movement of infected shrimp, and natural reservoirs, including other aquatic inhabitants and sea birds (Sanchez-Martinez et al., 2007). The virus has been recorded as latent infections in crabs and wild shrimp (Lo and Kou, 1998).

In 1992, a new disease with high mortalities of up to 100% (typically 40 to >90%) of the stock was recognized among shrimp, notably *L. vannamei,* which were farmed along the Taura river in Ecuador. It has been observed that the most frequent outbreaks of disease have occurred at salinities of <30 o/oo (Jimenez et al., 2000). The disease, for which the causal agent was a newly described icosahedral virus of 31 nm in diameter, was named as Taura syndrome, and rapidly became the most serious problem of shrimp farming in Ecuador (Bonami et al., 1997; Brock, 1997; Dhar et al., 2010). In Taiwan, mortalities of *L. vannamei* juveniles were recorded within 3 days of the onset of Taura syndrome during March 1999 (Yu and Song, 2000). Moribund juveniles were described by Yu and Song (2000) as lethargic, anorexic and with ataxic swimming. The musculature was opaque, and the shell was soft with a red colouration (Yu and Song, 2000). Extensive epithelial necrosis particularly in the appendages, body surface, gills and stomach was reported by Dhar et al. (2010). Taura syndrome was one of only three serious crustacean diseases included in the European Commission Council Directive 2006/88/EC of 2008 (Stentiford et al., 2009). Since 1992, the disease has spread throughout shrimp producing countries in the Americas, including Colombia, Peru, Mexico and the USA (Aranguren et al., 2013; Brock, 1997; Dhar et al., 2010; Hasson et al., 1999; Lightner, 1996a; Lightner et al., 1995), and in Asia, namely China, Indonesia, Korea, Malaysia, Taiwan and Thailand (Cheng et al., 2011; Do et al., 2006; Nielsen et al., 2005; Ruangsri et al., 2007; Tang and Lightner, 2005; Yu and Song, 2000) leading to significant economic losses. In Taiwan, 30 to 45 days after stocking post larvae obtained mostly from Ecuador, the majority (90%) of shrimp ponds were deserted as a result of mortalities that started 2–3 days after cessation of feeding (Tu et al., 1999). As an added complication, the presence of WSSV has been detected in shrimp otherwise infected with Taura syndrome (Phalitakul et al., 2006; Tsai et al., 2002). However, Taura syndrome resistant lines of *L. vannamei* have been developed, and survive extremely well (up to 100% survival) in laboratory challenges (Lightner et al., 2009).

Other diseases of significance include the following.

- Infectious hypodermal and haematopoietic necrosis (IHHN) is a serious pathogen of *Penaeus* spp. and *Litopenaeus* spp. causing up to 100% mortalities with the virus detected in all stages from eggs to adults (Motte et al., 2003). The causal agent, which is the smallest known shrimp virus, may be detected widely throughout infected animal tissues (Lightner, 1993, 2011). The disease is common in Asia, Central and South America, the Pacific islands, and the Middle East, occurring in farmed and wild shrimp (Bondad-Reantaso et al., 2001; Lightner, 1996b).
- Infectious myonecrosis has caused significant mortalities of 40–70% in juvenile and subadult *L. vannamei,* particularly those in water with low salinity, in Asia and Brazil (Andrade et al., 2007 Nunes et al., 2004; Saengchan et al., 2007).
- White tail disease of shrimp (also referred to as white muscle disease) has been responsible for large-scale mortalities in freshwater prawn (*Macrobrachium rosenbergii*) larvae and juveniles particularly in Asia (Qian et al., 2003; Sahul Hameed et al., 2004; Yoganandhan et al., 2006). The mortalities peak 5–6 days after the first indication of overt disease signs, with few post larvae surviving more than 15 days after the start of the outbreak; survivors grow normally into adults (Qian et al., 2003). Adult prawns are resistant to the disease, but may be carriers (Qian et al., 2003; Sahul Hameed et al., 2004).

- Yellow head disease of penaeids has been reported in East Africa, Asia, Australia and Mexico where mortalities may reach 100% within 3–5 days of the start of the outbreak (Chantanachookin et al., 1993).

Parasites

The caligid copepod salmon louse (*Lepeophtheirus salmonis*) has developed into a major constraint on the production of Atlantic salmon with losses affecting the growth of the salmon farming industry and its sustainability (Jones et al., 2015). In Norway, which is the world's largest producer of Atlantic salmon, the economic impact was estimated to be >NOK 3.4 billion in 2014 rising to >NOK 5 billion in 2015. This was equivalent to NOK 3836.28/t of salmon production (Iversen et al., 2015). However, higher estimates of NOK 7–8 billion per year have been presented (Rødseth, 2016). By extrapolating these figures, the global cost of sea lice would have been approximately NOK 9 billion in 2015, increasing in the following years (Brooker et al., 2018). The parasites attach to the salmon skin, and feed on the mucus/skin and blood (Mordue and Birkett, 2009). With a ready source of food, the lice proliferate causing profound skin erosion, immunosuppression and secondary infections (Hamre et al., 2009). The severe economic impact has led to much research activity devoted to determining suitable controls (Torrissen et al., 2013), and the costs were estimated as €430 million just in Norway during 2015 (Iversen et al., 2015). Essentially, the approaches have proceeded from therapeutic to preventative. Chemicals, including deltamethrin and hydrogen peroxide have been effective, but the benefit was negated by environmental issues concerning the release of the chemicals into the waterways around the salmon farms (Urbina et al., 2019). There has been success with the use of cleaner fish, notably wrasse and lumpsuckers, that consume the parasites from the surface of the salmon leading to a 100% reduction in lice numbers (Overton et al., 2019, 2020; Powell et al., 2018). Also, physical measures have been successful at reducing the likelihood of host-parasite contact. In particular, barriers/skirts around cages reduced infestation densities by up to 100%, increasing to 100% reduction if the cages were fully enclosed (Barrett et al., 2020).

With the expansion of the salmon and trout aquaculture industries in the last decade of the twentieth century, a new parasitic gill disease appeared initially in sea-caged Atlantic salmon in Tasmania (Munday, 1986), and spread to 14 countries across a wide geographical area (Oldham et al., 2016). The disease, which was coined amoebic gill disease (AGD), is caused principally in salmon and trout by *Neoparamoeba perurans* (Adams et al., 2012; Marcos-Lopez and Rodger, 2020; Powell et al., 2015). The greatest impact is on Atlantic salmon in seawater (Oldham et al., 2016). In Tasmania, the control of AGD was calculated to increase the cost of production of Atlantic salmon by ~20% resulting from mortalities, lost growth and treatments (Kube et al., 2012). In Scotland, the losses were worth ~US$81 million in 2011 (Shinn et al., 2015). Essentially, the parasites inhabit the gills where they feed on organic material and proliferate to such high populations that the outcome is fusion of the gill lamellae, gill necrosis and respiratory issues; inappetence and poor growth is a direct result (Hvas et al., 2017). Indeed, the mortalities, which result from asphyxiation, have been reported to reach as high as 82% of the stock in the absence of control measures (Steinum et al., 2008). A reduction in stocking density is helpful at controlling AGD (Douglas-Helders et al., 2004). Also, improvements in farm hygiene, including more regular net changes that improve water flow and increased oxygenation levels in the water were beneficial (Nowak, 2001; Rodger, 2014) possibly reducing biofouling communities that may be the reservoir for the parasite (Tan et al., 2002). The principal treatment has been bathing in freshwater, which helps but does not eliminate the disease (Parsons et al., 2001; Clark et al., 2003; Powell et al., 2015). Also, hydrogen peroxide has been used with some success (Powell et al., 2015).

Conclusion

Aquaculture continues to be affected by a diverse range of pathogens and parasites, which may have a serious impact on the sustainability and economic well-being of the industry. Great emphasis has been placed on developing effective disease control measures, and there has been a gradual move away from therapeutic to prophylactic approaches. This is commendable but mortalities continue, and new pathogens continue to emerge. Perhaps, there should not be any surprise when disease strikes a monoculture situation whereby large numbers of individuals are maintained in close proximity to each other and are exposed to the pathogens/parasites of wild species in the surrounding aquatic environment. Certainly, the movement of pathogens from wild to farmed aquatic animals is well documented, with examples including ISA and sea lice. However, it is unlikely that the transfer of pathogens/parasites from wild to farmed animals will cease unless there are profound changes to aquacultural infrastructures. Yet in terms of food security, aquaculture has been functioning for millennia, and production levels are increasing continuously. Despite the issues of disease, it is argued that aquaculture has a bright future. So, in terms of aquatic food security, what are the issues?

- Diseases may result in mortalities in all stages of production from eggs to adults. The economic impact will result also from the need for hygienic disposal of the corpses.
- Diseases may result in deformities or unsightly lesions thereby rendering the aquatic species virtually unsaleable.
- Disease may lead to a reduction in growth and feed conversion leading to higher costs for the producers.
- The presence of some pathogens (*Vibrio vulnificus* in bivalves and *V. parahaemolyticus* in shrimp) may have an impact on human health leading to embargoes on the sale of the infected animals for (human) consumption.
- The presence of some viral diseases, such as IHN and VHS, invoke slaughter policies, fallowing, and surveillance of surrounding sites. This has an inevitable effect on production.

References

Aamelfot, M., Dale, O.B. & Falk, K. (2014) Infectious salmon anaemia – Pathogenesis and tropism. *Journal of Fish Diseases*, 37, 291–307. https://doi.org/10.1111/jfd.12225.

Aarattuthodiyil, S., Griffin, M.J., Greenway, T.E., Khoo, L.H., Byars, T.S., Lewis, M., Steadman, J. & Wise, D.J. (2020) An orally delivered, live-attenuated *Edwardsiella ictaluri* vaccine efficiently protects channel catfish fingerlings against multiple *Edwardsiella ictaluri* isolates. *Journal of the World Aquaculture Society*, 51, 1354–1372. https://doi.org/10.1111/jwas.12693.

Adams, M.B., Crosbie, P.B. & Nowak, B.F. (2012) Preliminary success using hydrogen peroxide to treat Atlantic salmon, *Salmo salar* L., affected with experimentally induced amoebic gill disease (AGD). *Journal of Fish Diseases*, 35, 839–848. https://doi.org/10.1111/j.1365-2761.2012.01422.x.

Aguilera-Rivera, D., Escalante-Herrera, K., Gaxiola, G., Prieto-Dav, A., Rodriguez-Fuentes, G., Guerra-Castro, E., Hernandez-Lopez, J., Chavez-Sanchez, M.C. & Rodriguez-Canul, R. (2017) Immune response of the Pacific white shrimp, *Litopenaeus vannamei*, previously reared in biofloc and after an infection assay with *Vibrio harveyi*. *Journal of World Aquaculture Society*, 50, 119–136. https://doi.org/10.1111/jwas.12543.

Ahne, W. (1976) Untersuchungen über die Stabilität des karpfenpathogenen Virusstammes 10/3. *Fisch. Umwelt*, 2, 121–127.

Ahne, W., Bjorklund, H.V., Essbauer, S., Fijan, N., Kurath, G. & Winton, J.R. (2002) Spring viremia of carp (SVC). *Diseases of Aquatic Organisms*, 52, 261–272. https://doi.org/10.3354/dao052261.

Alexandrino, A.C., Ranzani-Paiva, M.J.T. & Romano, L.A. (1998) Identificación de viremia primaveral de la carpa (VPC) Carassius auratus en San Pablo, Brasil. *Revista Ceres*, 45, 125–137.

Altinok, I. & Grizzle, J.M. (2001) Effects of salinity on *Yersinia ruckeri* infection of rainbow trout and brown trout. *Journal of Aquatic Animal Health*, 13, 334–339. https://doi.org/10.1577/1548-8667(2001)013<0334:EOSOYR>2.0.CO;2.

Andrade, T.P.D., Srisuvan, T., Tang, K.F.J. & Lightner, D.V. (2007) Real-time reverse transcription polymerase chain reaction assay using TaqMan probe for detection and quantification of infectious myonecrosis virus (IMNV). *Aquaculture*, 264, 9–15. https://doi.org/10.1016/j.aquaculture.2006.11.030.

Angelidis, P., Karagiannis, D. & Crump, E.M. (2006) Efficacy of a *Listonella anguillarum*. *Diseases of Aquatic Organisms*, 71, 19–24. https://doi.org/10.3354/dao071019.

Aranguren, L.F., Salazar, M., Tang, K., Caraballo, X. & Lightner, D. (2013) Characterization of a new strain of Taura syndrome virus (TSV) from Colombian shrimp farms and the implication in the selection of TSV resistant lines. *Journal of Invertebrate Pathology*, 112, 68–73. https://doi.org/10.1016/j.jip.2012.08.009.

Aranguren, L.F., Han, J.E. & Tang, K.F.J. (2017) Enterocytozoon hepatopenaei (EHP) is a risk factor for acute hepatopancreatic necrosis disease (AHPND) and septic hepatopancreatic necrosis (SHPN) in the Pacific white shrimp Penaeus vannamei. *Aquaculture*, 471, 37–42. https://doi.org/10.1016/j.aquaculture.2016.12.038.

Arijo, S., Rico, R., Chabrillon, M., Diaz-Rosales, P., Martínez-Manzanares, E., Balebona, M.C., Magariños, B., Toranzo, A.E. & Moriñigo, M.A. (2005) Effectiveness of a divalent vaccine for sole, *Solea senegalensis* (Kaup), against *Vibrio harveyi* and *Photobacterium damselae* subsp. *piscicida*. *Journal of Fish Diseases*, 28, 33–38. https://doi.org/10.1111/j.1365-2761.2004.00597.x.

Ashraf, U., Lu, Y., Lin, L., Yuan, J., Wang, M. & Liu, X. (2016) Spring viraemia of carp virus: Recent advances. *Journal of General Virology*, 97, 1037–1051. https://doi.org/10.1099/jgv.0.000436.

Austin, B. & Austin, D.A. (2016). *Bacterial Fish Pathogens: Disease of Farmed and Wild Fish*, 6th edn. Springer: Dordrecht.

Balebona, M.C., Zorrilla, I., Moriñigo, M.A. & Borrego, J.J. (1998) Survey of bacterial pathogens affecting farmed gilt-head sea bream (*Sparus aurata* L.) in southwestern Spain from 1990 to 1996. *Aquaculture*, 166, 19–35. https://doi.org/10.1016/S0044-8486(98)00282-8.

Ball, H.J., Munro, A.L.S., Ellis, A., Elson, K.G.R., Hodgkiss, W. & McFarlane, I.S. (1971) Infectious pancreatic necrosis in rainbow trout in Scotland. *Nature*, 234, 417–418. https://doi.org/10.1038/234417a0] [PubMed: 4945908.

Barrett, L.T., Oppedal, F., Robinson, N. & Dempster, T. (2020) Prevention not cure: A review of methods to avoid sea lice infestations in salmon aquaculture. *Reviews in Aquaculture*, 12, 2527–2543. https://doi.org/10.1111/raq.12456.

Benediktsdóttir, E., Helgason, S. & Sigurjónsdóttir, H. (1998) *Vibrio* spp. isolated from salmonids with shallow skin lesions and reared at low temperature. *Journal of Fish Diseases*, 21, 19–28. https://doi.org/10.1046/j.1365-2761.1998.00065.x.

Benediktsdóttir, E., Verdonck, L., Spröer, C., Helgason, S. & Swings, J. (2000) Characterization of Vibrio viscosus and Vibrio wodanis isolated from different geographical locations: A proposal for reclassification of Vibrio viscosus as Moritella viscose comb. nov. *International Journal of Systematic and Evolutionary Microbiology*, 50, 479–488. https://doi.org/10.1099/00207713-50-2-479.

Bergmann, S.M., Sadowski, J., Kiełpiński, M., Bartłomiejczyk, M., Fichtner, D., Riebe, R., Lenk, M. & Kempter, J. (2010) Susceptibility of koi x crucian carp and koi x goldfish hybrids to koi herpesvirus (KHV) and the development of KHV disease (KHVD). *Journal of Fish Diseases*, 33, 267–272. https://doi.org/10.1111/j.1365-2761.2009.01127.x.

Birkbeck, T.H., Laidler, L.A., Grant, A.N. & Cox, D.I. (2002) Pasteurella skyensis sp. nov., isolated from Atlantic salmon (Salmo salar L.). *International Journal of Systematic and Evolutionary Microbiology*, 52, 699–704. https://doi.org/10.1099/00207713-52-3-699.

Birkbeck, T.H., Rennie, S., Hunter, D., Laidler, L.A. & Wadsworth, S. (2004) Infectivity of a Scottish isolate of Piscirickettsia salmonis for Atlantic salmon Salmo salar and immunity of salmon to this agent. *Diseases of Aquatic Organisms*, 60, 97–103. https://doi.org/10.3354/dao060097.

Bonami, J.R., Hasson, K.W., Mari, J., Poulos, B.T. & Lightner, D.V. (1997) Taura syndrome of marine penaeid shrimp: Characterization of the viral agent. *Journal of General Virology*, 78, 313–319. https://doi.org/10.1099/0022-1317-78-2-313.

Bondad-Reantaso, M.G., Mcgladdery, S.E., East, I. & Subasinghe, R.P. (2001) *Asia Diagnostic Guide to Aquatic Animal Diseases*. FAO Fisheries Technical Paper. Food and Agriculture Organization: Rome.

Bootland, L.M. & Leong, J.C. (1999) Infectious hematopoietic necrosis virus. In: *Fish Diseases and Disorders, Vol. 3: Viral, Bacterial and Fungal Infections* (edited by P. T. K. Woo & D. W. Bruno). CABI International: Wallingford.

Bovo, G., Håstein, T., Hill, B., Lapatra, S.E., Michel, C., Olesen, N.J., Shchelkunov, I., Storset, A., Wolffrom, T. & Midtlying, P.J. (2005) Hazard identification for vertical transfer of fish disease agents. Available at: www.crl-fish.eu/upload/sites/crl-fish/reports/links/fisheggtrade%20wp1.pdf.

Brock, J.A. (1997) Taura syndrome, a disease important to shrimp farms in the Americas. *World Journal of Microbiology and Biotechnology*, 13, 415–418. https://doi.org/10.1023/A:1018524216600.

Brooker, A.J., Skern-Mauritzen, R. & Bron, J.E. (2018) Production, mortality, and infectivity of planktonic larval sea lice, Lepeophtheirus salmonis (Krøyer, 1837): Current knowledge and implications for epidemiological modelling. *ICES Journal of Marine Science*, 75, 1214–1234. https://doi.org/10.1093/icesjms/fsy015.

Bruno, D.W., Van West, P. & Beakes, G.W. (2011) Saprolegnia and other oomycetes. In: *Fish Diseases and Disorders: Viral, Bacterial and Fungal Infections* (edited by D. W. Bruno & P. T. K. Woo). CABI International: Wallingford, pp. 669–720.

Cao, H.-P., Wen, L.-F., He, S., Lu, L.-Q., Yang, X.-L. & Chen, B.-Y. (2015) Vibrio cholerae: A causal agent for white feces syndrome in freshwater cultured whiteleg shrimp (Penaaeus vannamei). *Israeli Journal of Aquaculture Bamidgeh*, 67, article no.: 1212.

Chantanachookin, C., Boonyaratpalin, S., Kasornchandra, J., Direkbusarakom, S., Ekpanithanpong, U., Supamataya, K., Sriurairatana, S. & Flegel, T.W. (1993) Histology and ultrastructure reveal a new granulosis-like virus in Penaeus monodon affected by yellow-head disease. *Diseases of Aquatic Organisms*, 17, 145–157. https://doi.org/10.3354/dao017145.

Cheng, L.-T., Lin, W.H., Wang, P.C., Tsai, M.A., Ho, P.Y., Hsu, J.P., Chern, R.S. & Chen, S.C. (2011) Epidemiology and phylogenetic analysis of Taura syndrome virus in cultured Pacific white shrimp litopenaeus vannamei B. in Taiwan. *Diseases of Aquatic Organisms*, 97, 17–23. https://doi.org/10.3354/dao02407.

Chettri, J.K., Deshmukh, S., Holten-Andersen, L., Jafaar, R.M., Dalsgaard, I. & Buchmann, K. (2013) Comparative evaluation of administration methods for a vaccine protecting rainbow trout against Yersinia ruckeri O1 biotype 2 infections. *Veterinary Immunology and Immunopathology*, 154, 42–47. https://doi.org/10.1016/j.vetimm.2013.04.001.

Choudhury, T.G., Maiti, B., Venugopal, M.N. & Karunasagar, I. (2019) Influence of some environmental variables and addition of r-lysozyme on efficacy of Vibrio harveyi phage for therapy. *Journal of Biosciences*, 44, 8. https://doi.org/10.1007/s12038-018-9830-x.

Clark, G., Powell, M. & Nowak, B. (2003) Effects of commercial freshwater bathing on reinfection of Atlantic salmon, Salmo salar, with amoebic gill disease. *Aquaculture*, 219, 135–142. https://doi.org/10.1016/S0044-8486(03)00020-6.

Crumlish, M., Dung, T.T., Turnbull, J.F., Ngoc, N.T.N. & Ferguson, H.W. (2002) Identification of Edwardsiella ictaluri from diseased freshwater catfish, Pangasius hypophthalmus (Sauvage), cultured in the Mekong Delta, Vietnam. *Journal of Fish Diseases*, 25, 733–736. https://doi.org/10.1046/j.1365-2761.2002.00412.x.

Cvitanich, J.D., Garate N., O. & Smith, C.E. (1991) The isolation of a rickettsia-like organism causing disease and mortality in Chilean salmonids and its confirmation by Koch's postulate. *Journal of Fish Diseases*, 14, 121–145. https://doi.org/10.1111/j.1365-2761.1991.tb00584.x.

Dang, T.L., Pham, A.T. & Phan, T.V. (2018) Acute hepatopancreatic necrosis disease (AHPND) in Vietnam. *Asian Fisheries Science*, 31S. https://doi.org/10.33997/j.afs.2018.31.S1.020.

Dey, B.K., Dugassa, G.H., Hinzano, S.M. & Bossier, P. (2020) Causative agent, diagnosis and management of white spot disease in shrimp: A review. *Reviews in Aquaculture*, 12, 822–865. https://doi.org/10.1111/raq.12352.

Dhar, A.K., Lakshman, D.K., Amundsen, K., Robles-Sikisaka, R., Kaizer, K.N., Roy, S., Hasson, K.W. & Allnutt, F.C. (2010) Characterization of a Taura syndrome virus isolate originating from the 2004 Texas epizootic in cultured shrimp. *Archives of Virology*, 155, 315–327. https://doi.org/10.1007/s00705-009-0584-5.

Dixon, P., Paley, R., Alegria-Moran, R. & Oidtmann, B. (2016) Epidemiological characteristics of infectious hematopoietic necrosis virus (IHNV): A review. *Veterinary Research*, 47, article no.: 63. https://doi.org/10.1186/s13567-016-0341-1.

Do, J.W., Cha, S.J., Lee, N.S., Kim, Y.C., Kim, J.W., Kim, J.D. & Park, J.W. (2006) Taura syndrome virus from Penaeus vannamei shrimp cultured in Korea. *Diseases of Aquatic Organisms*, 70, 171–174. https://doi.org/10.3354/dao070171.

Dong, X., Wang, H., Zou, P., Chen, J., Liu, Z., Wang, X. & Huang, J. (2017) Complete genome sequence of Vibrio campbellii strain 20130629003S01 isolated from shrimp with acute hepatopancreatic necrosis disease. *Gut Pathogens*, 9, 31. https://doi.org/10.1186/s13099-017-0180-2.

Dopazo, C.P. (2020) The infectious pancreatic necrosis virus (IPNV) and its virulence determinants: What is known and what should be known. *Pathogens*, 9, article no.: 94. https://doi.org/10.3390/pathogens9020094.

Douglas-Helders, G.M., Weir, I.J., O'Brien, D.P., Carson, J. & Nowak, B.F. (2004) Effects of husbandry on prevalence of amoebic gill disease and performance of reared Atlantic salmon (Salmo salar L.). *Aquaculture*, 241, 21–30. https://doi.org/10.1016/j.aquaculture.2004.07.026.

Dumetz, F., Duchaud, E., LaPatra, S.E., Le Marrec, C., Claverol, S., Urdaci, M.C. & Le Hénaff, M. (2006) A protective immune response is generated in rainbow trout by an OmpH-like surface antigen (P18) of Flavobacterium psychrophilum. *Applied and Environmental Microbiology*, 72, 4845–4852. https://doi.org/10.1128/AEM.00279-06] [PubMed: 16820479.

Egidius, E., Andersen, K., Clausen, E. & Raa, J. (1981) Cold water vibriosis or 'Hitra disease' in Norwegian salmonid farming. *Journal of Fish Diseases*, 4, 353–354. https://doi.org/10.1111/j.1365-2761.1981.tb01143.x.

Egusa, S. (1983) Disease problems in Japanese yellowtail, Seriola quinqueradiata, culture: A review. In: *Diseases of Commercially Important Marine Fish and Shellfish* (edited by J. E. Stewart). Conseil Int Ernational pour l'Exploration de la Mer: Copenhagen, pp. 10–18.

Escobar, L.E., Escobar-Dodero, J. & Phelps, N.B.D. (2018) Infectious disease in fish: Global risk of viral hemorrhagic septicemia virus. *Reviews in Fish Biology and Fisheries*, 28, 637–655. https://doi.org/10.1007/s11160-018-9524-3.

Fijan, N. (1988) Vaccination against spring viraemia of carp. In: *Fish Vaccination* (edited by A. E. Ellis). Academic Press: London, pp. 204–215.

Flegel, T.W. (2012) Historic emergence, impact and current status of shrimp pathogens in Asia. *Journal of Invertebrate Pathology*, 110, 166–173. https://doi.org/10.1016/j.jip.2012.03.004.

Flegel, T.W. & Sritunyaluucksana, K. (2018) Recent research on acute hepatopancreatic necrosis disease (AHPND) and Enterocytozoon hepatopenaei in Thailand. *Asian Fisheries Science*, 31S. https://doi.org/10.33997/j.afs.2018.31.S1.018.

Fredriksen, B.N., Olsen, R.H., Furevik, A., Souhoka, R.A., Gauthier, D. & Brudeseth, B. (2013) Efficacy of a divalent and a multivalent water-in-oil formulated vaccine against a highly virulent strain of Flavobacterium psychrophilum after intramuscular challenge of rainbow trout (Oncorhynchus mykiss). *Vaccine*, 31, 1994–1998. https://doi.org/10.1016/j.vaccine.2013.01.016.

Fryer, J.L., Lannan, C.N., Giovannoni, S.J. & Wood, N.D. (1992) Piscirickettsia salmonis gen. nov., the causative agent of an epizootic disease in salmonid fishes. *International Journal of Systematic Bacteriology*, 42, 120–126. https://doi.org/10.1099/00207713-42-1-120] [PubMed: 1371057.

Garver, K.A., Dwilow, A.G., Richard, J., Booth, T.F., Beniac, D.R. & Souter, B.W. (2007) First detection and confirmation of spring viraemia of carp virus in common carp, Cyprinus carpio L., from Hamilton Harbour, Lake Ontario, Canada. *Journal of Fish Diseases*, 30, 665–671. https://doi.org/10.1111/j.1365-2761.2007.00851.x] [PubMed: 17958610.

Garver, K. & Wade, J. (2017). *Characterization of Infectious Hematopoietic Necrosis Virus (IHNV)*. Can sci. advisory Secretariat res Doc. 73, pp. 1–32.

Ghosh, B., Bridle, A.R., Nowak, B.F. & Cain, K.D. (2015) Assessment of immune response and protection against bacterial coldwater disease induced by a live-attenuated vaccine delivered orally or intraperitoneally to rainbow trout, Oncorhynchus mykiss (Walbaum). *Aquaculture*, 446, 242–249. https://doi.org/10.1016/j.aquaculture.2015.04.035.

Goodwin, A.E. (2002) First report of spring viremia of carp virus (SVCV) in North America. *Journal of Aquatic Animal Health*, 14, 161–164. https://doi.org/10.1577/1548-8667(2002)014<0161:FROSVO>2.0.CO;2.

Greger, E. & Goodrich, T. (1999) Vaccine development for winter ulcer disease, Vibrio viscosus, in Atlantic salmon, Salmo salar L. *Journal of Fish Diseases*, 22, 193–199. https://doi.org/10.1046/j.1365-2761.1999.00163.x.

Guðmundsdóttir, S., Helgason, S., Sigurjónsdóttir, H., Matthíasdóttir, S., Jónsdóttir, H., Laxdal, B. & Benediktsdóttir, E. (2000) Measures applied to control Renibacterium salmoninarum infection in Atlantic salmon: A retrospective study of two sea ranches in Iceland. *Aquaculture*, 186, 193–203. https://doi.org/10.1016/S0044-8486(99)00375-0.

Hamre, L.A., Glover, K.A. & Nilsen, F. (2009) Establishment and characterization of salmon louse (Lepeophtheirus salmonis (Krøyer 1837)) laboratory strains. *Parasitology International*, 58, 451–460. https://doi.org/10.1016/j.parint.2009.08.009] [PubMed: 19732850.

Han, J.E., Tang, K.F.J., Aranguren, L.F. & Piamsomboon, P. (2017) Characterization and pathogenicity of acute hepatopancreatic necrosis disease natural mutants, pirAB(vp) (–) V. parahaemolyticus, and pirAB(vp) (+) V. campbellii strains. *Aquaculture*, 470, 84–90. https://doi.org/10.1016/j.aquaculture.2016.12.022.

Hanif, A., Bakopoulos, V., Leonardos, I. & Dimitriadis, G.J. (2005) The effect of sea bream (Sparus aurata) broodstock and larval vaccination on the susceptibility by Photobcterium damsela subsp. piscicida and on the humoral immune parameters. *Fish and Shellfish Immunology*, 19, 345–361. https://doi.org/10.1016/j.fsi.2004.12.009] [PubMed: 15863015.

Hansen, G.H., Bergh, O., Michaelsen, J. & Knappskog, D. (1992) Flexibacter ovolyticus sp. nov., a pathogen of eggs and larvae of Atlantic halibut, Hippoglossus hippoglossus L. *International Journal of Systematic Bacteriology*, 42, 451–458. https://doi.org/10.1099/00207713-42-3-451.

Hassan, E.S., Mahmoud, M.M., Kawato, Y., Nagai, T., Kawaguchi, O., Iida, Y., Yuasa, K. & Nakai, T. (2012) Subclinical Edwardsiella ictaluri infection of wild ayu Plecoglossus altivelis. *Fish Pathology*, 47, 64–73. https://doi.org/10.3147/jsfp.47.64.

Hasson, K.W., Lightner, D.V., Mari, J., Bonami, J., Poulos, B.T., Mohney, L.L., Redman, R.M. & Brock, J.A. (1999) The geographic distribution of Taura Syndrome Virus (TSV) in the Americas: Determination by histopathology and in situ hybridization using TSV-specific cDNA probes. *Aquaculture*, 171, 13–26. https://doi.org/10.1016/S0044-8486(98)00429-3.

Hawke, J.P. (1979) A bacterium associated with disease of pond cultured channel catfish, Ictalurus punctatus. *Journal of the Fisheries Research Board of Canada*, 36, 1508–1512. https://doi.org/10.1139/f79-219.

Hawley, L.M. & Garver, K.A. (2008) Stability of viral hemorrhagic septicemia virus (VHSV) in freshwater and seawater at various temperatures. *Diseases of Aquatic Organisms*, 82, 171–178. https://doi.org/10.3354/dao01998.

Hoare, R., Ngo, T.P.H., Bartie, K.L. & Adams, A. (2017) Efficacy of a polyvalent immersion vaccine against Flavobacterium psychrophilum and evaluation of immune response to vaccination in rainbow trout fry (Oncorhynchus mykiss L.). *Veterinary Research*, 48, 43. https://doi.org/10.1186/s13567-017-0448-z] [PubMed: 28821298.

Hoferer, M., Akimkin, V., Skrypski, J., Schütze, H. & Sting, R. (2019) Improvement of a diagnostic procedure in surveillance of the listed fish diseases IHN and VHS. *Journal of Fish Diseases*, 42, 559–572. https://doi.org/10.1111/jfd.12968] [PubMed: 30779211.

Hong, X.-P., Lu, L.-Q. & Xu, D. (2016) Progress in research on acute hepatopancreatic necrosis disease (AHPND). *Aquaculture International*, 24, 577–593. https://doi.org/10.1007/s10499-015-9948-x.

Hussein, M.M.A., Hatai, K. & Nomura, T. (2001) Saprolegniosis in salmonids and their eggs in Japan. *Journal of Wildlife Diseases*, 37, 204–207. https://doi.org/10.7589/0090-3558-37.1.204.

Hvas, M., Karlsbakk, E., Mæhle, S., Wright, D.W. & Oppedal, F. (2017) The gill parasite Paramoeba perurans compromises aerobic scope, swimming capacity and ion balance in Atlantic salmon. *Conservation Physiology*, 5, 1–12. https://doi.org/10.1093/conphys/cox066.

Iberahim, N.A., Trusch, F. & Van West, P. (2018) Aphanomyces invadens, the causal agent of epizootic ulcerative syndrome, is a global threat to wild and farmed fish. *Fungal Biology Reviews*, 32, 118–130. https://doi.org/10.1016/j.fbr.2018.05.002.

Inoue, K., Yamano, K., Maeno, Y., Nakajima, K., Matsuoka, M., Wada, Y. & Sorimachi, M. (1992) Iridovirus infection of cultured red sea bream, Pagrus major. *Fish Pathology*, 27, 19–27. https://doi.org/10.3147/jsfp.27.19.

Iversen, A., Hermansen, Ø., Andreassen, O., Brandvik, R.K., Marthinussen, A. & Nystøyl, R. (2015) Kostnadsdrivere I lakseoppdrett (Cost Drivers in salmon Farming). No. 41, 1–43. *Nofima* [in Norwegian].

Jakob, E., Barker, D.E. & Garver, K.A. (2011) Vector potential of the salmon louse Lepeophtheirus salmonis in the transmission of infectious haematopoietic necrosis virus (IHNV). *Diseases of Aquatic Organisms*, 97, 155–165. https://doi.org/10.3354/dao02414.

Jimenez, R., Barniol, R., De Barniol, L. & Machuca, M. (2000) Periodic occurrence of epithelial viral necrosis outbreaks in Penaeus vannamei in Ecuador. *Diseases of Aquatic Organisms*, 42, 91–99. https://doi.org/10.3354/dao042091.

Jones, A.C., Mead, A., Kaiser, M.J., Austen, M.C.V., Adrian, A.W., Auchterlonie, N.A., Black, K.D., Blow, L.R., Bury, C., Brown, J.H., Burnell, G.M., Connolly, E., Dingwall, A., Derrick, S., Eno, N.C., Gautier, D.J.H., Green, K.A., Gubbins, M., Hart, P.R., Holmyard, J.M., Immink, A.J., Jarrad, D.L., Katoh, E., Langley, J.C.R., Lee, D.O., Le Vay, L., Leftwich, C.P., Mitchell, M., Moore, A., Murray, A.G., McLaren, E.M.R., Norbury, H., Parker, D., Parry, S.O., Purchase, D., Rahman, A., Sanver, F., Siggs, M., Simpson, S.D., Slaski, R.J., Smith, K., Syvret, M.L.Q., Tibbott, C., Thomas, P.C., Turnbull, J., Whiteley, R., Whittles, M., Wilcockson, M.J., Wilson, J., Dicks, L.V. & Sutherland, W.J. (2015) Prioritization of knowledge needs for sustainable aquaculture: A national and global perspective. *Fish and Fisheries*, 16, 668–683. https://doi.org/10.1111/faf.12086.

Jun, L., Jeong, J., Kim, J., Nam, J., Shin, K., Kim, J., Kang, J. & Jeong, H. (2009) Influence of temperature shifts on the onset and development of red sea bream iridoviral disease in rock bream Oplegnathus fasciatus. *Diseases of Aquatic Organisms*, 84, 201–208. https://doi.org/10.3354/dao02041.

Jung, S., Miyazaki, T., Miyata, M., Danayadol, Y. & Tanaka, S. (1997) Pathogenicity of Iridovirus from Japan and Thailand for the red sea bream Pagrus major in Japan, and histopathology of experimentally infected fish. *Fisheries Science*, 63, 735–740. https://doi.org/10.2331/fishsci.63.735.

Joseph, T.C., Murugadas, V., Reghunathan, D., Shaheer, P., Akhilnath, P.G. & Lalitha, K.V. (2015) Isolation and characterization of Vibrio cholerae O139 associated with mass mortality in Penaeus monodon and experimental challenge in postlarvae of three species of shrimp. *Aquaculture*, 442, 44–47. https://doi.org/10.1016/j.aquaculture.2015.02.020.

Kamilya, D. & Baruah, A. (2014) Epizootic ulcerative syndrome (EUS) in fish: History and current status of understanding. *Reviews in Fish Biology and Fisheries*, 24, 369–380. https://doi.org/10.1007/s11160-013-9335-5.

Kanai, K. (2017) Pseudotuberculosis. *Fish Pathology*, 52, 53–56. https://doi.org/10.3147/jsfp.52.53.

Karlsen, C., Espelid, S., Willassen, N.-P. & Paulsen, S.M. (2011) Identification and cloning of immunoprotective Aliivibrio salmonicida Pallike protein present in profiled outer membrane and secreted subproteome. *Diseases of Aquatic Organisms*, 93, 215–223. https://doi.org/10.3354/dao02302.

Karreman, G.A. (2006) Elements of an aquatic animal health program-infectious hematopoietic necrosis in farmed Atlantic salmon in British Columbia. In: *Aquaculture Biosecurity* (edited by A. D. Scarfe, C. S. Lee & P. J. O'Bryen), pp. 155–163.

Kawato, Y., Kiryu, I., Kawamura, Y. & Nakajima, K. (2017) Red Sea bream iridoviral disease in hatchery-reared devil stinger Inimicus japonicus. *Fish Pathology*, 52, 206–209. https://doi.org/10.3147/jsfp.52.206.

Kibenge, F.S.B., Munir, K., Kibenge, M.J.T., Joseph, T. & Moneke, E. (2004) Infectious salmon anemia virus: Causative agent, pathogenesis and immunity. *Animal Health Research Reviews*, 5, 65–78. https://doi.org/10.1079/ahr200461.

Kitancharoen, N., Yamamoto, A. & Hatai, K. (1997) Fungicidal effects of hydrogen peroxide on fungal infection of rainbow trout eggs. *Mycoscience*, 38, 375–378. https://doi.org/10.1007/BF02461675.

Kordon, A.O., Kalindamar, S., Majors, K., Abdelhamed, H., Tan, W., Karsi, A. & Pinchuk, L.M. (2020) Live attenuated Edwardsiella ictaluri vaccines enhance the protective innate immune responses of channel catfish B cells. *Developmental and Comparative Immunology*, 109, 103711. https://doi.org/10.1016/j.dci.2020.103711] [PubMed: 32311387.

Kube, P.D., Taylor, R.S. & Elliott, N.G. (2012) Genetic variation in parasite resistance of Atlantic salmon to amoebic gill disease over multiple infections. *Aquaculture*, 364–365, 165–172. https://doi.org/10.1016/j.aquaculture.2012.08.026.

Kumar, B.K., Deekshit, V.K., Raj, J.R.M., Rai, P., Shivanagowda, B.M., Karunasagar, I. & Karunasagar, I. (2014) Diversity of Vibrio parahaemolyticus associated with disease outbreak among cultured litopenaeus vannamei (Pacific white shrimp) in India. *Aquaculture*, 433, 247–251. https://doi.org/10.1016/j.aquaculture.2014.06.016.

Kumar I, P., Sarkar, P., V, S.R., V, M., Guru, A., Arshad, A., Elumalai, P. & Arockiaraj, J. (2020) Pathogenicity and pathobiology of epizootic ulcerative syndrome (EUS) causing fungus Aphanomyces invadens and its immunological response in fish. *Reviews in Fisheries Science and Aquaculture*, 28, 358–375. https://doi.org/10.1080/23308249.2020.1753167.

Kumar, R., Ng, T.H. & Wang, H.-C. (2020) Acute hepatopancreatic necrosis disease in penaeid shrimp. *Reviews in Aquaculture*, 12, 1867–1880. https://doi.org/10.1111/raq.12414.

Lal, T.M., Sano, M. & Ransangan, J. (2017) Isolation and characterization of large marine bacteriophage (Myoviridae), VhKM4 infecting Vibrio harveyi. *Journal of Aquatic Animal Health*, 29, 26–30. https://doi.org/10.1080/08997659.2016.1249578.

Leal, C.A.G., Tavares, G.C. & Figueiredo, H.C.P. (2014) Outbreaks and genetic diversity of Francizella noatunensis subsp. orientalis isolated from farm-raised Nile tilapia (Oreochromis niloticus) in Brazil. *Genetics and Molecular Research*, 13, 5704–5712. https://doi.org/10.4238/2014.July.25.26.

Lightner, D.V. (1993) Diseases of penaeid shrimp. In: *CRC Handbook of Mariculture: Crustacean Aquaculture.* (edited by J.P. McVey). CRC Press: Boca Raton, FL.

Lightner, D.V. (1996a) Epizootiology, distribution and the impact on international trade of two penaeid shrimp viruses in the Americas. *Revue Scientifique et Technique*, 15, 579–601. https://doi.org/10.20506/rst.15.2.944.

Lightner, D.V. (1996b). *A Handbook of Shrimp Pathology and Diagnostic Procedures for Diseases of Cultured Penaeid Shrimp*. World Aquaculture Society: Baton Rouge, LA.

Lightner, D.V. (2011) Status of shrimp diseases and advances in shrimp health management. In: *Diseases in Asian Aquaculture* VII (edited by M. G. Bondad-Reantaso, J. B. Jones, F. Corsin & T. Aoki). Fish Health Section, Asian Fisheries Society: Selangor, Malaysia, pp. 121–134.

Lightner, D.V., Redman, R.M., Hasson, K.W. & Pantoja, C.R. (1995) Taura syndrome in Penaeus vannamei (Crustacea: Decapoda): Gross signs, histopathology and ultrastructure. *Diseases of Aquatic Organisms*, 21, 53–59. https://doi.org/10.3354/dao021053.

Lightner, D.V., Redman, R.M., Arce, S. & Moss, S.M. (2009) Specific pathogen-free shrimp stocks in shrimp farming facilities as a novel method for disease control in crustaceans. In: *Shellfish Safety and Quality* (edited by S. Shumway & G. Rodrick). Woodhead Publishing: London, pp. 384–424.

Lipp, E.K. & Rose, J.B. (1997) The role of seafood in foodborne diseases in the United States of America. *Revue Scientifique et Technique*, 16, 620–640. https://doi.org/10.20506/rst.16.2.1048.

Lo, C.F. & Kou, G.H. (1998) Virus-associated white spot syndrome of shrimp in Taiwan: A review. *Fish Pathology*, 33, 365–371. https://doi.org/10.3147/jsfp.33.365.

López-León, P., Luna-González, A., Escamilla-Montes, R., Flores-Miranda, MdC., Fierro-Coronado, J.A., Álvarez-Ruiz, P. & Diarte-Plata, G. (2016) Isolation and characterization of infectious Vibrio parahaemolyticus, the causative agent of AHPND, from the whiteleg shrimp (litopenaeus vannamei). *Latin American Journal of Aquatic Research*, 44, 470–479. https://doi.org/10.3856/vol44-issue3-fulltext-5.

Lopez-Tellez, N.A., Corbala-Bermejo, J.A., Bustamante-Unzuera, M.L., Silva-Ledesma, L.P., Vidal-Martinez, V.M. & Rodriguez-Canul, R. (2019) History, impact and status of infectious diseases of the Pacific white shrimp Penaaeus vannamei (Boone, 1831) cultivated in Mexico. *J. Wld Aquacult. Soc.*, 51, 334–345.

Long, A., Call, D.R. & Cain, K.D. (2014) Investigation of the link between broodstock infection, vertical transmission, and prevalence of Flavobacterium psychrophilum in eggs and progeny of rainbow trout and coho salmon. *Journal of Aquatic Animal Health*, 26, 66–77. https://doi.org/10.1080/08997659.2014.886632.

Lönnström, L.G., Hoffrén, M.L. & Wiklund, T. (2008) Flavobacterium psychrophilum associated with mortality of farmed perch, Perca fluviatilis L. *Journal of Fish Diseases*, 31, 793–797. https://doi.org/10.1111/j.1365-2761.2008.00967.x.

MacKinnon, B., Jones, P., Hawkins, L., Dohoo, I., Stryhn, H., Vanderstichel, R. & St-Hilaire, S. (2019) The epidemiology of skin ulcers in saltwater reared Atlantic salmon (Salmo salar) in Atlantic Canada. *Aquaculture*, 501, 230–238. https://doi.org/10.1016/j.aquaculture.2018.11.035.

Madsen, L. & Dalsgaard, I. (2008) Water recirculation and good management: Potential methods to avoid disease outbreaks with Flavobacterium psychrophilum. *Journal of Fish Diseases*, 31, 799–810. https://doi.org/10.1111/j.1365-2761.2008.00971.x] [PubMed: 19238756.

Magariños, B., Romalde, J.L., Bandín, I., Fouz, B. & Toranzo, A.E. (1992) Phenotypic, antigenic, and molecular characterization of Pasteurella piscicida strains isolated from fish. *Applied and Environmental Microbiology*, 58, 3316–3322. https://doi.org/10.1128/aem.58.10.3316-3322.1992] [PubMed: 1444366.

Maj-Paluch, J., Borzym, E., Matras, M., Stachnik, M. & Reichert, M. (2019) Phylogenetic analysis of spring viremia of carp virus (SVCV) identified in Poland. *Medycyna Weterynaryjna*, 75, 6159–2019. https://doi.org/10.21521/mw.6159.

Marana, M.H., Skov, J., Chettri, J.K., Krossøy, B., Dalsgaard, I., Kania, P.W. & Buchmann, K. (2017) Positive correlation between Aeromonas salmonicida vaccine antigen concentration and protection in vaccinated rainbow trout Oncorhynchus mykiss evaluated by a tail fin infection model. *Journal of Fish Diseases*, 40, 507–516. https://doi.org/10.1111/jfd.12527.

Marcos-López, M. & Rodger, H.D. (2020) Amoebic gill disease and host response in Atlantic salmon (Salmo salar L.): A review. *Parasite Immunology*, 42, article no.: e12766. https://doi.org/10.1111/pim.12766.

Mastan, S.A. & Qureshi, T.A. (2001) Role of bacteria in the epizootic ulcerative syndrome (EUS) of fishes. *Journal of Environmental Biology*, 22, 187–192.

Matsuoka, S., Inouye, K. & Nakajima, K. (1996) Cultured fish species affected by red sea bream iridoviral disease from 11991 to 1995. *Fish Pathology*, 31, 233–234. https://doi.org/10.3147/jsfp.31.233.

McClure, C.A., Hammell, K.L. & Dohoo, I.R. (2005) Risk factors for outbreaks of infectious salmon anemia in farmed Atlantic salmon, Salmo salar. *Preventive Veterinary Medicine*, 72, 263–280. https://doi.org/10.1016/j.prevetmed.2005.07.010.

Meyer, F.P. & Bullock, G.L. (1973) Edwardsiella tarda, a new pathogen of channel catfish (Ictalurus punctatus). *Applied Microbiology*, 25, 155–156. https://doi.org/10.1128/am.25.1.155-156.1973.

Mladineo, I., Miletić, I. & Bočina, I. (2006) Photobacterium damselae subsp. piscicida outbreak in cage-reared Atlantic bluefin tuna Thunnus thynnus. *Journal of Aquatic Animal Health*, 18, 51–54. https://doi.org/10.1577/H05-012.1.

Morales-Covarrubias, M.S., Ruiz-Luna, A., Moura-Lemus, A.P., Solís Montiel, V.T. & Conroy, G. (2011) Prevalence of diseases in cultured white shrimp (Penaeus vannamei) in eight regions of Latin America. *Rev. Científica FCV-LUZ*, 2, 434–446.

Morales-Covarrubias, M.S., Tlahuel-Vargas, L., Martínez-Rodríguez, I.E., Lozano-Olvera, R. & Palacios-Arriaga, J.M. (2012) Necrotizing hepatobacterium (NHPB) infection in Penaeus vannamei with florfenicol and oxytetracycline: A comparative experimental study. *Rev. Científic, FCV-LUZ*, 22, 72–80.

Mordue Luntz, A.J. & Birkett, M.A. (2009) A review of host finding behavior in the parasitic sea louse, Lepeophtheirus salmonis (Caligidae: Copepoda). *Journal of Fish Diseases*, 32, 3–13. https://doi.org/10.1111/j.1365-2761.2008.01004.x.

Motte, E., Yugcha, E., Luzardo, J., Castro, F., Leclercq, G., Rodríguez, J., Miranda, P., Borja, O., Serrano, J., Terreros, M., Montalvo, K., Narváez, A., Tenorio, N., Cedeño, V., Mialhe, E. & Boulo, V. (2003) Prevention of IHHNV vertical transmission in the white shrimp litopenaeus vannamei. *Aquaculture*, 219, 57–70. https://doi.org/10.1016/S0044-8486(02)00631-2.

Munday, B.L. (1986) Diseases of salmonids. In: *Proceedings of the Workshop on Diseases of Australian Fish and Shellfish* (edited by J. D. Humphrey & J. S. Langdon). Benalla, Victoria, Australia.

Muthukrishnan, S., Defoirdt, T., Ina-Salwany, M.Y., Yusoff, F.M., Shariff, M., Ismail, S.I. & Natrah, I. (2019) Vibrio parahaemolyticus and Vibrio harveyi causing acute hepatopancreatic necrosis disease (AHPND) in Penaeus vannamei (Boone, 1931) isolated from Malaysian shrimp ponds. *Aquaculture*, 511, 734227. https://doi.org/10.1016/j.aquaculture.2019.734227.

Nakajima, K. & Maeno, Y. (1998) Pathogenicity of red sea bream Iridovirus and other fish iridoviruses to red sea bream. *Fish Pathology*, 33, 143–144. https://doi.org/10.3147/jsfp.33.143.

Nelapati, S., Nelapati, K. & Chinnam, B.K. (2012) Vibrio parahaemolyticus – An emerging foodborne pathogen – A review. *Veterinary World*, 5, 48–63.

Nelson, E.J. & Ghiorse, W.C. (1999) Isolation and identification of Pseudoalteromonas piscicida strain Cura-d associated with diseased damselfish (Pomacentridae) eggs. *Journal of Fish Diseases*, 22, 253–260. https://doi.org/10.1046/j.1365-2761.1999.00168.x.

Nielsen, L., Sang-Oum, W., Cheevadhanarak, S. & Flegel, T.W. (2005) Taura syndrome virus (TSV) in Thailand and its relationship to TSV in China and the Americas. *Diseases of Aquatic Organisms*, 63, 101–106. https://doi.org/10.3354/dao063101] [PubMed: 15819424.

Nilsen, H., Olsen, A.B., Vaagnes, Ø, Hellberg, H., Bottolfsen, K., Skjelstad, H. & Colquhoun, D.J. (2011a) Systemic Flavobacterium psychrophilum infection in rainbow trout, Oncorhynchus mykiss (Walbaum), farmed in fresh and brackish water in Norway. *Journal of Fish Diseases*, 34, 403–408. https://doi.org/10.1111/j.1365-2761.2011.01249.x] [PubMed: 21401645.

Nilsen, H., Johansen, R., Colquhoun, D.J., Kaada, I., Bottolfsen, K., Vågnes, Ø. & Olsen, A.B. (2011b) Flavobacterium psychrophilum associated with septicaemia and necrotic myositis in Atlantic salmon Salmo salar: A case report. *Diseases of Aquatic Organisms*, 97, 37–46. https://doi.org/10.3354/dao02390.

Nowak, B. (2001) Qualitative evaluation of risk factors for amoebic gill disease in cultured Atlantic salmon. In: *Risk Analysis in Aquatic Animal Health. Proceedings of the an International Conference Held in Paris, France*, 8–10 February (edited by C. J. Rodgers). World Health Organization for Animal Health: Paris.

Nunan, L.M., Pantoja, C.R., Gomez-Jimenez, S. & Lightner, D.V. (2013) Candidatus Hepatobacter penaei,' an intracellular pathogenic enteric bacterium in the hepatopancreas of the marine shrimp Penaeus vannamei (Crustacea: Decapoda). *Applied and Environmental Microbiology*, 79, 1407–1409. https://doi.org/10.1128/AEM.02425-12.

Nunan, L., Lightner, D., Pantoja, C. & Gomez-Jimenez, S. (2014) Detection of acute hepatopancreatic necrosis disease (AHPND) in Mexico. *Diseases of Aquatic Organisms*, 111, 81–86. https://doi.org/10.3354/dao02776.

Nunes, A.J.P., Cunha-Martins, P. & Vasconselos-Gesteira, T.C. (2004) Carcinicultura ameaçada. *Rev. Panoram Aquic*, 83, 37–51.

Oldham, T., Rodger, H. & Nowak, B.F. (2016) Incidence and distribution of amoebic gill disease (AGD) – An epidemiological review. *Aquaculture*, 457, 35–42. https://doi.org/10.1016/j.aquaculture.2016.02.013.

Ortega, C., Cañas-Lopez, L., Irgang, R., Fajardo, R., Poblete-Morales, M., Valladares-Carranza, B., Tapia-Cammas, D. & Avendaño-Herrera, R. (2019) First detection of spring viraemia of carp virus in common carp (Cyprinus carpio L.) affected by a septicaemic disease in Mexico. *Journal of Fish Diseases*, 42, 667–675. https://doi.org/10.1111/jfd.12969.

Overton, K., Dempster, T., Oppedal, F., Kristiansen, T.S., Gismervik, K. & Stien, L.H. (2019) Salmon lice treatments and salmon mortality in Norwegian aquaculture: A review. *Reviews in Aquaculture*, 11, 1398–1417. https://doi.org/10.1111/raq.12299.

Overton, K., Barrett, L.T., Oppedal, F., Kristiansen, T.S. & Dempster, T. (2020) Sea lice removal by cleaner fish in salmon aquaculture: A review of the evidence base. *Aquaculture Environment Interactions*, 12, 31–44. https://doi.org/10.3354/aei00345.

Özçelik, H., Taştan, Y., Terzi, E. & Sönmez, A.Y. (2020) Use of onion (Allium cepa) and garlic (Allium sativum) wastes for the prevention of fungal disease (Saprolegnia parasitica) on eggs of rainbow trout (Oncorhynchus mykiss). *Journal of Fish Diseases*, 43, 1325–1330. https://doi.org/10.1111/jfd.13229.

Pajdak-Czaus, J., Platt-Samoraj, A., Szweda, W., Siwicki, A.K. & Terech-Majewska, E. (2019) Yersinia ruckeri – Threat not only to rainbow trout. *Aquaculture Research*, 50, 3083–3096. https://doi.org/10.1111/are.14274.

Parsons, H., Nowak, B., Fisk, D. & Powell, M. (2001) Effectiveness of commercial freshwater bathing as a treatment against amoebic gill disease in Atlantic salmon. *Aquaculture*, 195, 205–210. https://doi.org/10.1016/S0044-8486(00)00567-6.

Phalitakul, S., Wongtawatchai, J., Sarikaputi, M. & Viseshakul, N. (2006) The molecular detection of Taura syndrome virus emerging with white spot syndrome virus in penaeid shrimps of Thailand. *Aquaculture*, 260, 77–85. https://doi.org/10.1016/j.aquaculture.2006.05.040.

Plumb, J.A. & Sanchez, D.J. (1983) Susceptibility of five species of fish to Edwardsiella ictaluri. *Journal of Fish Diseases*, 6, 261–266. https://doi.org/10.1111/j.1365-2761.1983.tb00075.x.

Powell, A., Treasurer, J.W., Pooley, C.L., Keay, A.J., Lloyd, R., Imsland, A.K. & Garcia de Leaniz, C. (2018) Use of lumpfish for sea-lice control in salmon farming: Challenges and opportunities. *Reviews in Aquaculture*, 10, 683–702. https://doi.org/10.1111/raq.12194.

Powell, M.D., Reynolds, P. & Kristensen, T. (2015) Freshwater treatment of amoebic gill disease and sea-lice in seawater salmon production: Considerations of water chemistry and fish welfare in Norway. *Aquaculture*, 448, 18–28. https://doi.org/10.1016/j.aquaculture.2015.05.027.

Pradham, P.K., Rathore, G., Sood, N., Swaminathan, T.R., Yadav, M.K., Verma, D.K., Chaudhary, D.K., Abidi, R., Punia, P. & Jena, J.K. (2014) Emergence of epizootic ulcerative syndrome: Large-scale mortalities of cultured and wild fish species in Uttar Pradesh, India. *Current Science*, 106, 1711–1718.

Prayitno, S.B. & Latchford, J.W. (1995) Experimental infections of crustaceans with luminous bacteria related to Photobacterium and Vibrio—Effect of salinity and pH on infectiosity. *Aquaculture*, 132, 105–112. https://doi.org/10.1016/0044-8486(94)00374-W.

Pujalte, M.J., Sitjà-Bobadilla, A., Macián, M.C., Álvarez-Pellitero, P. & Garay, E. (2007) Occurrence and virulence of Pseudoalteromonas spp. in cultured gilthead sea bream (Sparus aurata L.) and European sea bass (Dicentrarchus labrax L.). Molecular and phenotypic characterization of P. Undina strain U58. *Aquaculture*, 271, 47–53. https://doi.org/10.1016/j.aquaculture.2007.06.015.

Pulpipat, T., Maekawa, S., Wang, P.C. & Chen, S.C. (2020) Immune response and protective efficacy of a formalin-killed Francisella noatunensis subsp. orientali vaccine evaluated through intraperitoneal and immersion challenge methods in Oreochromis niloticus. *Vaccines*, 8. https://doi.org/10.3390/vaccines8020163] [PubMed: 32260212.

Qian, D., Shi, Z., Zhang, S., Cao, Z., Liu, W., Li, L., Xie, Y., Cambournac, I. & Bonami, J.R. (2003) Extra small virus-like particles (XSV) and nodavirus associated with whitish muscle disease in the giant freshwater prawn, Macrobrachium rosenbergii. *Journal of Fish Diseases*, 26, 521–527. https://doi.org/10.1046/j.1365-2761.2003.00486.x.

Ramírez-Paredes, J.G., Mendoza-Roldan, M.A., Lopez-Jimena, B., Shahin, K., Metselaar, M., Thompson, K.D., Penman, D.J., Richards, R.H. & Adams, A. (2019) Whole cell inactivated autogenous vaccine effectively protects red Nile tilapia (Oreochromis niloticus) against francisellosis via intraperitoneal injection. *Journal of Fish Diseases*, 42, 1191–1200. https://doi.org/10.1111/jfd.13041] [PubMed: 31184398.

Raszl, S.M., Froelich, B.A., Vieira, C.R.W., Blackwood, A.D. & Noble, R.T. (2016) Vibrio parahaemolyticus and Vibrio vulnificus in South America: Water, seafood and human infections. *Journal of Applied Microbiology*, 121, 1201–1222. https://doi.org/10.1111/jam.13246.

Restrepo, L., Bayot, B., Arciniegas, S., Bajaña, L., Betancourt, I., Panchana, F. & Reyes Muñoz, A.R. (2018) PirVP genes causing AHPND identified in a new Vibrio species (Vibrio punensis) within the commensal Orientalis clade. *Scientific Reports*, 8, 13080. https://doi.org/10.1038/s41598-018-30903-x.

Richards, R.H. & Pickering, A.D. (1978) Frequency and distribution patterns of Saprolegnia infection in wild and hatchery-reared brown trout Salmo trutta L., char Salvelinus alpinus (L.). *Journal of Fish Diseases*, 1, 69–82. https://doi.org/10.1111/j.1365-2761.1978.tb00006.x.

Rimstad, E. & Markussen, T. (2020) Infectious salmon anaemia virus – Molecular biology and pathogenesis of infection. *Journal of Applied Microbiology*, 129, 85–97. https://doi.org/10.1111/jam.14567.

Roberts, R.J. & Pearson, M.D. (2005) Infectious pancreatic necrosis in Atlantic salmon, Salmo salar L. *Journal of Fish Diseases*, 28, 383–390. https://doi.org/10.1111/j.1365-2761.2005.00642.x.

Robertson, P.A.W., Calderon, J., Carrera, L., Stark, J.R., Zherdmant, M. & Austin, B. (1998) Experimental Vibrio harveyi infections in Penaeus vannamei larvae. *Diseases of Aquatic Organisms*, 32, 151–155. https://doi.org/10.3354/dao032151.

Rodger, H.D. (2014) Amoebic gill disease (AGD) in farmed salmon (Salmo salar) in Europe. *Fish Vet*, J14, 16–27.

Rødseth, G.I. (2016). *Beregninger pa° Fem Milliarder I Lusekostnader Er Bare Deler Av Sannheten. iLaks*.no. Retrieved from https://ilaks.no/ beregninger-pa-fem-milliarder-i-lusekostnader-er-bare-deler-avsannheten/ [in Norwegian].

Ruane, N., Geoghegan, F. & Cinneide, M.O. (2007) Infectious pancreatic necrosis virus and its impact on the Irish salmon aquaculture and wild fish sectors. *Marine Environment and Health Series (Dublin)*, 30, 1–48.

Ruangsri, J., Tanmark, N., Penprapai, N. & Supamattaya, K. (2007) Epizootic and pathogenesis of Taura syndrome virus (TSV) in black tiger shrimp (Penaeus monodon) cultured in southern Thailand. *Songklanakarin Journal of Science and Technology*, 29, 1263–1274.

Rucker, R.R., Whipple, W.J., Parvin, J.R. & Evans, C.A. (1953) A contagious disease of sockeye salmon possibly of virus origin. *United States Fish and Wildlife Service Fishery Bulletin*, 54, 35–46.

Saengchan, S., Phewsaiya, K., Briggs, M. & Flegel, T.W. (2007) Outbreaks of infectious myonecrosis virus (IMNV) in Indonesia confirmed by genome sequencing and use of an alternative RT-PCR detection method. *Aquaculture*, 226, 32–38.

Sahul Hameed, A.S., Yoganandhan, K., Sri Widada, J. & Bonami, J.R. (2004) Studies on the occurrence and RT- PCR detection of Macrobrachium rosenbergii nodavirus and extra small virus-like particles associated with white tail disease of Macrobrachium rosenbergii in India. *Aquaculture*, 238, 127–133. https://doi.org/10.1016/j.aquaculture.2004.06.009.

Salonius, K., Siderakis, C., MacKinnon, A.M., & Griffiths, S.G. (2005). Use of *Arthrobacter davidanieli* as a live vaccine against *Renibacterium salmoninarum* and *Piscirickettsia salmonis* in salmonids. in P. J. Midtlyng, editor. *Progress in fish vaccinology*. Developments in biological standardization, volume 121, Karger, Basel, Switzerland. Pages 189–197

Salte, R., Rørvik, K.-A., Reed, E. & Norberg, K. (1994) Winter ulcers of the skin in Atlantic salmon, Salmo salar L.: Pathogenesis and possible aetiology. *Journal of Fish Diseases*, 17, 661–665. https://doi.org/10.1111/j.1365-2761.1994.tb00265.x.

Sánchez-Martínez, J.G., Aguirre-Guzmán, G. & Mejía-Ruíz, H. (2007) White spot syndrome virus in cultured shrimp: A review. *Aquaculture Research*, 38, 1339–1354. https://doi.org/10.1111/j.1365-2109.2007.01827.x.

Sano, M., Ito, T., Kurita, J., Yanai, T., Watanabe, N., Miwa, S. & Iida, T. (2004) First detection of koi herpesvirus in cultured common carp Cyprinus carpio in Japan. *Fish Pathology*, 39, 165–167. https://doi.org/10.3147/jsfp.39.165.

Santos, Y., García-Marquez, S., Pereira, P.G., Pazos, F., Riaza, A., Silva, R., El Morabit, A. & Ubeira, F.M. (2005) Efficacy of furunculosis vaccines in turbot, Scophthalmus maximus (L.): Evaluation of immersion, oral and injection delivery. *Journal of Fish Diseases*, 28, 165–172. https://doi.org/10.1111/j.1365-2761.2005.00610.x.

Schaefer, J.W., Alvarado, V., Enriquez, R. & Monras, M. (1991) Salmon farming in Chile. *Tierärztliche Umschau*, 45, 449.

Selvin, J., Huxley, A.J. & Lipton, A.P. (2005) Pathogenicity, antibiogram and biochemical characteristics of luminescent Vibrio harveyi associated with 'Black Shell Disease' of Penaeus monodon. *Fishery Technology*, 42, 191–196.

Senapin, S., Dong, H.T., Meemetta, W., Siriphongphaew, A., Charoensapsri, W., Santimanawong, W., Turner, W.A., Rodkhum, C., Withyachumnarnkul, B. & Vanichviriyakit, R. (2016) Hahella-chejuensis is the aetiological agent of a novel red egg disease in tilapia (Oreochromis spp.) hatcheries in Thailand. *Aquaculture*, 454, 1–7. https://doi.org/10.1016/j.aquaculture.2015.12.013.

Shahin, K., Shinn, A.P., Metselaar, M., Ramirez-Paredes, J.G., Monaghan, S.J., Thompson, K.D., Hoare, R. & Adams, A. (2019) Efficacy of an inactivated whole-cell injection vaccine for nile tilapia, Oreochromis niloticus (L), against multiple isolates of Francisella noatunensis subsp. orientalis from diverse geographical regions. *Fish and Shellfish Immunology*, 89, 217–227. https://doi.org/10.1016/j.fsi.2019.03.071.

Shao, L. & Zhao, J.R. (2017) Isolation of a highly pathogenic spring viraemia of carp virus strain from grass carp (Ctenopharyngodon idella) in late summer, China, 2016. *Virus Research*, 238, 183–192. https://doi.org/10.1016/j.virusres.2017.06.025.

Shchelkanov, M.Y., Shulgina, M.A., Stepan'kov, A.P., Lvov, N., Kakareka, N.N., Shestopalov, A.M., Galkina, I.V. & Shevchenko, O.G. (2017) Infectious salmon anaemia. *S Russ Ecol. Dev.*, 12, 120–134.

Shin, S., Kulatunga, D.C.M., Dananjaya, S.H.S., Nikapitiya, C., Lee, J. & De Zoysa, M. (2017) Saprolegnia parasitica isolated from rainbow trout in Korea: Characterization, anti-Saprolegnia activity and host pathogen interaction in zebrafish disease model. *Mycobiology*, 45, 297–311. https://doi.org/10.5941/MYCO.2017.45.4.297.

Shinn, A.P., Pratoomyot, J., Bron, J.E., Paladini, G., Brooker, E.E. & Brooker, A.J. (2015) Economic costs of protistan and metazoan parasites to global mariculture. *Parasitology*, 142, 196–270. https://doi.org/10.1017/S0031182014001437.

Shinn, A.P., Pratoomyot, J., Griffiths, D., Trong, T.Q., Vu, N.T., Jiravanichpaisal, P. & Briggs, M. (2018) Asian shrimp production and the economic cost of disease. *Asian Fish Sci.*, 31S, 29–58. https://doi.org/10.33997/j.afs.2018.31.S1.003.

Shors, S.T. & Winston, V. (1989) Detection of infectious hematopoietic necrosis virus in an invertebrate (Callibaetis sp.). *American Journal of Veterinary Research*, 50, 1307–1309.

Skall, H.F., Olesen, N.J. & Mellergaard, S. (2005) Viral haemorrhagic septicaemia virus in marine fish and its implication for fish farming – A review. *Journal of Fish Diseases*, 28, 509–529. https://doi.org/10.1111/j.1365-2761.2005.00654.x.

Songe, M.M., Willems, A., Wiik-Nielsen, J., Thoen, E., Evensen, Ø., van West, P. & Skaar, I. (2016) Saprolegnia diclina IIIA and S. parasitica employ different infection strategies when colonizing eggs of Atlantic salmon, Salmo salar L. *Journal of Fish Diseases*, 39, 343–352. https://doi.org/10.1111/jfd.12368.

Soto, E., Hawke, J.P., Fernandez, D. & Morales, J.A. (2009) Francisella sp., an emerging pathogen of tilapia, Oreochromis niloticus (L.). *Journal of Fish Diseases*, 32, 713–722. https://doi.org/10.1111/j.1365-2761.2009.01070.x] [PubMed: 19515205.

Soto, E., Endris, R.G. & Hawke, J.P. (2010) In vitro and in vivo efficacy of florfenicol for treatment of Francisella asiatica infection of tilapia. *Antimicrobial Agents and Chemotherapy*, 54, 4664–4670. https://doi.org/10.1128/AAC.00206-10.

Soto, E., Wiles, J., Elzer, P., Macaluso, K. & Hawke, J.P. (2011) Attenuated Francizella asiatica iglc mutant induces protective immunity to francisellosis in tilapia. *Vaccine*, 29, 593–598. https://doi.org/10.1016/j.vaccine.2010.06.040.

Soto, E., McGovern-Hopkins, K., Klinger-Bowen, R., Fox, B.K., Brock, J., Antonio, N., van der Waal, Zv, Rushton, S., Mill, A. & Tamaru, C.S. (2013) Prevalence of Francisella noatunensis subsp. orientalis in cultured tilapia on the island of Oahu, Hawaii. *Journal of Aquatic Animal Health*, 25, 104–109. https://doi.org/10.1080/08997659.2013.781554.

Soto-Rodriguez, S.A., Gomez-Gil, B., Lozano-Olvera, R., Betancourt-Lozano, M. & Morales-Covarrubias, M.S. (2015) Field and experimental evidence of Vibrio parahaemolyticus as the causative agent of acute hepatopancreatic necrosis disease of cultured shrimp litopenaeus vannamei. In: *Applied and Environmental Microbiology*, 81, 1689–1699. https://doi.org/10.1128/AEM.03610-14.

Stalin, N. & Srinivasan, P. (2017) Efficacy of potential phage cocktails against Vibrio harveyi and closely related Vibrio species isolated from shrimp aquaculture environment in the south east coast of India. *Veterinary Microbiology*, 207, 83–96. https://doi.org/10.1016/j.vetmic.2017.06.006.

Steinum, T., Kvellestad, A., Rønneberg, L.B., Nilsen, H., Asheim, A., Fjell, K., Nygård, S.M.R., Olsen, A.B. and Dale, O.B. (2008), First cases of amoebic gill disease (AGD) in Norwegian seawater farmed Atlantic salmon, Salmo salar L., and phylogeny of the causative amoeba using 18S cDNA sequences. *Journal of Fish Diseases*, 31: 205-214. https://doi.org/10.1111/j.1365-2761.2007.00893.x.

Stentiford, G.D., Bonami, J.-R. & Alday-Sanz, V. (2009) A critical review of susceptibility of crustaceans to Taura syndrome, yellowhead disease and white spot disease and implications of inclusion of these diseases in European legislation. *Aquaculture*, 291, 1–17. https://doi.org/10.1016/j.aquaculture.2009.02.042.

Sumithra, T.G., Kumar, T.V.A., Swaminathan, T.R., Anusree, V.N., Amala, P.V., Reshma, K.J., Kishor, T.G., Kumar, R.R., Sharma, S.R., Kripa, V., Prema, D. & Sanil, N.K. (2020) Epizootics of epizootic ulcerative syndrome among estuarine fishes of Kerala, India, under post-flood conditions. *Diseases of Aquatic Organisms*, 139, 1–13. https://doi.org/10.3354/dao03465.

Suzuki, K., Mizuno, S., Katsumata, Y., Misaka, N., Miyamoto, M. & Sasaki, Y. (2018) Asymptomatic infection of Renibacterium salmoninarum in hatchery-reared juvenile chum salmon Oncorhynchus keta resulted in mass mortalities after long-term rearing. *Fish Pathology*, 53, 40–43. https://doi.org/10.3147/jsfp.53.40.

Takano, T., Matsuyama, T., Sakai, T. & Nakayasu, C. (2011) Protective efficacy of a formalin-killed vaccine against atypical Edwardsiella tarda infections in red sea bream Pagrus major. *Fish Pathology*, 46, 120–122. https://doi.org/10.3147/jsfp.46.120.

Tan, C.K.F., Nowak, B.F. & Hodson, S.L. (2002) Biofouling as a reservoir of Neoparamoeba pemaquidensis (page, 1970), the causative agent of amoebic gill disease in Atlantic salmon. *Aquaculture*, 210, 49–58. https://doi.org/10.1016/S0044-8486(01)00858-4.

Tang, K.F.J. & Lightner, D.V. (2005) Phylogenetic analysis of Taura syndrome virus isolates collected between 1993 and 2004 and virulence comparison between two isolates representing different genetic variants. *Virus Research*, 112, 69–76. https://doi.org/10.1016/j.virusres.2005.03.023.

Tang, K.F.J., Bondad-Reantaso, M.G., Arthur, J.R., MacKinnon, B., Hao, B., Alday-Sanz, V., Liang, Y. & Dong, X. (2020). Shrimp acute hepatopancreatic necrosis disease strategy manual. *FAO Fisheries and Aquaculture Circular No. 1190*. Rome. FAO. https://doi.org/10.4060/cb2119en.

Taylor, N.G.H., Peeler, E.J., Denham, K.L., Crane, C.N., Thrush, M.A., Dixon, P.F., Stone, D.M., Way, K. & Oidtmann, B.C. (2013) Spring viraemia of carp (SVC) in the UK: The road to freedom. *Preventive Veterinary Medicine*, 111, 156–164. https://doi.org/10.1016/j.prevetmed.2013.03.004.

Torrissen, O., Jones, S., Asche, F., Guttormsen, A., Skilbrei, O.T., Nilsen, F., Horsberg, T.E. & Jackson, D. (2013) Salmon lice – Impact on wild salmonids and salmon aquaculture. *Journal of Fish Diseases*, 36, 171–194. https://doi.org/10.1111/jfd.12061.

Tsai, J.M., Shiau, L.J., Lee, H.H., Chan, P.W.Y. & Lin, C.Y. (2002) Simultaneous detection of white spot syndrome virus (WSSV) and Taura syndrome virus (TSV) by multiplex reverse transcription-polymerase chain reaction (RT-PCR) in Pacific white shrimp Penaeus vannamei. *Diseases of Aquatic Organisms*, 50, 9–12. https://doi.org/10.3354/dao050009.

Tu, C., Huang, H.T., Chuang, S.H., Hsu, J.P., Kuo, S.T., Li, N.J., Hsu, T.L., Li, M.C. & Lin, S.Y. (1999) Taura syndrome in Pacific white shrimp Penaeus vannamei cultured in Taiwan. *Diseases of Aquatic Organisms*, 38, 159–161. https://doi.org/10.3354/dao038159.

Urbina, M.A., Cumillaf, J.P., Paschke, K. & Gebauer, P. (2019) Effects of pharmaceutials used to treat salmon lice on nontarget species: Evidence from a systematic review. *Science of the Total Environment*, 649, 1124–1136. https://doi.org/10.1016/j.scitotenv.2018.08.334.

Villumsen, K.R., Neumann, L., Ohtani, M., Strøm, H.K. & Raida, M.K. (2014) Oral and anal vaccination confers full protection against enteric redmouth disease (ERM) in rainbow trout. *PLOS ONE*, 9, e93845. https://doi.org/10.1371/journal.pone.0093845.

Vishwanath, T.S., Mohan, C.V. & Shankar, K.M. (1998) Epizootic ulcerative syndrome (EUS), associated with a fungal pathogen, in Indian fishes: Histopathology – 'a cause for invasiveness'. *Aquaculture*, 165, 1–9. https://doi.org/10.1016/S0044-8486(98)00227-0.

Wang, R.X., Feng, J., Su, Y.L., Ye, L.T. & Wang, J.Y. (2013) Studies on the isolation of Photobacterium damselae subsp. piscicida from diseased golden pompano (Trachinotus ovatus Linnaeus) and antibacterial agents sensitivity. *Veterinary Microbiology*, 162, 957–963. https://doi.org/10.1016/j.vetmic.2012.09.020.

WHO (2006). *Risk Assessment of Choleragenic Vibrio cholerae O1 and O139 in Warm Water Shrimp for International Trade: Interpretative Summary and Technical Report. Microbiological Risk Assessment Series No. 9*. WHO/Food and Agriculture Organization: Rome, pp. 1–90.

Wllloughby, L.G. (1970) Mycological aspects of a disease of young perch in Windermere. *Journal of Fish Biology*, 2, 113–116. https://doi.org/10.1111/j.1095-8649.1970.tb03265.x.

Willoughby, L.G. (1989) Continued defence of salmonid fish against Saprolegnia fungus, after its establishment. *Journal of Fish Diseases*, 12, 63–67. https://doi.org/10.1111/j.1365-2761.1989.tb01292.x.

Willoughby, L.G., Roberts, R.J. & Chinabut, S. (1995) Aphanomyces invaderis sp. nov., the fungal pathogen of freshwater tropical fish affected by epizootic ulcerative syndrome. *Journal of Fish Diseases*, 18, 273–276. https://doi.org/10.1111/j.1365-2761.1995.tb00302.x.

Winton, J.R. (1991) Recent advances in the detection and control of infectious hematopoietic necrosis virus (IHNV) in aquaculture. *Annual Review of Fish Diseases*, 1, 83–93. https://doi.org/10.1016/0959-8030(91)90024-E.

Wolf, K. (1988) Infectious hematopoietic necrosis. In: *Fish Viruses and Fish Viral Diseases*. Cornell University Press: Ithaca, pp. 83–114.

Wood, E.M., Snieszko, S.F. & Yasutake, W.T. (1955) Infectious pancreatic necrosis in brook trout. *AMA Archives of Pathology*, 60, 26–28.

Xiao, J., Liu, L., Ke, Y., Li, X., Liu, Y., Pan, Y., Yan, S. & Wang, Y. (2017) Shrimp AHPND-causing plasmids encoding the PirAB toxins as mediated by pirAB-Tn903 are prevalent in various Vibrio species. *Scientific Reports*, 7, 42177. https://doi.org/10.1038/srep42177.

Yoshimizu, M. (2016) Bacterial kidney disease of salmonids. *Fish Pathology*, 51, 49–53. https://doi.org/10.3147/jsfp.51.49.

Yoganandhan, K., Leartvibhas, M., Sriwongpuk, S. & Limsuwan, C. (2006) White tail disease of the giant freshwater prawn Macrobrachium rosenbergii in Thailand. *Diseases of Aquatic Organisms*, 69, 255–258. https://doi.org/10.3354/dao069255.

Yu, C.I. & Song, Y.L. (2000) Outbreaks of Taura syndrome in Pacific white shrimp Penaeus vannamei cultured in Taiwan. *Fish Pathology*, 35, 21–24. https://doi.org/10.3147/jsfp.35.21.

Zhao, Y.-R., Yue, Z.-Q., Tan, L.-Y., Ma, Y., Li, Q., Liang, C.-Z. & Li, B.-F. (2011) Preliminary investigation and analysis of red sea bream iridoviral disease (RSIVD) in some coastal areas of Shandong Province. *Yuye Kexue Jinzgan*, 32, 31–36.

Zheng, L.P., Geng, Y., Yu, Z.H., Wang, K.Y., Ou, Y.P., Chen, D.F., Huang, X.L., Deng, L.J., Gan, W.X., Fang, J., Zhong, Z.J. & Lai, W.M. (2018) First report of spring viremia of carp virus in Percocypris pingi in China. *Aquaculture*, 493, 214–218. https://doi.org/10.1016/j.aquaculture.2018.04.056.

Zhou, J., Fang, W., Yang, X., Zhou, S., Hu, L., Li, X., Qi, X., Su, H. & Xie, L. (2012) A nonluminescent and highly virulent Vibrio harveyi strain is associated with 'Bacterial White Tail Disease' of litopenaeus vannamei shrimp. *PLOS ONE*, 7, e29961. https://doi.org/10.1371/journal.pone.0029961.

Zorrilla, I., Balebona, M.C., Mori·igo, M.A., Sarasquete, C. & Borrego, J.J. (1999) Isolation and characterization of the causative agent of pasteurellosis, Photobacterium damsela ssp. piscicida, from sole, Solea senegalensis (Kaup). *Journal of Fish Diseases*, 22, 167–172. https://doi.org/10.1046/j.1365-2761.1999.00157.x.

6

Global aquatic food production

Ram C. Bhujel

Background

SDG goals and concerns

In 2015, the United Nations set 17 Sustainable Development Goals (SDGs) to be achieved by 2030. Goal 2.1 states that by 2030 we will end hunger and ensure access by all people to safe, nutritious and sufficient food all year round. However, a report by FAO (2018) and other UN agencies showed that 17 million more people became hungry in 2017 reaching 821 million compared to 804 million in 2016. More alarmingly, it reported that 24% more people faced severe food insecurity in 2017 (769 million) compared to 2015 (619 million). Similarly, Goal 2.2 aims at ending all forms of malnutrition by 2030. A report by FAO (2018) and others has shown that little progress has been made in 5 years. For example, they showed that 22.2%, i.e. 151 million, of children under 5 are still stunted in 2017 as compared to 25% in 2012. At the same time, the obesity problem increased slightly in adults (13.2% from 11.7%) as well as in children (5.6% from 5.4%) between 2012 and 2017. The degree of food insecurity varies among vulnerable groups and regions. The percentage of women of reproductive age with anaemia, a symptom of iron shortage and of hidden hunger, increased to 33.8% in 2017 from 30.3% in 2012. Variations exist among the regions, and between urban and rural areas. These facts indicate that the food and nutrition SDG will be difficult to achieve by 2030 unless more efforts are made. The countries facing these challenges need to be serious and their policymakers and planners to implement appropriate plans effectively.

Availability and access to aquatic food

Aquatic foods could play a significant role in achieving the sustainable development goal, especially Goal 2.2, if the appropriate policies are formulated, practical plans are made and effectively implemented. This can be better elucidated using the four pillars of food security: availability, accessibility, use and stability as elaborated by Charlton (2016). In order to achieve the goal, we need to enhance aquatic food production (availability), to make aquatic produce cheaper and more easily accessible to all (accessibility), to help people use them wisely (utilization) and to create an environment that will stabilize market prices and other inputs.

Margaret Crumlish and Rachel Norman (eds) **Aquatic Food Security**
DOI: 10.52517/9781789183368.006, © 5m Books Ltd, 2023

Communities living in low lying flood plains and close to lakes, rivers and the sea normally are familiar with aquatic animals and have easy access to them; they enjoy a wide variety of aquatic animals as sources of protein. Whereas communities living in dry land zones normally consume terrestrial animal meat as their main source of protein. For them aquatic food is either unknown or is a rare alternative or dietary substitute for terrestrial meat sources. Choice of food largely depends on knowledge about health benefits, availability and affordability. More interestingly, people are slow to or may not easily accept new types of food. Community norms, values, cultures, customs, religions, tastes and social status play a very important role in food choices. These factors are often so strong that many people do not go against them even when they know seafood items to have excellent health benefits. For example, the inclusion of seafood items or a large and expensive fish at a wedding party or other significant social event may only be for recognition of high social rank. In some religions killing animals and eating their meat is forbidden and adherents take protein from vegetable sources. Hunger and malnutrition are complex problems to solve. Rigorous consideration is needed in selecting appropriate response strategies. I suggest the following three major interventions.

- Educate consumers about nutrition and food options, their role in human health, their sources and their proper utilization.
- Expand the production of nutritious food items and increase efficiency of production systems.
- Improve the access of those on lower incomes to nutritious food items by enhancing the efficiency of marketing systems and even creating new markets to reach new, particularly, remote, rural areas.

This chapter has been devoted to cover all of these aspects hoping that it would contribute to achieving the goal by 2030.

Aquatic food demand

About 20% of total animal protein comes from aquatic food for the total population (Chan et al., 2017; FAO, 2018). More importantly, demand for aquatic food is increasing rapidly due mainly to the increasing population. The world's population is currently increasing by approximately 1 billion each decade, which means at least 22 million tonnes (mt) of additional aquatic food needs to be supplied by the end of each subsequent decade. According to the FAO there will be an annual shortage of about 50 mt seafood by 2030 (Table 6.1). Seafood supply needs to be doubled by 2050 to meet the expected demand.

Awareness about global warming and other environmental threats is increasing. Environmental concerns are becoming an increasingly important consideration for consumers when choosing food items to grow and consume. Demand for aquatic food is going to see further uplift as more people become aware that it is more efficient in terms of resource use and less polluting to the environment. For example, fish normally can convert feed almost three times more efficiently than pork (i.e. feed conversion ratio (FCR)). Fish production needs 1000 times less land and water use per kilogram produced and the industry has a lower carbon footprint and overall lower production of green-house gases (Table 6.2). Beef and lamb production has 3–10 times more impact than other animal-based foods, such as fish produced in ponds or non-circulatory water systems (Clark and Tilman, 2017). There are many recommending we turn to aquaculture in order to feed the predicted 9 billion people who will be living on the Earth by 2050 (Béné et al., 2015).

Table 6.1 Fish supply and predicted deficits.

Continents	Supply by 2030 (mt)	Requirement by 2030 (mt)	Supply deficit (mt)
Africa	11.7	18.7	–7.0
Asia	156.5	186.3	–29.8
Europe	18.6	23.4	–4.8
Latin America & Caribbean	16.2	18.3	–2.1
North America	6.2	12.9	–6.6
Oceana	1.5	1.8	–0.3
World total	210.7	261.2	–50.6

Source: FAO (2018).

Table 6.2 Comparative inputs and results of different animal protein sources.

Parameter	Farmed finfish	Chicken	Pork	Beef	Lamb
Land use (m^2/kg protein)[1]	1	80	130	1020	–
Water use (l/kg)[2]	500	4,325	5,988	15,415	8,763
Feed conversion ratio[3]	1.1	2.2	3.0	4–10	–
Protein retention (%)[4]	31	21	18	15	–
Carbon footprint (kg CO_2e/100 g protein)[5]	5.98	5.7	7.61	19.85	49.89

Sources: [1]Clark & Tilman (2017), [2]Armstrong (2017), [3,4]GAA (2018) and [5]EWG (2011).

In many countries aquatic food is a lot cheaper than red meat. Increasing numbers of people change their food habits from red meat to white meat, especially seafood; a process that is happening in many developed countries including Canada and the USA. People in developed countries are demanding more aquatic food. However, production of sufficient aquatic food to fulfil demand is difficult for the countries where water is scarce and input costs are high.

Most developing countries produce more than they can consume and therefore surpluses for export are important sources of revenue. Aquatic food is now produced and processed on an industrial scale and transported to different markets around the world. For example, Norwegian salmon can be bought in Asian countries including the recently established Timor-Leste. Similarly, Vietnamese pangasius is found all over Europe and the USA.

Demand for aquatic food is increasing in almost every country. China exhibits the fastest growing aquatic food demand, driven mainly by increased income levels and awareness about its health benefits. If China's more than 1 billion people start consuming extra fish then the demand is likely to skew aquatic food markets away from the USA and EU to China. Demand should also increase in South Asia especially in India and Pakistan following the introduction of new species such as tilapia and pangasius, which have no small bones in the muscle. Similarly to China, increased awareness about the health benefits and an emerging middle class should also contribute to higher demand.

Aquatic food production

Aquatic food is the largest source of high-quality animal protein (Béné et al., 2015). An estimated 178.5 mt of aquatic animals and plants (approximately 96.48 mt from catch and 82.14 mt from culture) was produced in 2018; out of which about 156.4 mt was used as human food (Table 6.3, Figure 6.1, FAO, 2020). This figure was more than 2 times that of beef, 1.2 times that of broiler chicken, 1.4 times that of pork and almost 10 times that of mutton or lamb. Annual per capita consumption of aquatic food increased from 13.4 kg (average over 1986–1995) to 20.5 kg in 2018. Consumption and the demand for aquatic food has increased drastically and will continue to do so. The increase in the supply comes mainly from farming which rose from an average annual production of 14.9 mt during 1986 to 1995 to 82.1 mt in 2018, a 551% increment. Aquatic food is produced in every country; however, countries in tropical regions have specific climatic, geographical and socio-economic advantages. Rich northern countries produce a lot less than they need. Therefore, aquatic food is transported and traded. Of the 2018 production of 156.4 mt, 37.6% was traded internationally with a value of US$164 billion (Table 6.3, FAO, 2020). Trade flows mainly from developing countries to developed- and high-income countries contributing to food security in the purchasing nations.

Total global aquatic food production from wild catch has been more or less stable since the 1980s. Some fishing grounds have completely collapsed, e.g. Monterey Bay in California, USA (Parrish, 2000). In most cases fishers are catching smaller and smaller fish to get the same volume. Most countries in the west started facing lower catches from around 1950s and 1960s. Aquaculture was thought to be the main solution to reduced wild catches. Developed countries, such as France, Germany, Japan, started farming of aquatic species and initially saw aquaculture production growth until the 1980s when production stalled and then started to drop due mainly to high cost of inputs, labour and environmental problems. Only a few cases remain commercially successful, e.g. Salmon in Canada, Norway and Scotland. At the same time, farming started to boom in developing countries, especially in Asia and Latin America. The developed world adopted a strategy of funding through aid and investment to support developing countries in producing cheap and good quality aquatic food which they then import.

The status and the scope of the main sources of aquatic foods, i.e. capture and culture, are now discussed briefly.

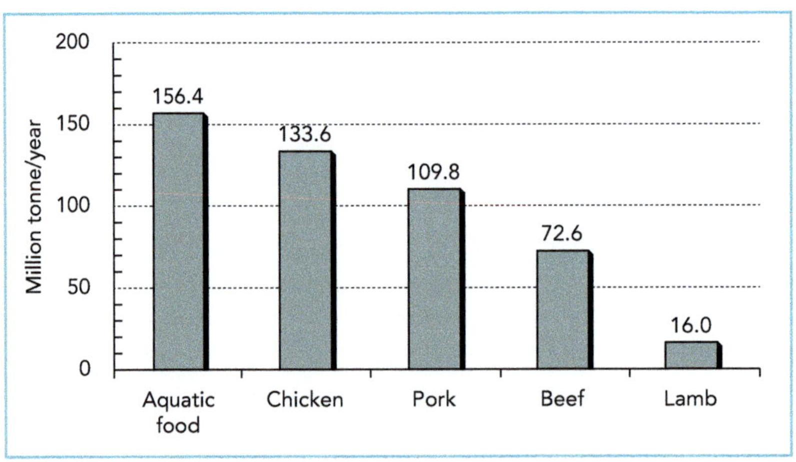

Figure 6.1 Sources of animal protein for human food and nutrition security.

Source: data from https://www.fao.org/3/ca9229en/ca9229en.pdf, https://www.fao.org/3/cc0461en/cc0461en.pdf, https://www.fao.org/fishery/statistics-query/en/global_production/global_production_quantity.

Table 6.3 Production of aquatic food from capture and culture (mt)

	2012	2013	2014	2016	2017	2018
Capture	86.9	91.4	89.8	89.6	93.1	96.4
Aquaculture	14.9	34.2	59.7	76.5	79.5	82.1
Total aquatic food production	101.8	125.6	149.5	166.1	172.6	178.5
Non-food use	30.0	27.1	20.3	17.9	19.7	22.1
Total human consumption	71.8	98.5	129.2	148.2	152.9	156.4
Per capita consumption (kg)	13.4	15.9	18.4	19.9	20.3	20.5
Export volume	34.9	46.7	56.7	59.5	64.9	67.1
Export value (US$)	37.0	59.6	117.1	142.6	156.0	164.1

Note: * 10-year average.
Source: FAO (2020).

Aquatic food from capture

Seafood was the common name of the food items mainly captured from the sea or marine environments, especially anchovies, crabs, mackerel, sardine, sea bass, squid, shrimps, tuna, and sharks. Aquatic food animals are captured from inland as well as marine waters. Figure 6.2 shows the relative importance of the species that are caught from the wild (FAO, 2020).

Anchovies

Anchovies are small pelagic seawater fish commonly found in the surface waters of the Atlantic, Indian and Pacific oceans. Large stocks are found in the seas off Peru and northern Chile. They belong to the family Engraulidae, comprising 17 genera and 144 species. They are small, ranging from 2 to 40 cm in length. They are an oily fish, rich in omega-3 and, therefore, are nutritious human food. Oily fish are also good sources of vitamins A, B and D, and many minerals. They are mostly caught by artisanal fishermen, consumed by their families and also supplied to the local markets and also processing factories where they are dried, fried, salted, or fermented, and packaged in cans. Ready-to-eat processed products are distributed worldwide as a source of good animal protein. They are often consumed alone as snacks or used as ingredients for other main dishes. They play an important role in human nutrition especially in Asia and Europe. Large fishing boats and fleets have been used to produce fishmeal and fish oils in large volumes for aquaculture and livestock industry. Peru became the leading producer of fishmeal in the 1960s and had a total landing of around 8 mt, i.e. approx. 40% of the global total volume (IFFO, 2009). Emphasis was given to shifting anchovies towards human consumption (Bórquez and Hernández, 2009). An annual average of 6.5 mt over the 2005–2014 period was produced for direct human consumption, later production declined to 4.3 mt 2015 and 3.2 mt in 2016 but again increased to over 7 mt in 2018 (Figure 6.2). Exploitation for direct human consumption resulted in a sharp decline in catch. As a solution, a quota has been imposed by policy makers in Peru, i.e. a 300,000 tonnes limit was set for human consumption in 2017 (Mereghetti, 2017).

Alaska pollock

Alaska pollock (*Gadus chalcogrammus*) is also known as walleye pollock and is one of the world's largest fisheries. Pollock is widely distributed in the North Pacific Ocean with the largest concentrations in the eastern Bering Sea. The USA has good regulation and wild-caught Alaska pollock is considered one of

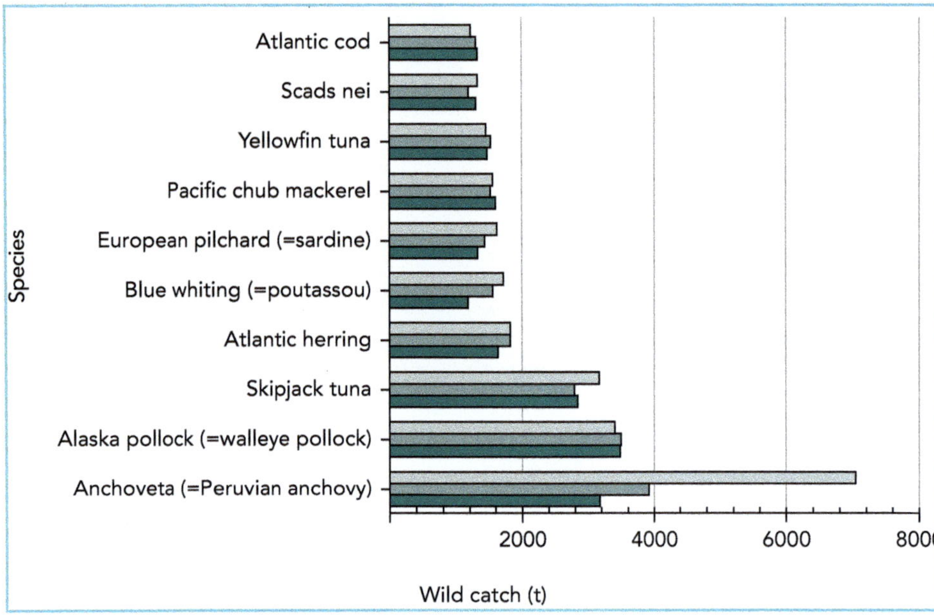

Figure 6.2. Annual production of major species caught from the wild, 2016–2018.

Source: data from https://www.fao.org/3/ca9229en/ca9229en.pdf, https://www.fao.org/3/cc0461en/cc0461en.pdf, https://www.fao.org/fishery/statistics-query/en/global_production/global_production_quantity.

the best choices of seafood as it is sustainably managed and responsibly harvested. Scientists estimate the female spawning biomass to be above the target level. The regulations for the pollock fishery aim to conserve the spawning population to ensure pollock can successfully reproduce and keep the population size at healthy levels. Figure 6.2 shows that catch reached 3–4 mt per year over the period 2016–2018.

Tuna

Tuna is one of the most important species caught in the wild and they are highly prized. There are several species of tuna under the genus *Thunnus*. The most popular are Atlantic bluefin or giant bluefin (*Thunnus thynnus*), Pacific bluefin (*T. orientalis*), southern bluefin (*T. maccoyii*) and yellow fin (*T. albacares*). The Japanese are way ahead in terms of wild catch and consumption. Bluefin tuna is the most preferred species. Tuna sushi (wrapped with vinegar rice), sashimi (fresh raw slices) and surimi (made from fish paste) are very popular in Japan and worldwide. Whereas Thailand has been the top supplier of canned tuna for which smaller tunas are caught. Considering the declining numbers of tuna in the ocean various efforts have been made to restore their population. There were some increments in the production of skipjack and yellow fin tuna between 2016 and 2018 (Figure 6.2). However, smaller tunas are often caught to meet increasing demand. Attempts have been made to breed and commercially grow tuna but the cost of production is prohibitively high; the FCR is as high as 18:1 for bluefin tuna (Yan and van Beijnen, 2019). Its future is uncertain unless there is a break-through in the management of their wild population or in farming methods.

Mackerel

Mackerel are another group of different species of oily pelagic fish belonging to the family Scombridae, which has more than 30 species including tuna. The oil content varies from 5% to 25% across the seasons. As oil content increases, water content decreases. They are distributed mostly along the coasts and offshores of temperate and tropical seas. Mackerel prefer shallow waters and often migrate in large schools along the coast to find suitable spawning grounds. Mackerel have stripes or bands on the back and the name 'mackerel' means 'marked' or 'spotted'. Atlantic mackerel (*Scomber scombrus*) is one of the important species. It is abundant in the north east Atlantic from Norway to Morocco and the Canaries, and in the Mediterranean and Black seas. In the north west Atlantic it occurs from Labrador to North Carolina. Catch volume is approximately 5 mt per year. They are consumed fresh or are preserved (chilled, frozen, fermented, dried, fried, smoked, steamed or salted) and are found in cans or various packaging. Opinion is divided on whether mackerels should be produced for human consumption directly or they should be used in feed for livestock and aquaculture (Bórquez and Hernández, 2009).

Sardines

Sardines, also known as pilchards, are small forage fish that are widely consumed by humans as they are rich in omega-3 fatty acids. They are consumed fresh, often grilled, and are pickled or smoked and packaged mostly in cans. Sardines are commercially fished for direct human consumption, and also for processing. They are canned, dried, salted and smoked. They are also used to produce fishmeal and fish oil. Fishmeal is mostly used to feed animals. Oil is used to manufacture products such as paint, varnish and linoleum. Sardine fishing has existed thousands of years in Croatia, Morocco, Peru and other countries in Europe. The Peruvian fishing industry is huge and produces over 9 mt annually but with little for direct human consumption, i.e. a little over 100,000 mt tonnes (about 2% of catch). It is promoted for the sake of human health, environment and economic benefits by the government and NGOs. The wild catch has been reduced since 2000 to about 3 mt per year through regulatory action.

Herring

Herring is another forage fish belonging to the Clupeidae family. They are found in large schools in shallow waters of the temperate North Pacific and North Atlantic oceans, including the west coast of South America. They are also found in the Arabian Sea, Bay of Bengal and Indian Ocean. The genus *Clupea* is the most abundant and commercially important: 90% of all herring catch is from three species of *Clupea*. The most abundant is the Atlantic herring which accounts for about 50% of catch. Herring has been a staple food since 3000 BC. They are eaten raw, and often salted, smoked, pickled or fermented. Some herrings may contain mercury and also polychlorinated biphenyl (PCB) and dioxin which cause cancer.

Atlantic cod

Unlike anchovies, sardines and herring, cod, haddock and flatfish are white, demersal fish and are widely consumed. The Atlantic cod (*Gadus morhua*) is the primary source of cod liver oil. The livers account for about 13% of total weight. The cod is a benthopelagic fish of the family Gadidae. According to FishBase (www.fishbase.org), they are distributed around the North Atlantic and Arctic (Ungava Bay in Canada along the North American coast to Cape Hatteras), off North Carolina in the western Atlantic, along both east and west coasts of Greenland, around Iceland, and from the Barents Sea, including the region around Bear Island, down the European coast to the Bay of Biscay.

Shrimp and prawn

Shrimp is a crustacean. There are 40,000 species of crustaceans found in fresh and saline waters. There are over 300 species of shrimps cultured or captured for human consumption. They are widely distributed from shallow to deep seas. They mostly inhabit saline waters. *Macrobrachium rosenbergii* and a few other species are also found in freshwaters such as rivers, lakes and swamps, and are often called the giant river prawn or giant freshwater prawn. Among saline water species, the Asian tiger shrimp (*Penaeus monodon*) and Pacific white shrimp or white leg shrimp (*Litopenaeus vannamei*) are the major ones. Both shrimp and prawn are highly value and one of the most commonly traded aquatic food products. Almost one-half of the total volume for human consumption still comes from the sea. Among the crustaceans, shrimp is the most popular and highly commercialized food item and is well established among the consumers all over the world. It is grown mostly in tropical countries and coastal areas. Wild catch has increased from about 0.5 mt in 1950 to 4 mt per year in recent years due to high demand which is never met. Wild catch is taken using simple hand nets to huge trawlers. Broodstock caught from the wild are generally considered the best source of genetic materials for aquaculture and wild stock replenishment.

Crab

Crab are highly developed and the most diverse crustaceans within the order Decapoda. Most commercially significant crab belong to the Brachyura (true crabs) or Anomura (hermit crabs and king crabs) infraorders. They are generally characterized by a pair of claws and a wide and flat body. They are widely dispersed and found in shallow tropical seas to deep ocean trenches, estuarine and freshwaters. Some tropical species spend the majority of their life on land, only returning to the water to reproduce. Most crab species are highly fished but they are still available abundantly because they have high fecundity. A gravid female may produce 50,000 to 1 million eggs per year. Most of them carry multiple batches of fertilized eggs attached to the underside of the female, which are extruded at different times. Global production of different crabs has surpassed 1 mt per year (Figure 6.3). Traps are used to catch crab as they are simple to operate, useful for deep water, labour-efficient, and the crabs are less likely to be injured which is important as they are sold at higher prices live. Trawling, tangle netting, dredges, trotlines and drop/ring nets are also used to harvest crabs. The majority of the large crab fisheries operate in tropical and northern hemisphere temperate and Arctic waters. By volume Gazami crab has the highest catches as it has a wide distribution throughout the western Pacific, found especially in shallow inshore waters in coastal bays. The catch may also come from stocking of crablets into the natural waters particularly off the Japanese coast. However, the most important commercial crab species are all fast-growing 'swimming crabs,' found in shallow tropical or temperate waters and bays.

Crabs are everywhere in all oceans, fresh water, and on land. About 10,000 species have been described so far. Most crabs live in the sea and they are still abundant. However, in some fishing grounds they are limited, and breeding and farming have been tried due to the expectation that populations in the wild may decline any time.

Other shellfish

Squid, octopus and cuttlefish look similar to most people. They are highly nutritious, they have no bones and all the parts except inner skeleton are edible. The best-known octopus is the common octopus, *Octopus vulgaris*, a medium-sized animal that is widely distributed in tropical and temperate seas throughout the world. It lives in holes or crevices along the rocky bottom and is secretive and

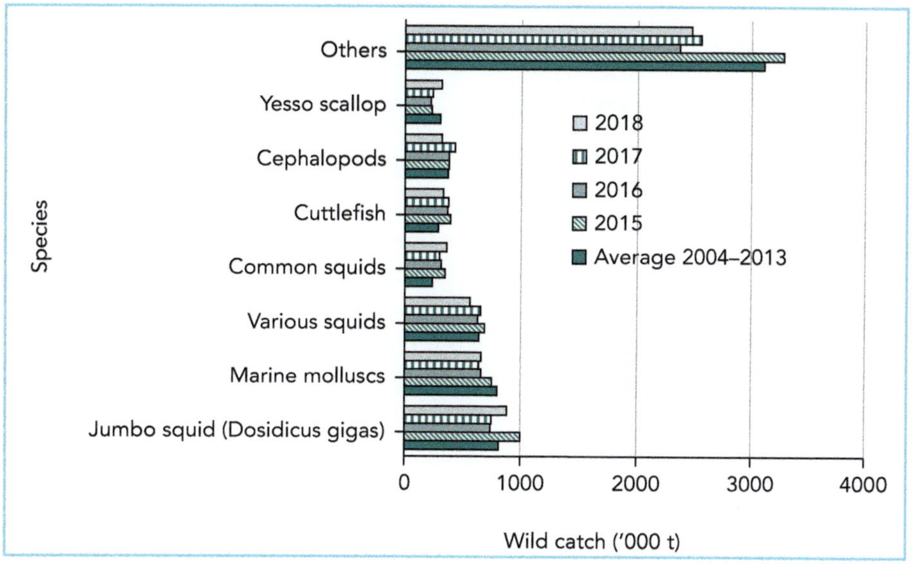

Figure 6.3 Production of molluscs and other bivalves from wild catch.

Source: data from https://www.fao.org/3/ca9229en/ca9229en.pdf, https://www.fao.org/3/cc0461en/cc0461en.pdf, https://www.fao.org/fishery/statistics-query/en/global_production/global_production_quantity.

retiring by nature. It feeds mainly on crabs and other crustaceans. This species is thought to be the most intelligent of all invertebrate animals. Marine molluscs were among the earliest human foods. Figure 6.3 shows the production trends since 2010. Prehistoric heaps of discarded mollusc shells have been found in coastal areas throughout the world especially in China, Japan, Peru, Brazil, Portugal and Denmark indicating that molluscs were the source of animal protein for people since long ago.

Oysters are bivalve molluscs found in the temperate and warm coastal waters of all oceans. They are considered high-value food items eaten either raw, cooked or smoked. There are two main families, i.e. Ostreidae (true oysters) or Aviculidae (pearl oysters). The true oysters (family Ostreidae) include species of *Ostrea*, *Crassostrea* and *Pycnodonte* genuses. Common *Ostrea* species include the European flat, or edible, oyster (*O. edulis*), the Olympia oyster (*O. lurida*) and *O. frons*. *Crassostrea* species include the Portuguese oyster (*C. angulata*), the North American, or Virginia, oyster (*C. virginica*) and the Japanese oyster (*C. gigas*) Pearl oysters (family Aviculidae) are mostly of the genus *Meleagrina*, sometimes called *Pinctada* or *Margaritifera*.

Mussels are an important aquatic food item found worldwide. They have about 1000 species found in streams, lakes, and ponds over most of the world. They are mostly common in cool seas. Many species are dark blue or dark greenish brown on the outside, but all are pearly inside. Mussels burrow into soft mud or wood but mostly attach themselves to solid objects often in dense clusters or to one another by proteinaceous threads called byssus threads. Mussels are collected from deep water by means of dredges or rakes. Some species are raised commercially, e.g. the blue mussel (*Mytilus edulis*), and are very important as food in Europe and elsewhere. The blue or purple coloured *M. edulis* has been cultivated in Europe since the 13th century.

Lobsters are very good food for humans and have high commercial value. Some species of lobsters, especially of true and spiny lobsters, are of the highest value aquatic food items. Therefore, they are eaten by the most affluent. Almost all are sourced from the wild. Major lobster supplying countries are Canada, the USA, Australia, Brazil, Myanmar and Vietnam. Most lobsters are nocturnal and live in the deep-sea, i.e. bottom-dwellers. Lobsters scavenge for dead animals but also eat live fish, small molluscs and other bottom-dwelling invertebrates and seaweed. Breeding technology has not been developed so far. Such technology might be necessary in the future to meet growing demand. Limited farming has been tried in Vietnam collecting juveniles from the wild (Towers, 2014). There are several species of lobsters; with, *Panulirus ornatus*, *P. homarus* and *P. polyphagus* are the major ones. The culture period varies with the species: *P. ornatus* takes about 20–24 months to reach the commercial size of 800–1000 g/piece, whereas the other species require 12–15 months to reach 200–300 g/piece.

Other species

There are several hundred species that are caught from the wild and consumed by human as sources of food. For example, abalone, cockles, clams, scallops, sea cucumber, sea horses, and others that play important roles in supplying animal protein and other nutrients especially minerals. Many of them are even high-value species and have niche markets. Some communities consume many aquatic species while others very few. For example, shrimp may not be accepted by some communities. Acceptability of new species depends on their deeply rooted culture, tradition and taste. Social media and easy communications around the globe have helped achieve a better and more widespread understanding of food security problems. Knowing some aquatic species are available in the surrounding environment, accepting them for consumption and managing their population, realizing their benefits to human health and the environment could enhance food security to a great extent. Only about 300 (0.2%) aquatic animal species have been used for consumption, culture and commercial purposes and only about one-half of them have been traded (FAO, 2020). There is great potential to explore new species because a huge number of species exist in nature. About 85,000 species of molluscs, 40,000 species of crustaceans and 30,000 species of finfishes exist on earth. According to Science Daily (2011), the latest findings showed that there are 8.7 million animal species on the earth. Out of which 6.5 million are on land and 2.2 million are in the oceans. However, only 212,000 species of aquatic animals (i.e. less than 10%) have been identified so far (Bouchet, 2006). There are plenty of potential species but more research is needed. Even if a species has no value where it is abundant, it can have higher demand in other places. Exploring opportunities for different markets is important. Many aquatic species have been brought from their original habitat to other places for farming or restocking into the natural waters.

Future outlook

Fishers only catch the fish that they know they can sell somewhere. Initially, fishing became a very lucrative business as there were no restrictions and fishers could catch plenty of fish in a single day as they were abundant in nature. When some of the fishing grounds collapsed, restrictions, such as complete stoppage of fishing for certain breeding seasons or ban on small mesh size of fishing nets, were brought in. Aquatic food catch has stagnated over the last decade and remains around 90 mt per year (Figure 6.4, Table 6.3). Catch may remain at similar level even until 2030 (Figure 6.2). Whereas aquatic food demand is increasing because of increased human population and also shifting food habits from red meat to seafood. Except for a few cases, wild stock and the catch from the sea cannot increase to meet the demand. Some scholars are arguing that actual catches are a lot higher and the peak was 130 mt in 1996 from when it has declined (Pauly and Zeller, 2016). Moreover, there are alarming research findings and projections that fisheries may collapse by 2048; all the fish species we catch will be gone unless critical measures are taken (Worm et al., 2006).

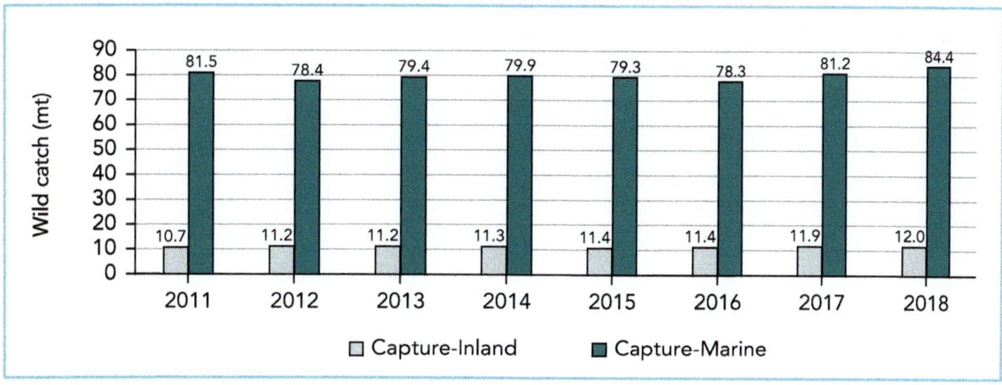

Figure 6.4 Wild catch from inland and marine water.

Source: data from https://www.fao.org/3/ca9229en/ca9229en.pdf, https://www.fao.org/3/cc0461en/cc0461en.pdf, https://www.fao.org/fishery/statistics-query/en/global_production/global_production_quantity.

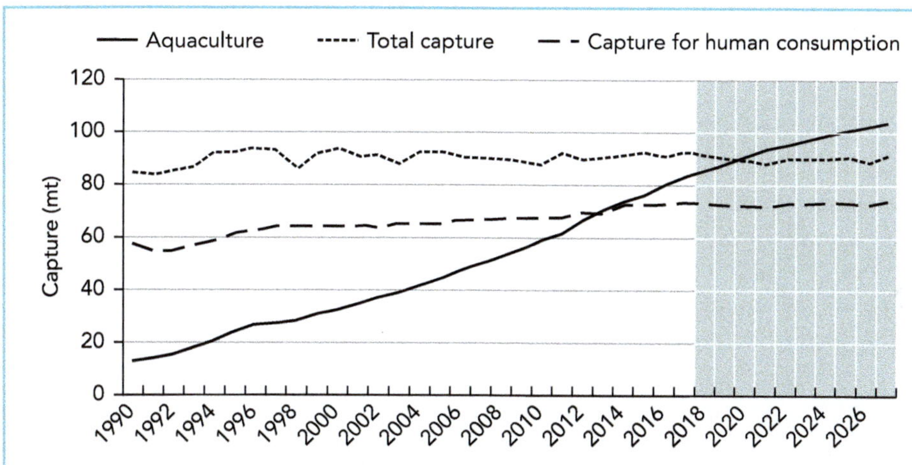

Figure 6.5 Aquatic food production trend from capture and aquaculture.

Source: data from https://www.fao.org/3/ca9229en/ca9229en.pdf, https://www.fao.org/3/cc0461en/cc0461en.pdf, https://www.fao.org/fishery/statistics-query/en/global_production/global_production_quantity.

Annual aquatic food production for human consumption from capture is stabilizing at around 90–100 mt out of which about 80–85 mt is from marine sources and 10–15 mt from inland or freshwater (Figures 6.4, 6.5 and Table 6.3). Both of these two natural sources are reaching their maximum limits. There is no indication that wild catch can meet the ever-increasing demand due to population increase and shifting food habits from terrestrial animal meat to aquatic foods.

Efforts are also being made to minimize losses as a large part of catch is wasted or used as animal food to increase the percentage available as human food (Table 6.3). Recently, due to awareness and some actions, wastage has been reduced to some extent from 23% to 20% which means the catch used as human food has improved from 77% to 80%. However, data have shown production cannot go higher than this. The volume of aquatic food production has to be increased dramatically if the projected demand by 2030 for food and nutrition security is to be met. Therefore, alternative means such as aquaculture has been aggressively promoted in many countries. Results are promising (Table 6.3, Figure 6.5).

Aquatic food from culture

The farming of fish and shellfish is known to have taken place in ancient Chinese civilizations and the Roman Empire. The oldest written history of aquaculture dates back to 475 BC with the 'Treatise in Fish Culture' by Fan Li. From India, around 321–300 BC, there are mentions in Kautilya's *Arthashastra* (Economics) (Kutty, 1999; Silas, 2003) and some indications are also found in King Someswara's *Manasoltara* (1127 AD). In Europe, fish farming began by the middle of the 11th century evident through the construction of ponds. The culture of aquatic animals has a long tradition in most parts of Asia especially in Southeast Asia where people live in flood plains and were familiar with the aquatic animals. Primitive forms of aquaculture might have started with small hanging nets or cages and with pens to hold the animals which were fed kitchen wastes (Pantulu, 1979). Some other scholars think farming might have started in rice paddies. Rice cultivation started around 10,000 years ago. Fish were abundant even in rice fields especially during summer when they start migrating upstream for breeding. People may have learned to trap fish for consumption and to keep them a live in rice paddies by blocking the inlets and outlets, in a similar way to tidal coastal areas where people used 'trap and hold' methods, which are still practised in various parts of the world.

Modern aquaculture started in the 1950s when some of the fishing grounds collapsed or showed signs of collapsing. Farming the common carp (*Cyprinus carpio*) was the 'first wave'. Aquaculture started in industrialized nations such as France, Germany, Japan and others. It grew until the 1990s, but growth could not be sustained further due mainly to high cost of inputs and labour. Many countries tried aquaculture but it was the developing countries of Asia and Latin America that had the most success. However, it was not until the 1980s that significant growth occurred. A 'second wave' during the period 1950–1980 saw a lot of research and promotional activities with other species, such as Indian major carps and also Chinese carps. At the same time, Mozambique tilapia (*Oreochromis mozambicus*) was also promoted; however, it became a nuisance due to excessive breeding. Later it was replaced by Nile tilapia (*Oreochromis niloticus*) during the mid 1980s. Since then aquaculture has expanded to many areas of the world as a 'third wave'. Many new species were introduced to many countries and promoted heavily. Following successes in breeding, nursing and farming of other species, such as catfish, salmon, shrimp and others, farming started to boom. As a combined result, aquaculture started showing great progress especially during the 1980s and 1990s during which annual growth rates were 10.8% and 9.5%, respectively. Growth continued at slower rate, 5.8% during 2001–2016, which was still faster than other major food production sectors. In 2020, almost 90% of total aquatic food (all species) is produced in Asia (Table 6.4) where molluscs and other aquatic animals production is well over 90% of the global total (FAO, 2022). Table 6.5 shows the top 10 aquatic food producing countries and their contribution. Eight out of ten are in Asia. China is by far the largest producer of aquatic food followed by Indonesia, India and Vietnam. Real growth occurred after 1990 when commercialization started, especially with the expansion of cage culture in lakes, rivers and reservoirs, and in ponds on land. At present, the largest expansion is expected in India, Latin America and the Caribbean, and Southeast Asia.

Farmers grow the fish that they know they can sell. Choosing the right species for production is critical: carp, tilapia, salmon, pangasius, tuna and crustaceans are probably the most popular species

Table 6.4 Production (mt) of major aquatic food group and the relative contribution by Asia.

Category	Production ('000 mt)		Asia's contribution (%)
	World	Asia	
Finfish	54,279	47,400	87.3
Molluscs	17,511	16,083	91.8
Crustacea	9,387	8,414	89.6
Other aquatic animals	919	915	99.6
Total	**82,095**	**72,812**	**88.7**

Source: data from https://www.fao.org/3/ca9229en/ca9229en.pdf, https://www.fao.org/3/cc0461en/cc0461en.pdf, https://www.fao.org/fishery/statistics-query/en/global_production/global_production_quantity.

Table 6.5 Top 10 aquatic food (from catch and culture) producing countries, 2017.

Rank	Country	Total aquatic food production (mt)	Contribution (%)
1	China	76.15	44.6
2	Indonesia	20.88	12.2
3	India	9.60	5.6
4	Vietnam	6.33	3.7
5	USA	5.41	3.2
6	Myanmar	5.05	3.0
7	Japan	4.77	2.8
8	Philippines	4.69	2.7
9	Russia	4.40	2.6
10	Chile	3.89	2.3
	Others	29.43	17.3
	Total	**170.6**	**100**

Source: https://worldatlas.com/articles/leading-countries-in-fishing-and-aquaculture-output.html.

and are widely available. Production has contributed greatly to global food and nutrition security. Aquaculure also creates employment and enhances economic growth. The fastest growth rates have been seen with tilapia and shrimp (Kobayashi et al., 2015). There are many other species that play important roles in specific countries and communities, such as catfish in Nigeria and Thailand, groupers and sea bass in Thailand and Malaysia, milk fish in the Philippines, mullet and hilsa in Bangladesh and trout in cold-water areas. A brief discussion follows of the major aquatic species and the characteristics that have made them successful or considered to have potential in the future.

Carps

Carps, the freshwater cyprinids native to Asia and Europe, have long been a staple food for humans; whether they are captured wild or cultured in ponds. Farming of common carp (*Cyprinus carpio*) might have begun about 3500 BC in China in ponds at silk farms utilizing silkworm pupae, their faeces and the decayed leaves of mulberry plants as food. As farming started expanding other carp species emerged as more important, such as grass carp (*Ctenopharyngodon idella*), silver carp (*Hypophthalmichthys molitrix*) and bighead carp (*Aristichthys nobilis*) as they feed mainly on grasses,

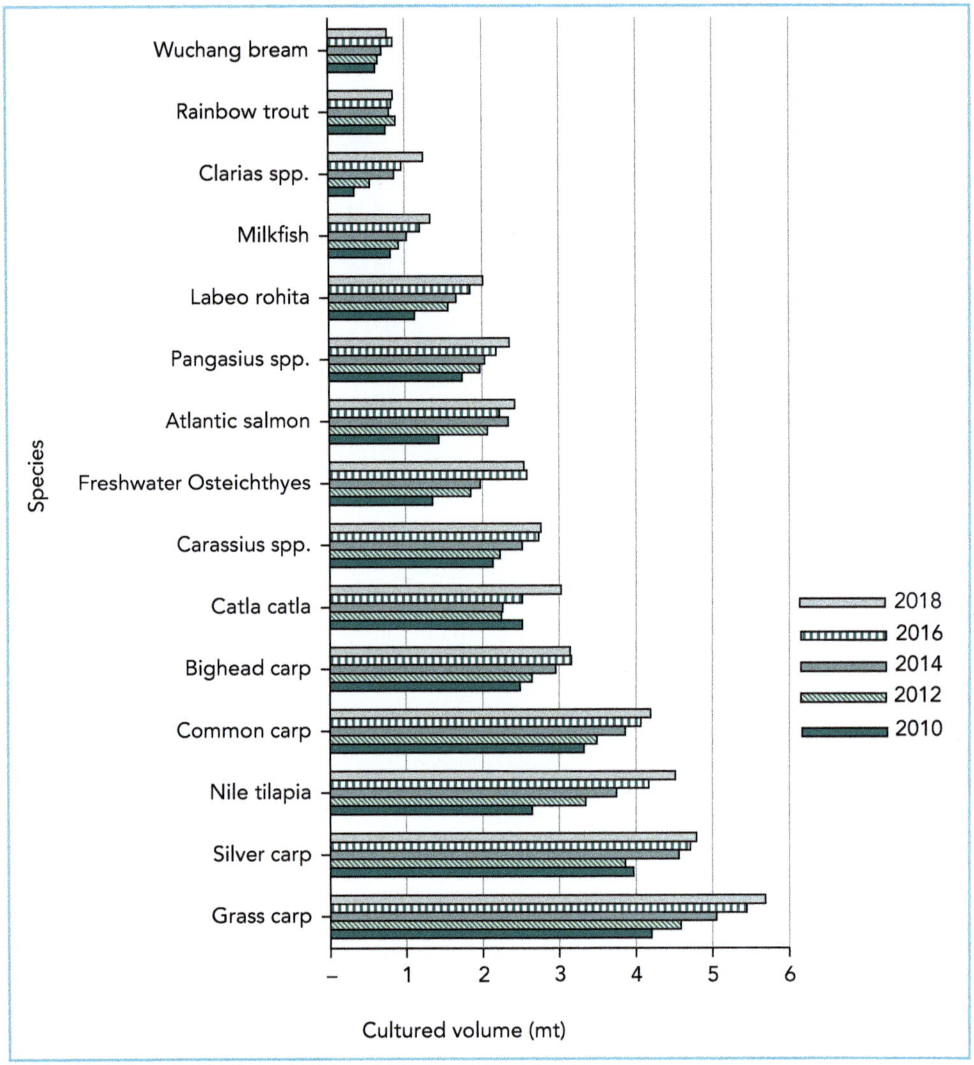

Figure 6.6 Major species of finfish cultured by volume, 2010–2018 (mt/year).

Source: data from https://www.fao.org/3/ca9229en/ca9229en.pdf, https://www.fao.org/3/cc0461en/cc0461en.pdf, https://www.fao.org/fishery/statistics-query/en/global_production/global_production_quantity.

phyto-planktons and zooplanktons, respectively. Similarly, evidence of three species of carp, namely, rohu (*Labeo rohita*), mrigal (*Cirrihunus mrigala*) and catla (*Catla catla*), with similar feeding habits, has been found in 4th-century BC India.

In general, carp are grown in green water with limited or no feeding. They are very important in terms of nutritional security and are produced in large volumes especially in China and the Indian subcontinent. They serve as staple food for many and also contribute to nutrition and health of hundreds of millions of people living in rural communities. Carp production was heavily promoted all over the developing world by development agencies such as the UNDP and the FAO with a view to supporting poor families in Bangladesh, Myanmar, Nepal and Pakistan. A number of training and consultancy services were provided. Facilities were built for breeding and nursing at many

Table 6.6 Aquaculture production of major carps and other top 15 species in 2020 (mt).

Major carp species	2014	2015	2016
Grass carp (*Ctenopharyngodon idella*)	5,539	5,822	6,068
Silver carp (*Hypophthalmichthys molitrix*)	4,968	5,125	5,301
Common carp (*Cyprinus carpio*)	4,161	4,328	4,557
Bighead Carp (*Aristichthys nobilis*)	3,255	3,402	3,527
Crucian carp (*Carassius spp.*)	2,769	2,913	3,006
Catla (*Catla catla*)	2,770	2,764	2,961
Roho labeo (*Labeo rohita*)	1,670	1,785	1,843
Wuchang bream (*Megalobrama amblycephala*)	783	796	826
Black carp (*Mylopharyngodon piceus*)	557	596	632

Source: FAO (2022).

government stations. However, because of the presence of small Y-bones in the muscle many people are reluctant to eat carp. More importantly, carp are not suitable for value addition processing, such as fillet preparation. As a result, many countries producing carps facing marketing problems. There is some trading of carp, mostly frozen, between countries in the Indian sub-continent. However, trade between carp producing countries and developed countries is minimal.

Tilapias

Tilapias are native cichlids to the African continent and are found mainly in the freshwater of shallow streams, rivers, ponds and lakes. There are nearly a hundred species of tilapia. Among them Nile tilapia (*Oreochromis niloticus*) and its hybrids (*O. mossambinus, O. aureus*) are commercially important species which are grown in over 100 countries in low-cost to highly intensive systems (Bhujel, 2014; Bhujel and Suharman, 2021). Their fast growth, short production cycle and frequent breeding behaviour, i.e. once a month, mean they are favoured by farmers (Bhujel, 2014; Suresh and Bhujel, 2018). More importantly, they feed mostly on plankton or other sources low in food chain unlike many other carnivorous species which feed on other live animals or require high-protein diets to grow. Most tilapias are grown in Asia and Latin America, but are exported to Europe, USA and back to African countries. The USA is the leading tilapia importer in the world where it became the fourth-most consumed aquatic food species in 2012 within a decade of imports starting because of its low price, mild taste and ease of preparation. Global production is steadily increasing and doubling or tripling each decade (Figure 6.7). Recent research and technological development in various aspects of farming such as the production of monosex fry, broodstock development and management and feeding have contributed to the expansion (Bhujel, 2014; Bhujel and Perera, 2017).

In various parts of the world tilapia is considered a poor person's fish; however, it contains similar or even higher levels in many nutrients compared to other species. For example, 100 g of tilapia contains 128 kcal calories, 26.15 g protein, 2.65 g fat (but no saturated fats), 56 mg sodium, 0.69 mg iron, and 0 g carbohydrate compared to 206 kcal calories, 22.1 g protein, 12.35 g fat, 61 mg sodium, 0.34 mg iron and 0 g carbohydrate in salmon. Therefore, it is suitable for those who want the diet low in caloric, sodium, fat and cholesterol diets (Yacoubian, 2023). Additionally, tilapia is rich in niacin, vitamin B_{12}, potassium, selenium and phosphorus and all the amino-acids (Yacoubian, 2023). Recently, tilapia fin has been used to make a soup

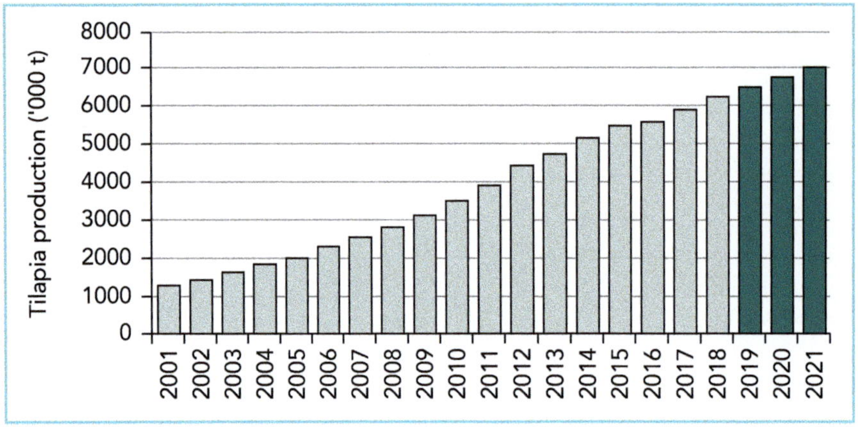

Figure 6.7 Farmed tilapia production (2001–2018) and estimation for 2019–2021.

Source: data from https://www.fao.org/3/ca9229en/ca9229en.pdf, https://www.fao.org/3/cc0461en/cc0461en.pdf, https://www.fao.org/fishery/statistics-query/en/global_production/global_production_quantity.

in an effort to replace shark fin soup in Taiwan. Its low level of omega-3 fatty acids has been considered a concern, however, bearing in mind that the main purpose of producing and supplying the fish is for protein, especially in rural areas and developing countries, this disadvantage is more than outweighed by the advantages. It is common that all freshwater fish species have low omega-3 fatty acid levels and are higher in omega-6 fatty acids. Interestingly, tilapia grown in green water, which is common practice in developing countries, has a relatively higher level of omega-3 fatty acids compared to those grown in cages with complete feeds (Karapanagiotidis et al., 2006).

Salmon

Salmon farming is supported by well-developed technology, particularly cage culture in the sea using large cages (20–50 m diameter and 15 m deep), cost-effective feeds and processing. It has a good reputation as a source of omega-3 and carotenoids. It is high-value seafood, but is still traded worldwide including to least developed countries (LDCs). It is farmed in many countries, such as Norway, Chile, Scotland (UK), Canada and the Faroe Islands (Denmark) (Table 6.7), and considerable salmon is still caught in the wild. Many people think wild-caught salmon is better than farmed. However, wild salmon cannot fulfil demand. During the 18th century, wild fish populations declined due to increased human population and pollution caused by the industrial revolution. Owing to scarcity in the wild, salmon became a luxury product. Farming became necessary. Much research and investment was put into the development of farming in Europe. Since then, breeding and hatchery methods were developed to produce smolts and young fish. However, successful growing of larger salmon only came in the late 1960s following the adoption of cages. Salmon farming became a success story in Europe during the 1970s and 1980s. Many farms appeared in fjords and bays in Norway and Scotland. Although the cost of inputs and labour is high in Europe, production of salmon has been made very successful mainly due to mechanization and automation. Salmon farming in sea cages is the most advanced and commercially successful form of aquaculture in the world. However, its farming is restricted to saline waters and cold temperate zones and is not widely distributed.

Table 6.7 Atlantic salmon production, 2016–2018 mt.

Country	Production (t)		
	2016	2017	2018
Norway	1,220,000	1,250,000	1,300,000
Chile	480,000	480,000	480,000
Scotland, UK	175,000	175,000	175,000
Canada	135,000	135,000	135,000
Faroe Islands, Denmark	86,000	90,000	89,000
Australia	65,000	65,000	65,000
Ireland	30,000	30,000	30,000
USA	20,000	20,000	20,000
Others	24,000	24,000	24,000

Source: Tveteras et al., (2019), FAO (2020).

Pangasius

Pangasius cultivation was one of the fastest growing industries during the early decades of this century. Vietnam led the industry where production increased from 50,000 t in 1998 to 1.2 mt in 2008 mainly due to a booming export market. Initially, it was exported to the USA, later to Eastern Europe and then all over the world. Pangasius has a white meat with a mild flavor. Favourable government policy, abundant water resource along the Mekong river and its tributaries, the presence of a processing industry, product certification and the emergence of a feed industry all helped the pangasius cultivation to grow very fast. About 30,000–40,000 farmers grow pangasius in Vietnam alone and over 200,000 people, mostly women, are involved in the processing industry supporting livelihoods of many people along its value chain. Farming has expanded to several other countries, such as Bangladesh, India, Indonesia, and Malaysia (Table 6.8), due to its high productivity and profitability. Average production has reached 400 t/ha per 6-month cropping (Phan et al., 2009).

African catfish

A number of catfish species are found in Asia and Africa. African catfish (*Clarias gariepinus*) has made a significant contribution to food security in the African countries of Nigeria, Kenya, among others. It has also been introduced to Asia to bring its fast-growing trait to Asian hybrids, such as *C. macrocephalus* and others. Catfish are mostly sold live in local markets and also preserved sun-dried and

Table 6.8 Pangasius production, 2016–2018. Source: Tveteras et al. (2019), FAO (2020).

Country	Production (t)		
	2016	2017	2018
Vietnam	1,188,680	1,251,680	1,334,680
India	447,500	462,500	640,000
Bangladesh	422,182	435,666	454,312
Indonesia	100,000	105,000	110,000
Malaysia	100,000	105,000	110,000

smoked for long storage and transportation. As catfish are airbreathers, they can be grown in a small area at high densities. Farming has expanded even in rural villages among poor farmers with small land holdings. They are often found to be grown in ditches or even in drums. They also can consume the offal of chicken or other animals; they are easy to grow as scavengers in backyard gardens. They are often found in foul waters. More recently, commercial farmers grow them in water recirculatory tank systems using floating pellets in addition to earth ponds.

Asian sea bass

Sea bass (*Lates calcarifer*) is white-fleshed fish native to Southeast Asia. It is also called Baramundi in Australia where some farming in cages is carried out. Sea bass is a carnivorous species, which is often used to control other fish, such as tilapia fry. It is seawater fish and requires 28–32 ppt to breed but will grow in all types of saline, brackish and even fresh waters. It is grown in the brackish water of river mouths in Thailand and other Asian countries by hundreds of farmers and supplied to the local restaurant trade. It is also grown in freshwater ponds due to pollution in brackish waters and disease problems. It has good markets in most countries and has a high value to the restaurant trade. All sea bass start as males when they are young and convert to being female when they become mature after reaching about 3 kg size. Therefore, it is very difficult to maintain broods and to be successful in breeding. Few farmers know its specialized breeding techniques and they are reluctant to share their experience as they fear their business will be affected by competition. An important stage of seed production is nursing the fry because they are carnivorous and can eat each other if they do not get suitable natural food in the culture system. Therefore, suitable feeding methods using a combination or a gradual shifting from one live food to another, such as artemia or moina, before starting pellet feeding is critical. Frequent grading every 1 or 2 days to maintain uniformity of fry and fingerlings size in each compartment is a tedious but very important practice. Owing to complexities in mass-scale seed production, farming has not taken off compared to the other species, such as catfish, tilapia or salmon. Concerted efforts in training and investment are needed if large areas of unused coastal zones of the tropics are to be utilized to contribute to food and nutrition security.

Trout

Trout is a freshwater fish, which requires very clear and cold water, the temperature of which ranges from 14°C to 18°C. Breeding and hatching require even lower temperatures, 9–14°C (Rai et al., 2005). There are several species of trout but two dominate: Rainbow trout (*Onchorhynchus mykiss*) and brown (river, brook) trout (*Salmo trutta*). The first trout hatchery was built in Westphalia in 1741 by Stephan Ludwig Jacobi, a German multidisciplinary scientist, with a view to producing on a mass scale and re-stocking natural water bodies, where wild fish stock had started to deplete due to industrialization (EU, 2019). However, it took a century for his discovery to be implemented on a large scale. Rainbow trout led fish farming in Europe from the end of the 19th century. In the 1960s, this progress led to the commercial development of intensive rainbow trout farming, first in Denmark and then throughout Europe. Norwegian fjords, where the salmon industry has now boomed, were originally used to farm rainbow trout in floating cages. Nevertheless, trout is an important species in terms of food and nutrition security as it is grown worldwide in over 100 countries including across Asia. It is normally farmed in raceways in highland areas. Trout require specialized diets and care during the hatchery stage. Being high in omega-3 and having a tender white meat trout are attractive to consumers. Mass-scale and high-volume production has not taken off due to the specific environmental and the technological requirements. Most of the product are sold locally.

Crustaceans

Figure 6.8 shows the relative importance of farmed shrimp, prawns, crawfish, crabs and other crustaceans. Shrimp is the most important commercially farmed species and is one of the most globally and highly traded aquatic food items. The FAO database showed that global annual farmed production of shrimp was nearly 5 mt in 2018, valued over US$30 billion, compared to a mere 43,000 mt in 1977. Shrimp culture, especially that of whiteleg shrimp, boomed because of successes in breeding, feeding, nursing and grow-out technologies. Over 75% of production is in Asia and the Pacific countries, led by China and followed by India, Vietnam, Indonesia and Thailand. Although there are many types of shrimps cultured or caught in various parts of the world, two types of shrimp are the most popular ones for farming: *Penaeus monodon* and *P. vannamei*. Over one-half (53%) of the farmed shrimp production in 2018 was *P. vannamei* (FAO, 2020). Annually, an estimated nearly 6 mt is produced globally and 3 mt is traded internationally (FAO-Globefish, 2019; FAO, 2022).

Shrimp farming can be divided into four types (extensive, semi-intensive, intensive and super-intensive) with production levels per hectare of 100–200 kg, 1–4 t, 5–10 t and >10 t, respectively. Most shrimp farms are intensively managed using high stocking densities, high feeding, with aeration, and antibiotic use to exploit profit potential. These have resulted in occurrence of diseases caused by bacteria and viruses. For example, in mid-1990s, *P. monodon* was devastated by white spot syndrome virus and others forcing a shift in production to the Pacific whiteleg shrimp (*P. vannamei*) by early 2000, which accounted for 50% of total shrimp production (FAO-Globefish, 2019). After two decades, whiteleg shrimp has also started suffering from disease incidence, e.g. early mortality syndrome (EMS) or acute hepatopancreatic necrosis disease (AHPND), causing a 50% reduction in production since 2015, a reduction which still has not been recovered. It is believed that haphazard and over-use of antibiotic caused loss of immunity in the shrimp and the development of antibiotic resistant strains of bacteria (Pham et al., 2018). Much has been learned from these problems, including the

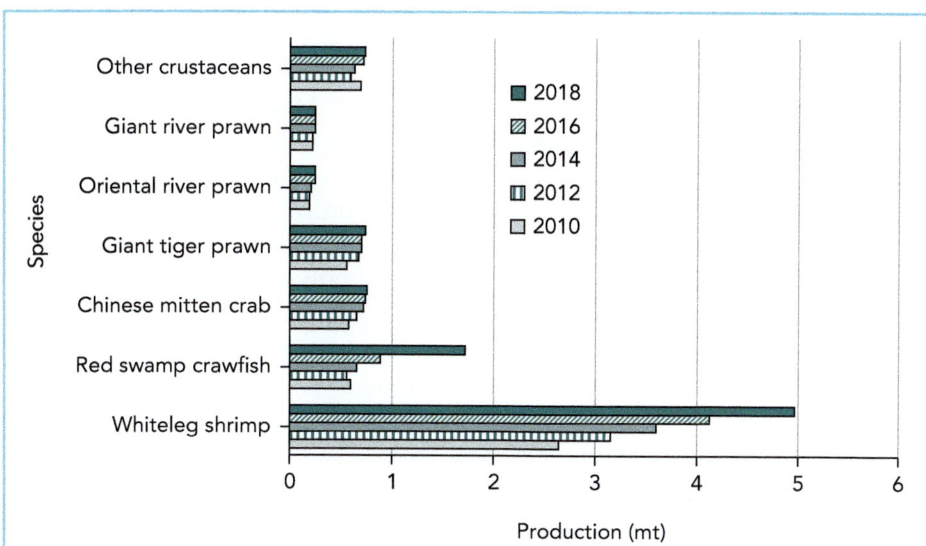

Figure 6.8 Crustacean production, 2010–2018.

Source: data from https://www.fao.org/3/ca9229en/ca9229en.pdf, https://www.fao.org/3/cc0461en/cc0461en.pdf, https://www.fao.org/fishery/statistics-query/en/global_production/global_production_quantity.

development of SPF (specific pathogen free) shrimp (*P. vannamei*), new technologies, such as biofloc, aqua-mimicry and so on (Avnimelech, 2015) and the onsite conversion of ammonia nitrogen to produce zooplanktons (copepods) by adding a carbohydrate source (e.g. rice bran). Similarly, aqua-mimicry uses fermented rice bran mixed with probiotics (e.g. *Bacillus* and *Lactobacillus*), which have benefitted shrimp farming (Chakravarty et al., 2018; Pinoargote et al., 2018). More importantly, considering the sustainability of shrimp farming itself and also that of environment, good aquaculture practices (GAP) and other certification schemes have become important and are required for export. Euope, the USA, Vietnam, China, Japan, Korea and Canada are the top export markets for shrimp in addition to the local markets in producer countries.

Freshwater prawn or river prawns have been farmed for some time and breeding, nursing and grow-out technologies are well-developed. However, its farming has not taken off due to low productivity. The FAO (2020) reported total production of two types of river prawns (*Macrobrachium nipponense* and *M. rosenbergii*) at nearly 1 million ton in 2018. Recently *M. rosenbergii* has attracted the attention of shrimp farmers in SouthEast Asia, especially after the rise of diseases in their shrimp. More recently, the production of neo-male or monosex technology described by Sagi and Aflalo (2005) has made prawn farming popular mainly because all-male culture makes possible to stock at higher densities.

Recently, crawfish (crayfish) is becoming popular in China as a delicacy. Production was reported to be over 1.7 mt in 2018 (FAO, 2020). Farmers are stocking and culturing in rice field systems. Crabs are still mostly caught in the wild. Low survival during larval rearing and nursing is the main constraint to farming (Boonyapakdee and Bhujel, 2019). Prawns, crawfish, and crabs fetch high prices; therefore, if practical technologies were to be developed to enhance productivity, farming would take off.

Other shellfish

In addition to commercially farmed aquatic finfish, there are many other aquatic species of crustaceans and molluscs that contribute to food and nutrition security especially supplying macro- and micro-nutrients. Marine molluscs were among the earliest human foods as evidenced by finds of prehistoric discarded mollusc shells in coastal areas worldwide. Molluscs are easy to keep live when caught and when more than could be immediately consumed were harvested the beginnings of aquatic farming were laid. Farming follows traditional methods and is very popular in coastal areas. However, it is often very localized in terms of production and catch from the wild is very limited below 1 mt annually (Figure 6.3). As the outer cover or the shells of molluscs or shellfish are made up of calcium carbonate or similar compounds, most offer good carbon sequestration or carbon deposition. However, their culture has not received adequate attention even when it is proven that they are good for the environment. Therefore, their farming needs to be promoted wherever is appropriate.

Oysters have been cultured successfully. The FAO database (Table 6.9) shows production from farming reached 5.7 mt in 2017, nearly a fivefold increase from 1.2 mt in 1987. Various species of oysters are cultured at the bottom or off the bottom in rafts and other structures on hanging ropes. They have been introduced to several countries for farming purposes. Mussels are another important group with global production 3.4 mt in 2017. About 1000 species of mussels exist in nature. Some of them are collected from deep water while others are raised commercially, e.g. the blue mussel (*Mytilus edulis*) in Europe and elsewhere. The blue or purple coloured *M. edulis* has been cultivated in Europe since the 13th century. Abalone, clams, cockles, scallops, squid (cuttlefishes and octopus) and many other aquatic animals play an important role in supplying animal protein and other

Table 6.9 Production of aquatic foods shellfishes and seaweeds (mt/year).

Year	Oysters	Mussels	Molluscs	Clams, cockles	Scallops	Squid	Abalone	Seaweed
1977	0.86	0.65	0.68	0.83	0.73	0.65	0.65	3.24
1987	1.27	0.92	1.10	1.46	1.11	0.92	0.92	4.64
1997	2.93	1.10	2.10	2.99	2.30	1.10	1.10	9.93
2007	4.40	1.60	2.60	5.86	3.10	1.60	1.90	16.59
2017	5.71	2.20	3.40	7.82	4.35	2.20	2.60	34.47

Source: FAO Database – online query at: http://www.fao.org/figis.

nutrients especially minerals. Production of clams and cockles was over 7.8 mt in 2017 rising from only 0.83 mt in 1977. Similarly, production of scallops has increased from 0.7 mt in 1977 to 4.35 mt in 2017. Many are high values species and have niche markets. Crab farming is being practised in some areas, but is still limited; the FAO database showed about 0.4 mt/year in 2017, which has doubled in a decade. Breeding of mud crab (*Scylla serrata*), Portunid crab (*Portunus pelagicus*) and other species has been successful.

Seaweed

Seaweed are red, green or brown marine algae. Their forms can be similar to broad leaves, fingers, sphericals or fruits. Seaweeds are high-value, nutrient-rich aquatic food items characterized by reasonable protein, low fat and low sodium levels and high calcium, phosphorous, magnesium, potassium, choline, sulphur and micronutrients (iodine, iron zinc, copper, selenium, manganese, boron and cobalt) content. They are also rich in vitamin A, B_{12} and K. The FAO database showed that total production of seaweeds was over 34.5 mt in 2017, out of which slightly over 1 mt is harvested from the wild (Ferdouse et al., 2018). The Asia-Pacific region is the largest producer. Seaweed grows naturally along seashores anchoring on rocks or other solid structures or to the sea bottom by root like 'holdfasts'. These root-like parts attach, but not to extract nutrients like the roots of the plants. Some seaweeds are used as fertilizers or as sources of polysaccharides and others are edible, and many are commercially important. There are several types of seaweeds. *Porphyra* spp. also called nori or a black seaweed, is often used in Japanese sushi. *Monostroma* spp. and *Enteromorpha* spp. are called aonori and are green tubular filaments. *Gracilaria* spp. is another species of seaweed, it is often called sea moss. It is used as a source of agar. It is harvested and sold as a salad vegetable in Hawaii (USA), the Philippines, Korea, Japan and China. Excessive demand cannot be met by collection from natural sources. It is being successfully cultivated in Hawaii and the West Indies. *Gracilaria* are also collected for food in Indonesia, Malaysia, the Philippines and Vietnam. It is considered an aphrodisiac and is also used to produce a non-alcoholic drink. Another popular seaweed variety is *Laminaria japonica*. Kombu is dried seaweed made from a mixture of *Laminaria* species. *Caulerpa lentillifera*, also called grapes or green caviar due to their spherical shape, can be grown in ponds with 30–35 ppt. Harvesting can be done within 2 months of transplanting, and then every 2 weeks. As they are grown on sandy and muddy bottoms, thorough washing is done in seawater. Seaweeds are inspected, sorted and placed in 100–200 g packages. They can stay fresh for a week or are frozen, which allows enough time for export.

Seaweed is a traditional food item in China, Japan and Korea and is also common in Indonesia, Malaysia and the Philippines. Seaweed consumption is growing in America and Europe, where it was not traditionally consumed, following migration of people from Asia. Seaweed has been cultivated onshore in east and southeast Asia, in tanks with the purpose of export to Japan. Markets are growing and it is already on the market in many countries around the world. People have tried to improve its image by rebranding seaweed as 'sea vegetables'. Current trends for consumers to embrace vegetarian, vegan and organically grown 'natural' foods should see the market for and seaweed farming grow rapidly. Farming of seaweed and also some macrophytes might take off considering they are green products and healthy superfoods (Ferdouse et al., 2018). The more than tripling of farmed seaweed production from 10.6 mt in 2000 to 35 mt in 2020, has already demonstrated its importance (FAO, 2022). According to Ferdouse et al. (2018), two countries, namely, China (47.4%) and Indonesia (38.4%) produce 86% of the total worldwide seaweed production. Other seaweed producing countries are the Philippines (5.3%), Korea (4.1%) and Japan (1.4%) and Malaysia (0.9%), which, with China and Indonesia, constitute almost 97.5% of the total global seaweed production and with Asia as a whole accounting for almost 99.7% of production in 2018. Wild catch contributes very little, i.e. 3.6% of the total. Seaweed production has the potential to expand to other countries that have long coastlines, such as India and several Africa and Latin America countries. Seaweed is environmentally advantageous and suitable for all diets including vegan. Concerted effort is needed to promote its importance and farming with a view to producing, supplying and consuming more seaweed. Governments, policymakers and researchers should plan more research and technology transfer so that it can fulfil its potential.

New species

There are a number of aquatic species still unidentified which may have potential commercially in the future. For an example, the sea cucumber was not widely recognized as a foodstuff about a decade ago, but now it is grown commercially. Sea cucumbers are found worldwide. The 1250 species of sea cucumbers are echinoderms from the class Holothuroidea, which live on the sea floor (Ross, 2019). They have a leathery skin and an elongated body with a branched gonad.

Sea cucumbers have been a delicacy in Asia for centuries. In the past, it used to be the wealthiest class with high-protein diets who could afford the luxury item (Lee and Polan, 2019). Demand exploded in the 1980s even among the middle class in China. Most of the sea cucumbers harvested are exported to Asian markets. Catching sea cucumber from the ocean floor is risky, results in a high price. According to a report by FAO (2020), wild-caught sea cucumber production increased from an average of 22,000 t during 2004–2013 to 48,000 t in 2018. They are heavily fished except for some populations in temperate waters. Farming has recently begun with fancy Japanese sea cucumber (*Apostichopus japonicus*) in Southeast Asia. Farmed production increased by 40% from 126,600 t in 2010 to 176,800 t in 2018 (FAO, 2020). Sea cucumbers are mostly dried and packaged in ornate boxes to give as gifts on special occasions. However, they are not easy to farm because of high larvae mortality. Even those that do survive take 2–6 years to grow to a marketable size.

Mantis shrimps are marine crustaceans of the order Stomatopoda. Similar to sea cucumbers they were not widely known. In the past mantis shrimps were commonly found in Cantonese, Japanese, Hawaiian, Mediterranean, Philippine and Vietnamese cuisines; however, nowadays they are now becoming popular species in restaurants although they are still caught from the wild. They are native of tropical and subtropical areas of the Indian and Pacific oceans. There are many types of mantis shrimp; namely, Japanese mantis shrimp (*Oratosquilla oratoria*), Malaysian, Mexican, Philippine and Thai varieties of mantis shrimps are also available. They are small, aggressive marine crustaceans with a hard shell similar to crabs, lobsters, crawfish, shrimp and krill.

Crawfish (crayfish) are species with potential, especially considering the devastating disease occurrences the shrimp industry has faced recently. Crawfish are freshwater crustaceans and are members of the infraorders Astacoidea and Parastacoidea. They are very similar to small lobsters and are often called freshwater or mountain lobsters. As mentioned above, they are already becoming popular in China and have been farmed successfully. Interest has been shown in many other countries. They have a profound potential and more effort and scientific studies are required to develop sustainable farming techniques to fulfil this.

Other species

There are many more species and a diversity of products available that play important roles in terms of food security for particular communities where these items are consumed. For example, jelly fish, frogs (American bull frog, European green frog, Asian bullfrog), soft shell turtles, crocodiles, sea urchines, seahorses, sea cucumbers, sand fishes, and so on. Some are also traded internationally, e.g. soft shell turtles.

Outlook of aquaculture

Large-scale farming of fish and aquatic animals started in the 1950s. Once the technologies developed, it grew exponentially. Total aquaculture production increased from 0.5 mt in 1950 to 87.5 mt in 2020, which is a huge increment. Production by volume and value from aquaculture is increasing at an average growth rate of about 3% per annum (FAO, 2022), while catch from marine and inland waters is stagnating. Since 1980s, production from almost all farmed species (Figure 6.9) has increased significantly with some species having skyrocketing growth, e.g. white leg shrimp and African catfish. Similarly, countries which realized the value of farming, and started promoting it in the early 1970s have seen their production increase staggeringly, such as China, Indonesia, Thailand and Vietnam (Figure 6.10).

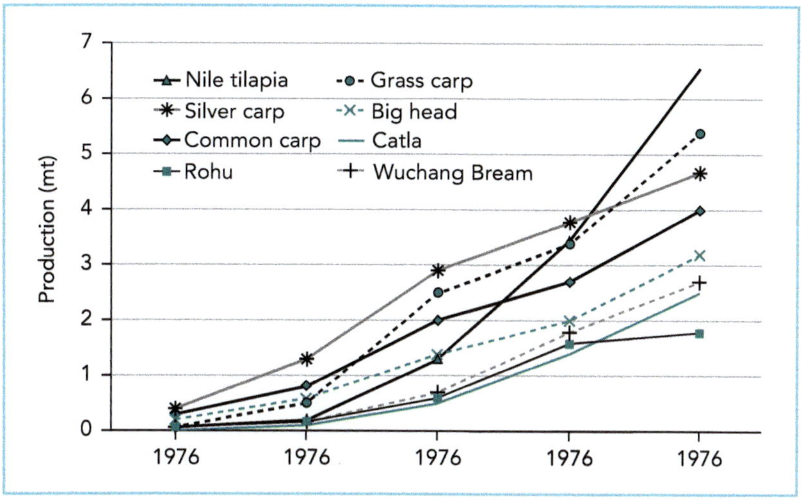

Figure 6.9 Production of commercially important species, 1976–2020.

Source: data from https://www.fao.org/3/ca9229en/ca9229en.pdf, https://www.fao.org/3/cc0461en/cc0461en.pdf, https://www.fao.org/fishery/statistics-query/en/global_production/global_production_quantity.

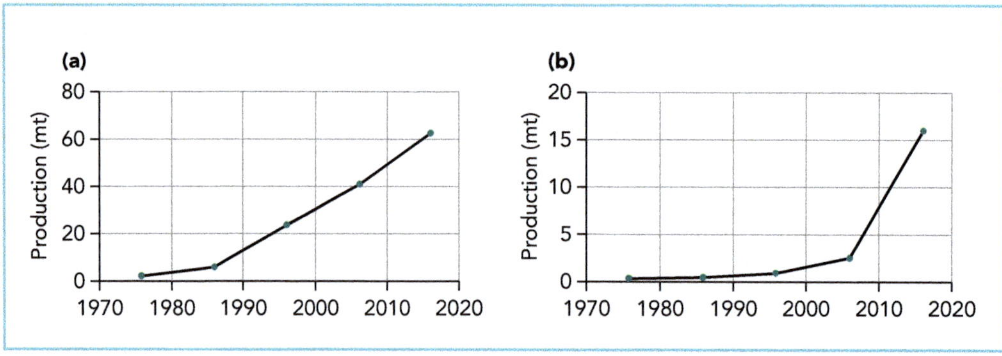

Figure 6.10 Growth of aquaculture production in (a) China and (b) Indonesia, 1973–2020.

Sources: data from https://www.fao.org/3/ca9229en/ca9229en.pdf, https://www.fao.org/3/cc0461en/cc0461en.pdf, https://www.fao.org/fishery/statistics-query/en/global_production/global_production_quantity.

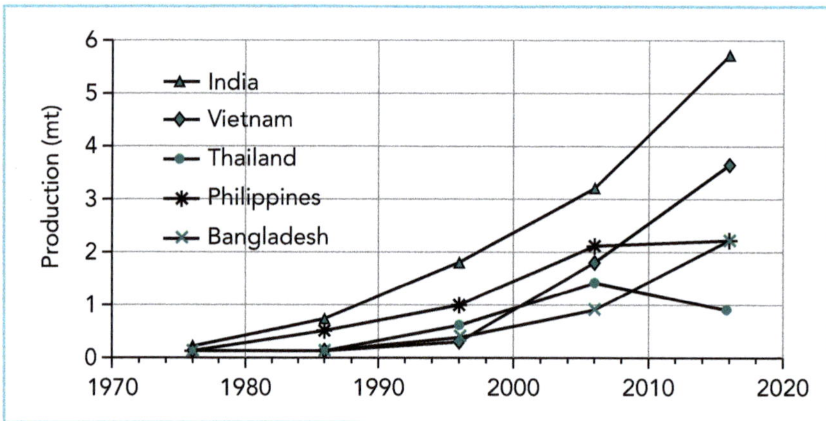

Figure 6.11 Growth of aquaculture production in selected Asian countries, 1973–2020.

Sources: data from https://www.fao.org/3/ca9229en/ca9229en.pdf, https://www.fao.org/3/cc0461en/cc0461en.pdf, https://www.fao.org/fishery/statistics-query/en/global_production/global_production_quantity.

It should be noted that in the countries where commercial aquaculture is growing rapidly, there have been lots of issues and problems with haphazard use of chemicals and antibiotics, over-fishing, use of low-paid, slave and child labourers, and environmental damage. Most southeast Asian countries are facing some ups and downs with some species such as shrimp farming. Many of these issues have been or are being addressed and as a result, in general, the aquatic food production sector is moving forward using either less intensive processes, such as polyculture shrimp and tilapia and organic shrimp farming, or using more controlled environments, such as aquaponics, bioflocs and water recirculatory systems.

Conclusion

Aquatic foods are critical to achieving food security for all people, be they rural communities or urban elites, in less developed or developed nations. Its role is even greater in solving the problem of malnutrition or micro-nutrient deficiencies (i.e. hidden hunger). Increasing production is important but at the same time physical as well as economic access is important. Therefore, the development of well-functioning and efficient markets and marketing channels is equally important to achieving aquatic food security as making production systems efficient and sustainable. Strong demand and market forces are driving technological innovation and production growth in aquatic food supply. Establishing different markets – local, regional and international – is essential to help solve the multifaceted problems related to food and nutritional security, employment and income. Fishing is one of the oldest employments of humankind and fish markets especially in coastal regions have existed for as long as people have made their homes there. Increasing awareness about the human health and environmental benefits of aquatic products is driving demand. Emerging new export markets are demanding more, especially the Middle East, China and India. As a result, aquatic products are becoming more valuable. Most governments have made suitable policies for the enhancement of aquaculture production in addition to managing fisheries resources. To meet the growing demand, production has to be significantly increased, which requires considerable investment in research, human resource and infrastructure development.

The aquatic food production sector generates income, creates employment, alleviates poverty, and improves food and nutrition security in developing countries directly at the production sites and indirectly across the value chains including transportation, processing, packaging and marketing. According to the FAO (2020), nearly 60 million people globally are engaged in primary fisheries and the aquaculture sector, out of which over 80% are in Asia, 10% in Africa and remainder in other countries. Over 18 million people are engaged in aquaculture; over 90% are in Asia where fish contributes more than 30% of animal protein in the diet of majority low-income, food-deficit countries. In capture fisheries, following the sector's rises and falls, the number of fishers goes up and down especially in large-scale fisheries, whereas the number of families dependent upon small-scale capture fishing is remaining more or less similar to 2000. In India alone, 14 million people are involved in the fisheries sector (NFDB, 2018). Similarly, in China, it could be a lot higher, although exact figures are difficult to determine. Income depends on the scale of operation and type of business. In Bangladesh, the aquatic food production sector provides millions of jobs. According to Das et al. (2018) and DoF (2017), there are over 18 million people engaged in this sector, i.e. 10% of its population, out of which nearly 14 million are employed in aquaculture.

Small-scale farmers and fishers usually struggle to survive and make a living. Well-managed larger aquaculture farmers are more likely to be well-off especially when export markets boom. For example, when shrimp (*P. monodon*) farming boomed in Thailand during mid-1980s and mid-1990s profits used to be over 100%. About 40,000 farmers were attracted into the shrimp farming business. When a viral disease devastated their crops, many of them lost their investment. Many farmers shifted to tilapia farming, which is carried out by over 300,000 families. The number of farmers shrimp farming has halved. They either stop completely or switch to polyculture to reduce risk.

Similarly, in Vietnam, pangasius farmers earned good income at the beginning and a number of farmers reached 40,000 and the processing industry created over 200,000 jobs, especially for women as filleting is done manually. However, when competition became tough, prices fell and profit margins went down. As a result, there was a decline in the number of farmers as they shifted to farm other species, such as tilapia.

Most aquatic food items are sold fresh or even live in local markets. More distant markets are served with frozen fillets or whole frozen items. Among the farmed species salmon, pangasius, shrimp and tilapia are the most popular species that are exported in considerable volume.

Only salmon is produced in the developed north, with a wider range of species produced in the south and developing countries. There are many varieties that are currently sold and consumed in local communities only, any of which may emerge with a large-scale future. Human population pressure is threatening many species such as anchovies, sardine, tuna, crabs, squid, shrimp, lobsters among others. Therefore, continuous research and development on domestication and technological advancement is needed, especially with regard to their breeding, feeding and overall management, to maintain and enlarge the role of aquatic food in meeting the SDGs by 2030 and achieving food security for the future.

References

Armstrong, M. (2017) How thirsty is our food? https://www.statista.com/chart/9483/how-thirsty-is-our-food/.

Avnimelech, Y. (2015) *Biofloc Technology – A Practical Guide Book*. World Aquaculture Society.

Béné, C., Barange, M., Subasinghe, R., Pinstrup-Andersen, P., Merino, G., Hemre, G.-I. & Williams, M. (2015) Feeding 9 billion by 2050 – Putting fish back on the menu. *Food Security*, 7, 261–274. https://doi.org/10.1007/s12571-015-0427-z.

Bhujel, R.C. (2014). *A Manual for Tilapia Business Management*. CABI Publishing: Wallingford.

Bhujel, R.C. & Perera, A. (2017) Shading of breeding hapas enhances reproductive performance of Nile tilapia (*Oreochromis niloticus*) and seed output. Journal of Aquaculture in the Tropics. *Journal of Aquaculture in the Tropics*, 32, 205–215.

Bhujel, R.C. and Suharman, I. (2021). Food and nutrition security in the context of COVID-19 and the potential role of tilapia aquaculture. The 9th International and National Seminar on Fisheries and Marine Science (ISFM), IOP Conf. Series: Earth and Environmental Science 695 (2021) 012023. doi:10.1088/1755-1315/695/1/012023.

Boonyapakdee, A. & Bhujel, R.C. (2019) Effects of nursing methods, astaxanthin supplementation and water quality on the survival of blue swimming crablets (*Portunus pelagicus*, Linnaeus, 1758). *Aquaculture Studies*, 19, 45–56. https://doi.org/10.4194/2618-6381-v19_1_05.

Bórquez, A.S. & Hernández, A.J. (2009) Status of and trends in the use of small pelagic fish species for reduction fisheries and for human consumption in Chile. In: *Fish as Feed Inputs for Aquaculture: Practices, Sustainability and Implications. FAO Fisheries and Aquaculture Technical Paper*, Vol. 518 (edited by M. R. Hasan & M. Halwart). Food and Agriculture Organization: Rome, pp. 289–324.

Bouchet, P. (2006) The magnitude of marine biodiversity. In *The Exploration of Marine Biodiversity: Scientific and Technological Challenges* (edited by C. Duarte). Funcación BBVA: Bibbao, pp. 31–62.

Chakravarty, S., Kumar, S. & Prakesh, S. (2018) Back to the basics: biomimicry in shrimp farming. *International Journal of Current Microbiology and Applied Sciences*, 7(5), 2172–2184. https://doi.org/10.20546/ijcmas.2018.705.253.

Chan, C.Y., Tran, N., Dao, C.D., Sulser, T.B., Phillips, M.J., Batka, M., Wiebe, K. & Preston, N. (2017) Fish to 2050 in the ASEAN Region. Penang, Malaysia: WorldFish and Washington DC, USA. International Food Policy Research Institute, Working paper.

Charlton, K.E. (2016) Food security, food systems and food sovereignty in the 21st century: A new paradigm required to meet Sustainable Development Goals. *Nutrition and Dietetics*, 73, 3–12. https://doi.org/10.1111/1747-0080.12264.

Clark, M. & Tilman, D. (2017) Comparative analysis of environmental impacts of agricultural production systems, agricultural input efficiency, and food choice. *Environmental Research Letters*, 12. https://doi.org/10.1088/1748-9326/aa6cd5.

Das, M., Islam, M.R., Akter, T., Kawser, A.Q.M.R. & Mondal, M.N. (2018) Present status, problems and prospect of fish farming at Gazipur Sadar upazila in Bangladesh. *Progressive Agriculture*, 29, 53–63. https://doi.org/10.3329/pa.v29i1.37480.

DoF (2017) *Yearbook of Fisheries Statistics of Bangladesh 2016–17. Fisheries Resources Survey System (FRSS).* Department of Fisheries: Bangladesh.

EWG (2011) Meat eaters guide to climate change+health. https://static.ewg.org/reports/2011/meateaters/pdf/methodology_ewg_meat_eaters_guide_to_health_and_climate_2011.pdf.

EU (2019) https://ec.europa.eu/fisheries/cfp/aquaculture/aquaculture_methods/history_en.

FAO (2018) The state of world fisheries and aquaculture, 2018. Meeting the Sustainable Development Goals. Food and Agriculture Organization: Rome.

FAO (2019) GlobeFish. *Information and Analysis on World Fish Trade.* www.fao.org/in-action/globefish/fishery-information/world-fish-market/en/.

FAO (2020) *Food Outlook – Biannual Report on Global Food Markets.* www.fao.org/3/ca9509en/CA9509EN.pdf. Food and Agriculture Organization: Rome.

FAO (2022) *The State of World Fisheries and Aquaculture – Towards Blue Transformation.* FAO: Rome. https://www.fao.org/3/cc0461en/cc0461en.pdf.

Ferdouse, F., Yang, Z., Holdt, S.L., Murúa, P. & Smith, R. (2018) *The Global Status of Seaweed Production, Trade and Utilization.* FAO-Globefish Research Programme, Vol. 124. Rome.

GAA (2018) Four reasons why you should, and will be, eating more seafood. *Global Aquaculture Advocate.* https://www.aquaculturealliance.org/blog/four-reasons-why-seafood/.

IFFO (International Fishmeal and Fish Oil Organization) (2009) *Case Study: Peruvian Anchovy – Why Feed, not Food?* https://www.iffo.com/case-study-peruvian-anchovy-why-feed-not-food.

Karapanagiotidis, I.T., Bell, M.V., Little, D.C., Yakupitiyage, A. & Rakshit, S.K. (2006) Polyunsaturated fatty acid content of wild and farmed tilapias in Thailand: Effect of aquaculture practices and implications for human nutrition. *Journal of Agricultural and Food Chemistry*, 54, 4304–4310. https://doi.org/10.1021/jf0581877.

Kobayashi, M., Msangi, S., Batka, M., Vannuccini, S., Dey, M.M. & Anderson, J.L. (2015) Fish to 2030: The role and opportunity for aquaculture. *Aquaculture Economics and Management*, 19, 282–300. https://doi.org/10.1080/13657305.2015.994240.

Kutty, M.N. (1999) Aquaculture development in India from a global perspective. *Current Science*, 76, 333–341. https://www.jstor.org/stable/24101129.

Lee, N. & Polan, S. (2019) Sea cucumbers are so valuable that people are risking their lives diving for them. Business Insider. https://www.businessinsider.com/why-sea-cucumbers-so-expensive-seafood-2019–1.

Mereghetti, M. (2017) Peru launches anchovy quota for human consumption. https://www.undercurrentnews.com/2017/05/02/peru-launches-anchovy-quota-for-human-consumption/.

NFDB (National Fisheries Development Board) (2018). https://nfdb.gov.in/welcome/annual_report.

OECD-FAO (2020). Agricultural Outlook 2020-2029, OECD Agriculture statistics. https://www.fao.org/documents/card/en/c/ca8861en/.

Pantulu, V.R. (1979) Floating cage culture of fish in the lower Mekong River Basin. In: *Advances in Aquaculture* (edited by T. V. R. Pillay & W. A. Dill). Fishing News Books: Farnham, Surrey, pp. 423–427.

Parrish, R. (2000) Monterey sardine story. Pacific Fisheries Environmental Group, National Marine Service. https://swfsc.noaa.gov/publications/CR/2000/2000ParrR2.pdf.

Pauly, D. & Zeller, D. (2016) Catch reconstructions reveal that global marine fisheries catches are higher than reported and declining. *Nature Communications*, 7, 10244. https://doi.org/10.1038/ncomms10244.

Pham, T.T.H., Rossi, P., Dinh, H.D.K., Pham, N.T.A., Tran, P.A., Ho, T.T.K.M., Dinh, Q.T. & de Alencastro, L.F. (2018) Analysis of antibiotic multi-resistant bacteria and resistance genes in the effluent of an intensive shrimp farm (Long An, Vietnam). *Journal of Environmental Management*, 214, 149–156.

https://doi.org/10.1016/j.jenvman.2018.02.089.Phan, L.T., Bui, T.M., Nguyen, T.T.T., Gooley, G.J., Ingram, B.A., Nguyen, H.V., Nguyen, P.T. and De Silva, S.S., (2009). Current status of farming practices of striped catfish, Pangasianodon hypophthalmus, in the Mekong Delta, Vietnam. Aquaculture, 296: 227–236.

Pinoargote, G., Flores, G., Cooper, K. & Ravishankar, S. (2018) Effects on survival and bacterial community composition of the aquaculture water and gastrointestinal tract of shrimp (litopenaeus vannamei) exposed to probiotic treatments after an induced infection of acute hepatopancreatic necrosis disease. *Aquaculture Research*, 49, 3270–3288. https://doi.org/10.1111/are.13791.

Rai, A.K., Bhujel, R.C., Basnet, S.R. & Lamsal, G.P. (2005). *Rainbow Trout (Oncorhynchus mykiss) Culture in the Himalayan Kingdom of Nepal – A Success Story*. APAARI, FAO Regional Office for Asia and the Pacific: Bangkok. APAARI Publication no. 2005/1.

Ritchie, H., Rosado, P. and Roser, M. (2022). Environmental Impacts of Food Production. Published online at OurWorldInData.org. Retrieved on June 20, 2023, from: 'https://ourworldindata.org/environmental-impacts-of-food' [Online].

Ross, R. (2019) What is a sea cucumber? https://www.livescience.com/sea-cucumbers.html.

Sagi, A. & Aflalo, E.D. (2005) The androgenic gland and monosex culture of freshwater prawn Macrobrachium rosenbergii (De Man): A biotechnological perspective. *Aquaculture Research*, 36, 231–237. https://doi.org/10.1111/j.1365-2109.2005.01238.x.

Science Daily (2011) How many species on Earth? About 8.7 million, new estimate says. https://www.sciencedaily.com/releases/2011/08/110823180459.htm.

Silas, E.G. (2003) History and development of fisheries research in in India. *Journal of the Bombay Natural History Society*, 100(2&3), 502–520.

Suresh, V. & Bhujel, R.C. (2018) Tilapias. In: *Aquaculture: Farming of Aquatic Animals and Plants*, 3rd edn (edited by J. Lucas & P. C. Southgate). Wiley-Blackwell: Chichester, pp. 338–364.

Towers, L. (2014) Spiny lobster farming in viet nam and the role of probiotics during production. The FishSite. https://thefishsite.com/articles/spiny-lobster-farming-in-viet-nam-and-the-role-of-probiotics-during-production.

Tveteras, R., Nystoyl, R. & Jory, D.E. (2019) GOAL 2019: Global finfish production review and forecast. Global aquaculture advocate. https://www.aquaculturealliance.org/advocate/goal-2019-global-finfish-production-review-and-forecast/.

Tveteras, R., Nystoyl, R. & Jory, D.E. (2019) GOAL 2016: Global finfish production estimates and trends. Global aquaculture advocate. https://www.aquaculturealliance.org/wp-content/uploads/2017/06/Day1_RagnarTveteras.pdf.

Worm, B., Barbier, E.B., Beaumont, N., Duffy, J.E., Folke, C., Halpern, B.S., Jackson, J.B.C., Lotze, H.K., Micheli, F., Palumbi, S.R., Sala, E., Selkoe, K.A., Stachowicz, J.J. & Watson, R. (2006) Impacts of biodiversity loss on ocean ecosystem services. *Science (New York, NY)*, 314, 787–790. https://doi.org/10.1126/science.1132294.

Yan, G. & van Beijnen, J. (2019) Asian aquaculture: Trends for 2019. The FishSite. https://thefishsite.com/articles/asian-aquaculture-trends-for-2019.

Yacoubian, J. (2023). Salmon vs Tilapia – Health Impact and Nutrition Comparison. https://foodstruct.com/compare/fish-vs-tilapia (data based on USDA retrieved from: https://fdc.nal.usda.gov/fdc-app.html#/food-details/175177/nutrients.

7

The role of markets in global aquatic food security

Ram C. Bhujel

Trade in aquatic food products is strong. The annual trade value of aquaculture production i.e. 122 million tonnes (mt) in 2020 exceeded US$281.5 billion (FAO, 2022). Production takes place mostly in developing countries while the high-value markets are to be found in developed countries. The USA is the largest seafood importing country in the world and more than 90% of the aquatic food consumed in USA is imported (Asche, 2017). The European Union and Japan both import huge amounts of aquatic products from Asia. China, the largest producer which exported seafood worth US$19 billion in 2022 (Harkell, 2023), is itself starting to import and has become a major importer of salmonids, 70,000 t annually mostly from Chile, Denmark and Norway. Chinese farmers are trying to raise salmon and other temperate fish, like trout to reduce the volume imported. After China, Norway led the seafood export with the value of US$15 billion in 2022 (Holland, 2023). Seafood export of Vietnam reached US$11 billion in 2022 (Thanh, 2023). Aquatic food products are now the largest group in Indian agricultural exports. About 1.4 mt of 50 different fish and shellfish products were exported to 75 countries around the world in 2021/22 with an export value of nearly US$8 billion in value (Srinivas, 2022).

Unlike terrestrial animals, the large variety of aquatic organisms also have multiple product forms and are available live, fresh, whole frozen, fillet frozen, smoked, dried, fried, powdered, fermented, pickled and as pastes. Among them, live, fresh and chilled are the most preferred forms (nearly half by volume) for human consumption and command high prices. These popular forms are only practical for nearby markets within a few hours transportation time. Therefore, about 99% of these products are utilized in developing countries where most of the aquatic food items are sold fresh or even live at local markets. Markets further from fishing or farming areas need other product forms, such as fillet frozen and whole frozen. Frozen forms are the most popular (about one-third by volume) as they can be transported all over the world because they remain saleable for about 6 months. However, only about one-third of the total volume of frozen fish is consumed in developed countries.

Among the farmed species salmon, pangasius, shrimp and tilapia are the most popular and likely to be exported in any considerable volume. Salmon is the only species produced on a mass scale in the north or developed countries; the other species are produced in the south and developing countries. Notably, compared to the land species, there are many more aquatic species that are known, sold

Margaret Crumlish and Rachel Norman (eds) **Aquatic Food Security**
DOI: 10.52517/9781789183368.007, © 5m Books Ltd, 2023

and consumed in local communities. There is the potential for any among these only locally known species to emerge as the subjects of large-scale farming. Such untapped potential is one of the reasons aquaculture has an important part to play in achieving several of the Sustainable Development Goals (SDGs) by 2030.

Aquatic food markets

A market is a place where sellers and buyers meet to exchange goods and services. They can be permanent or temporary and occur on various frequencies: daily, weekly, monthly or even annually. Markets help the flow of goods and services from areas where they are produced in excess to the areas where they are scarce. Therefore, developing efficient markets helps to improve everyone's lives and livelihoods. Many people, especially those living in the cities and urban centres, rely on markets to access food items, especially aquatic foods. Almost all aquatic foods are either caught in water bodies (seas, rivers, lakes and reservoirs) or grown in the rural areas, which are often located far away from the populous cities. In the case of aquatic foods that are traded frozen worldwide the distance can be many thousands of miles.

Historical background

Fish markets have existed for a long time and are most often located along the coasts. Some are in inland centres on the trade routes. As fish can spoil quickly, the use of ice or other simple cooling methods became very important. When refrigeration and rapid transport became available in the 19th and 20th centuries, fish markets started to appear away from the coasts. However, modern trade has shifted away from traditional marketplaces to retail outlets, e.g. supermarkets. As a result, most fish markets now deal mainly with wholesale trade. The existing major fish retail markets continue to operate as much for traditional reasons as for commercial ones. As with any food, aquatic food market has three levels: auction, wholesale and retail markets.

Auction markets are a unique tradition of seafood markets. Those who do not have a tie up with a wholesaler are free to sell to any wholesaler operating in their port market. Auction is the direct transaction for negotiation between sellers (fishers) and buyers (middle-traders or wholesalers to retailers). These auctions are organized markets where goods are awarded to highest bidders or based on rules that determine who wins and the price the winner pays (Cassady, 1967). The price is decided on the spot, based on bidding. Payment to fishers is usually flexible and can be on cash and carry terms, made within a few days after the transaction or monthly depending on the agreement between both the parties. The auction method is used because price changes from time to time depending on the market situation and variety of products. The fishermen prefer it, as they do not want to be tied to traders on pre-fixed prices. There are two systems in auction: price incremental and decreasing. In the incremental auction, the price starts from low and goes up until the highest above which no one goes further, then the highest bidder wins. In the decreasing price or Dutch system, the auctioneer starts from a high price then reduces it gradually until a bidder is found, the first bidder is the winner. The decreasing price or Dutch system is popular in Europe whereas the incremental auction is practiced all over the world. Most auction markets are located together with wholesale and retail markets. Many of them are well-known historical markets known all over the world. Some of the major markets are listed below:

- Tsukiji Fish Market in Tokyo, Japan: it was the world's largest fish market (selling over 660,000 t a year). It closed in 2018.
- Scania Market, a historical annual market at Falsterbo, Sweden. A major fish market for herring that took place annually in Scania during the Middle Ages. From around 1200, it became one of the most important events for trade around the Baltic Sea and made Scania (the southernmost province of Sweden) into a major distribution centre for West European goods bound for eastern Scandinavia. The Scania Market continued to be an important trade centre for 250 years and was a cornerstone of the wealth of the Hanseatic League (a commercial and defensive confederation of merchant guilds and market towns in Central and Northern Europe).
- Sydney Fish Market, Sydney, Australia, is the world's third largest fish market for volume sold and second largest in terms of variety.
- Billingsgate Fish Market, London, UK.
- Fulton Fish Market, New York, USA.
- La Nueva Viga Market, Mexico City, Mexico is the world's second largest fish market. Selling from 250,000 t up to 550,000 t of seafood per year.
- Boston Seafood Market, Massachusetts, USA.
- Mercamadrid, Madrid, Spain is the world's fourth largest fish market. Selling about 220,000 t per year.
- Busan Cooperative Fish Market, Busan, South Korea.
- Feskekôrka, Gothenburg, Sweden.
- Maine Avenue Fish Market, Washington, DC, USA.

Modern day markets

At the macro level fish markets can be divided into the regions or countries that are briefly described in the following subsections.

EU

The EU is the biggest market for aquatic food products with the population of over 740 million. Per capita consumption is approximately 27 kg/year (FAO-GlobeFish, 2019) whereas pork consumption is 46 kg/year. Europeans know aquatic food is good for health, but per capita consumption has not gone up mainly because of the high price of seafood which has been driven by low production volumes within Europe and the high cost of transportation from Asia. Some efforts have been made to increase the production of aquatic foods within Europe, e.g. land-based salmon farming and, at the same time, to reduce the cost of transport and increase the shelf-life of the products (e.g. blast freezing). If these efforts are successfully or widely adopted, access to aquatic foods in Europe will increase considerably.

In Europe the market operates through retailers or food service operators who purchase from an agent or a European importer. According to the CBI (2016), the retail segment is dominated by a small number of large retailers who own the supermarket chains. These hypermarkets and supermarkets account for about 90% of all seafood sales, between them have over 600 different supermarket chains and over 420,000 non-specialist food retail stores across Europe. Several of these chains operate in multiple European countries, and the larger ones have annual turnovers in excess of €100 billion, such as:

- Carrefour (France, supermarkets, and branches worldwide)
- Metro AG (Germany and branches worldwide)
- Tesco (UK, supermarkets and local convenience stores and branches worldwide)
- Koninklijke Ahold N.V. (the Netherlands, supermarkets)
- Schwarz Group and Aldi GMBH (Germany & UK discount supermarkets)
- Rewe (Germany, operates Billa, BIPA, Merkur and Penny brands targeting different market segments).

To sell in these Europe markets requires certification by Global GAP and the Aquaculture Stewardship Council (ASC) for cultured seafood and the Marine Stewardship Council (MSC) for captured seafood. Aquatic food exporters can go either directly to agents, importers or retail and food service companies. The role of middlemen or agents in the supply chain is declining, as buyers want to have more control over the quality of the products they need direct relationships with suppliers, bypassing the agents. In some cases, these agents still perform their role as the facilitators of trading relationships between European buyers and exporters from developing countries. Within Europe, Spain, France, Germany, Italy, Sweden and UK are the largest importers of aquatic foods. They spent US$7.11 billion, US$6.18 billion, US$6.15 billion, US$5.60 billion, US$5.19 billion and US$4.2 billion in 2016, respectively (FAO, 2019).

USA

With a population of about 329 million, the USA imports 90% of the seafood consumed by Americans, spending over US$20 billion annually (NOAA, 2018). Per capita consumption of seafood in the USA is about 21 kg/year, which is almost the same as the world average. People consume beef more: 100 kg/capita. Although beef consumption is showing a per capita declining trend (100 kg/year in 1990 to about 80 kg/year in 2015). This suggests there may be more demand for seafood. According to NOAA (2018), annually, the USA imports fresh and frozen shrimp (0.5 mt), 0.43 mt of fresh and frozen fish fillet, canned products (0.25 mt), fresh and frozen whole fish (0.175 mt) and other products (0.195 mt). The USA have the Seafood Import Monitoring Program (SIMP) which is mandatory for foreign products and requires them to be accompanied by harvest and landing data. Importers need to maintain records for shrimp and abalone entering the USA. These requirements are to combat illegal, unreported and unregulated (IUU)-caught and/or misrepresented seafood from entering the USA. SIMP is a risk-based traceability programme requiring importers to record key data from the point of harvest to the point of entry into the USA for 13 imported fish and fish products identified as vulnerable to IUU fishing and/or seafood fraud. The 13 species included are abalone, Atlantic cod, blue crab (Atlantic), dolphinfish (Mahi Mahi), grouper, king crab (red), Pacific cod, red snapper, sea cucumber, sharks, shrimp, swordfish, and tunas (albacore, bigeye, skipjack, yellowfin and bluefin). In addition, USA Regulations for Fish and Fishery Products are provided by FAO-Globefish (2019).

Japan

Japan is the third largest seafood market. It imports more than any single European country and is, therefore, among countries second to the USA. The annual value of seafood imports is approximately US$14 billion per year. Seafood has been the staple food of the Japanese people throughout their history. They like to eat fish at every meal. Sashimi, sushi, surimi are the major preparations and are well-known worldwide. The Japanese traditional diet (washoku), contains mainly fish products and less animal fat and meat than the diet of many countries, which is likely to have a positive impact on longevity (Gabriel et al., 2018). Since the 9th century, grilled fish and sliced raw fish were widely popular. As Japan is an island nation seafood was easily accessible in the past but due to overfishing, seafood catch has declined sharply, and so therefore has consumption since the mid-1990s (Matsuyama, 2021). They have a seasonal consumption pattern following the peak seasons of various products. Seafood is also preferred for cultural and religious reasons. For Buddhists the killing of terrestrial animals was considered inhuman and the eating of 'four-legged creatures' was a taboo or considered unclean and as such was something to be avoided by personal choice especially during the 17th to 19th centuries. Japanese cuisine is also popular in many foreign countries and their food ingredients are often sourced from Japan. The Japanese consume 80% of the global bluefin tuna catch, which are declining in the wild along with many other species. Therefore, Japan imports different types of aquatic foods such as shrimp/prawn and frozen fillets of salmon, tuna and trout from many other countries, such as China, the USA, Chile and Thailand. Per capita consumption of seafood in Japan has fallen to 27 kg/year from its peak, which used to be 80 kg annually (FAO-Globefish, 2019). The big fall in consumption is mainly due to declines in catch, in shrimp production in Thailand and imported aquatic foods becoming relatively expensive; however, Japanese prefer aquatic foods, if the quality seafood can be supplied.

China

The Chinese seafood market is one of the emerging markets in the world. It is booming, and the demand for imported seafood is growing rapidly. China has a long history of eating seafood, especially for coastal inhabitants. Following economic reform in China, foreign seafood started to be imported during the 1990s. Initially, imported seafoods were considered luxury items but now seafood is accessible to almost all the Chinese. Consumers are more aware of the health benefits of omega-3 fatty acids. They are also aware of environmental pollution and food safety issues; imported seafood is normally considered more nutritious and safer due to water quality standards and stricter quality control in the production regions. Consumers prefer recognized brands, which give the impression of freshness and good quality. Since 2012, the seafood market has expanded along with the growth of the Chinese economy as a whole. For example, sales of chilled, frozen and shelf-stable seafood are showing a steady growth. In 2017 about 7.6 mt of seafood, including shrimp, salmon and crab, was imported from Russia, the USA, Canada, New Zealand and Norway (Zhu & Mikhailova, 2020). More recently, rock lobsters are becoming popular, the price of which might go as high as US$230/kg in some restaurants in Guangzhou and Shanghai (FAO-Globefish, 2019).

The seafood market is highly seasonal and festival-dependent especially during the Chinese New Year when more seafood is imported and purchased online, e.g. 38% of food sales in 2018 on the website Fresh.jd.com were of seafood for Chinese New Year. Valentine's Day (14 February) is immediately after Chinese New Year and rock lobsters are in high demand during that time (FAO-Globefish, 2019). Another important festival is the Chinese National Day (1 October) when seafood is in high demand. The whole country has a 7-day holiday at this time, and seafood producers may also prepare special gift boxes containing living or frozen seafood. Unlike consumers in other countries, Chinese consumers love peculiar items such as sea cucumber, fish balls, turtle and eel. Purchasing living aquatic foods from a Chinese traditional food market is common, especially for the older generation. Many retailers and restaurants present their aquatic foods in glass tanks or on ice, no matter if it is fish, crabs or turtles, in order to show the freshness of their products.

In China, aquatic food items are the most attractive items used in catering. Hot pot contributes 22% of the total revenue of China's catering industry, followed by buffets (12%) and barbecue and Japanese cuisine (5% each). The major share of imported salmon is not consumed at home, but goes to the catering market. China does not have a long history of consuming salmon, but Japanese cuisine has become very popular, especially sushi and sashimi, which are becoming widely accepted. Seafood is also sold via online platforms e.g. Taobao and JD, which are achieving great success. Fish dominates seafood consumption in China because of its rich variety and affordable prices. Another successful online market is Hema Xiansheng, a part of Alibaba which is a combination of a supermarket, restaurant, seafood market and mobile application. Hema supermarket is a blend of the online and offline shopping experience. Chinese consumers can order seafood on Hema's mobile app, which will be delivered within 2 hours. Consumers can also go to Hema supermarket to purchase groceries and pick up fresh seafood, which can be prepared and served in the dining area of Hema supermarkets. The Chinese seafood market offers huge opportunities for international seafood exporters, particularly if they pay attention to and take advantage of the peculiarities of the Chinese market.

Middle East

The Middle East is an emerging market for seafood. Most of these oil-rich countries used to consume more beef and wild caught seafood, but nowadays with the realization that seafood is better for health and of declining wild catches various attempts have been made to boost cultured aquatic food supply. For example, Oman had a US$1.29 billion programme running from 2013 to 2020 to establish 41 modern fish markets and fishing ports (Chiber, 2013). Some of the Middle Eastern countries already have high levels of seafood consumption, such as the UAE, Oman, Bahrain, Qatar and also Saudi Arabia. The aquatic food market is likely to reach over US$7 billion by 2021 according to the Straight from the Sea Trade Exhibition (SEAFEX) organizers. It is organized every year in Dubai World Trade Centre. More importantly, in these countries demand is also diversifying with the consumption of other seafood items such as shrimp, oysters, crabs and mussels instead of just fish. Unfortunately, among the Middle East countries, aquaculture has not developed and expanded due to shortages of water, human resources and other inputs. Among those practising aquaculture, Egypt is the leading country, producing tilapia and mullet. There is much opportunity for exporting countries, especially from South Asia, to supply in to the Middle Eastern market.

Markets in other countries

Seafood markets exist in almost every town in the world, with the size and frequency depending upon the population and location. Most countries have at least one centralized major aquatic food market, usually together with an agriculture market, in the capital city, e.g. the Manhattan Fish Market in Dhaka, Bangladesh. In Bangladesh there is a market in almost every village. In addition, other larger regional cities also have a well-known seafood market (Chittagong, Cox's Bazaar, Raipur and Khulna). In India, the Digha Mohana International Fish Market in West Bengal might be the largest one. Other large state level markets exist in Andhra Pradesh, Karnataka, Kerala, Gujarat, Tamilnadu and Maharashtra. The Jimbaran fish market in Bali and the fish market in Jakarta are the major fish markets in Indonesia. Kuala Lumpur, Penang and other cities in Malaysia have seafood markets. In the Philippines, there are markets spread all over each island, e.g. Bulungan Seafood Market, Dampa Seaside, Magsaysay Fish Market and Navotas Fish Complex within Metro Manila. Dumaguete Fish Terminal, Sutukil Seafood Market, Corong and Motiong Public Markets are other notable venues. In Thailand, there are several well-known fish/seafood markets such as the Ang Sila Seafood Market,

Saphan Pla Fish Markets, Shinsen Fish Market, Thalay Thai Market, Talat Thai Market, among others. Vietnam also has many markets where fresh and processed seafood items are traded. Main ones are Dong Ba Market, Mui Ne Fishing Village Market, Hoi An Fish Market, Hang Duong, Harbor Fish Market Phuoc, Tan Than Seafood market, Chu Lo Seafood Market and so on. Sydney fish market and harbor seafood market are popular in Australia. In Brazil, market of Sao Paulo and JM Pescado Fish Market of Ceara. Many other countries do have considerable sizes of seafood markets at the capital as well as at provincial or district level markets. When aquatic food is abundantly produced, these markets expand, and it may even create new markets. Therefore, selling aquatic food is not a problem.

Market types and characteristics

Markets are divided into three groups: primary, intermediate (wholesalers and retailers) and terminal (FAO, 2001).

Primary markets

Primary markets are those in which aquatic foods are sold fresh on landing or at production sites, nearby streets, road junctions or along the highways. They are usually near areas where fish are caught or cultured. A great variety of aquatic animals and a wide variation in sizes are available if they are wild caught with more or less uniform size and few species available if they are farmed. Markets frequency varies greatly daily and the size of the market depends on the local population and the location.

Most primary markets sell fresh but dead aquatic foods, but there is a significant, and growing, proportion that sell live aquatic animals. These markets are popularly called wet markets as they need to keep the fish in water. Live aquatic food markets are traditionally found in south east Asia. One or two decades ago, live fish or shellfish markets were limited to the Chinese mainland and Hong Kong, Singapore, Vietnam, but live aquatic animal markets can now be seen in other parts of Asia. There was a tradition of selling walking catfish, snakehead fish and sand goby as they are air breathers and can survive out of water for a considerable time. However, wet market provision is expanding to other aquatic species, especially prawn/shrimp, lobsters, sea bass and tilapia. The popularity of live sales is expanding especially as the use of chemicals as preservatives (e.g. formalin) becomes common. Consumers believe that if the animals are live, it means food items are fresh and free of chemicals.

Intermediate markets

Intermediate markets are predominately wholesale markets and are often centralized in one place usually near the national capital or other seafood producing location. Major cities in each country will have at least one wholesale market. They can be run by the government, the private sector or sometimes by producer cooperatives. Farmers sell in bulk to the wholesalers and do not need to sell their products themselves in smaller volumes to multiple purchases as in a primary market. This can be beneficial to farmers in that they have assurance and can sell in large volumes although farmers lose some margin. Wholesalers (or middlemen) are sometimes accused of ripping off producers, doing little work for their cut – simply providing communication, information, transportation and storage facilities. However, these services come at a cost, especially chilled storage with today's high energy prices.

Large retailers are also intermediate markets doing trade between wholesalers and individual consumers. They buy in bulk at a cheaper price per unit and sell to the customers in smaller units and at higher prices.

Terminal markets

Terminal markets are those from which final customers purchase for consumption at home or those from which commercial outlets, such as restaurants and hotels, buy. Some may process further and repackage in different forms such as sun-dried squid or other aquatic animals, smoked, canned, powdered, fermented, deep fried and so on. Owing to their long shelf life and ease of transport, these processed food items play a key role in ensuring food security in remote areas where they cannot be produced. Some of the terminal markets are traditional in nature, found in open spaces and often called dry seafood markets, while most items are distributed through well-organized and specialized retail outlets, such as the supermarkets and convenience stores. The growth in home delivery, both as ingredients for final home preparation and as fully prepared fast food, particularly in big cities, is one major change in consumer behaviour. Home delivery has always been popular for the time it saves, but the switch from eating out to staying at home was driven by the COVID-19 pandemic and lockdowns in many countries. As a result, food delivery businesses boomed (grab food). Most terminal markets, including restaurants, had to adapt and offer a delivery service to stay in business.

Market access and food safety and certification

To have market access, especially the international markets, exporters have to secure certification.

Marine Stewardship Council certification

The Marine Stewardship Council (MSC; www.msc.org) is an independent, international charity established in 1997 by the WWF and Unilever aiming at reversing the decline in wild aquatic animal stocks. It works based on FAO code of conduct for responsible fishing and eco labelling. It has eight global offices but is active in over 40 countries worldwide. MSC manages two international standards: fisheries and traceability partnering with governments. Fisheries bodies apply for certification on a voluntary basis. They are assessed against the MSC Standard by third-party independent certifiers supported by a team of experts. Members of the supply chain apply for 'chain of custody' and audits are performed by third-party independent certifiers. Aquatic foods from successfully certified fisheries are marketed with the MSC Eco-label once 'chain of custody' is completed.

Aquaculture Stewardship Council (ASC) certification

The ASC's (https://asc-aqua.org) mission is to transform aquaculture towards environmental sustainability and social responsibility using efficient market mechanisms that create value across the chain. ASC aims to be the world's leading certification and labelling program for responsibly farmed aquatic foods. The ASC's primary role is to manage the global standards for responsible aquaculture, which were developed by the WWF Aquaculture Dialogues in different parts of the world.

BAP

Global Aquaculture Alliance (GAA) is coordinating the development of Best Aquaculture Practices (BAP; www.bapcertification.org) certification standards for hatcheries, farms, processing facilities and feed mills. Its objective it to promote responsible practices across the aquaculture industry. It drives continued improvements via high standards that deliver significant benefits industry wide. The BAP standards cover aquaculture facilities for a variety of finfish and crustacean species, as well as mussels.

Good Aquaculture Practice (GAP)

GLOBALG.A.P. (www.globalgap.org) began as an initiative by retailers of the Euro-Retailer Produce Working Group (EUREPGAP) in 1997. British retailers working together with supermarkets in continental Europe became aware of consumers' growing concerns regarding product safety, environmental impact, and the health, safety and welfare of workers and animals. GLOBALG.A.P. is presently the world's leading farm assurance programme, translating consumer requirements into Good Agricultural/Aquaculture Practice in more than 100 countries. Some countries have their own GAPs, such as Vietnam's VietGAP. The Vietnamese Ministry of Agriculture and Rural Development has its own Criteria VietGAP issued on 28 January 2008 which are based on (1) standard production techniques, (2) food safety, including measures to ensure no chemical or physical contamination during harvesting, (3) environment and human health, including abuse of the labourers, and (4) product traceability. Vietnam made it compulsory for all commercial fish farms to obtain certification by VietGAP or other international certifications under Decree No. 36/2014 / ND-CP in 2014. The certification programme has encouraged the farming of other fishes varieties in Vietnam in addition to pangasius. Similarly, Thailand has its own ThaiGAP and hopefully many other countries (e.g. BanglaGAP etc.) will follow suit. The main objective of these GAPs is to promote aquaculture seafood, agriculture products and food safety in general and fruit and vegetables for consumption in the country in particular and for export.

Good Manufacturing Practices (GMP)

Food items which have been processed and packed according to Good Manufacturing Practices (GMP) are certified based on the six prerequisites, such as application of hygiene, safe and standard procedures in the equipment and facilities, during transportation and storage, personnel involved, sanitation and pest control measures, and recalls if necessary.

HACCP

A hazard analysis and critical control point (HACCP) system is a management tool applied for the protection of the food supply chain against microbiological, chemical and physical hazards monitored at critical points where there are risks of adulteration and contamination. It is a preventive system developed specifically to enhance food safety through which a hazard analysis is carried out to determine critical control points judged against the critical limits of hazards. Therefore, food items with HACCP certificates are considered to be safe to consume. Most of the food items sold nowadays in convenient stores and supermarkets have HACCP certificates. Online certification services specific to seafood is also available at: https://ehaccp.org/seafood-haccp/

Halal certification

Products targeted at Muslim consumers and for the markets in the Middle East and other Islamic countries require Halal certification. Certification is also beneficial if a company plans to export to western countries with substantial practising Muslim communities, such as in France, Germany, Spain and UK. Halal certification is a process that ensures the features and quality of the products according to the rules established by the Islamic Council that allow the use of the mark Halal. The Halal certificate is a document that guarantees the products meet the requirements of Islamic law and therefore are suitable for consumption. Halal is mainly applied to meat products as it certifies that the animals were slaughtered in a single cut, thoroughly bled, and their meats were not in contact with pork. Products that are Halal certified are often marked with a Halal symbol, or simply the letter 'M', similar to the use of the letter K to identify kosher products for Jewish consumers.

Organic certification

Interest in organic produce has grown steadily for decades, partly in response to haphazard use of chemicals and antibiotics, which have long-term effects on human health including increasing the incidence of cancer and other diseases and disorders. Although it is difficult to produce aquatic products without the use of any chemicals, some attempts have been made and several agencies have emerged to certify products, for example, Australian Certified Organic, Canada Organic, Biologique Canada, Naturland from Germany, USDA, the Soil Association in UK and Organic Agriculture Certification Thailand (ACT). Currently, in Europe alone, over 20 different standards for organic aquaculture exist. In Europe, most carp, salmonids (trout, char, salmon) and cod are bred and fed according to organic regulations. Organically farmed mussels and oysters are also available. When maintaining a completely organic process is not possible, because of difficulty in ensuring that all inputs including feed and feed ingredients are organic, products can be certified as bio-products, or carry green or eco-labelling. Though not as stringent as organic certification these alternatives certainly enhance the marketability of aquatic food products and should be sought when organic certification is not pratical.

Seafood Watch

Monterey Bay Aquarium in California USA maintains the Seafood Watch (www.seafoodwatch.org) list. It assesses seafood sources (both fished and farmed) against a number of stringent criteria and makes recommendations to customers in North America. They use a scoring system based on 10 aspects such as data records, sources of broodstock, seed, feeds, escapees, and so on. They categorize seafood sources into three groups: **Green** for recommended sources; **Yellow** for those under consideration or improving; and **Red** for those to be avoided.

Aquatic food marketing and trade

Agricultural trade has increased substantially since 1990 and nearly 15% of the food produced globally is traded internationally. FAO data (Table 7.1) shows that among protein sources, aquatic food has the highest percentage produced for trade. Out of 178 mt of aquatic food produced globally, more than one-third (34.5%) was traded in 2020 (FAO, 2022) compared to the proportions of the other meats, which were less than half, such as 15% for bovine meat, 11% for poultry meat, and 7% for porcine meat. Among the cereals, 23% of the wheat and only 9% of the rice produced was traded. In 2005, the volume of traded aquatic food was even higher (40%). The reduction in proportion is down to both an increase in production levels and an increase in internal consumption in producing countries, especially in Asia.

Aquatic products feature in the list of most traded food items, and most of the world's countries' trade reports. FAO (2020) data showed that the majority of aquatic foods exported moved from developing countries to developed countries and the value of the trade is almost doubling each decade. Of the 178 mt of total estimated aquatic food production, nearly 90% (over 157.9 mt) was assumed to be for direct human consumption. This has increased significantly in recent decades following policy direction to reduce caught fish destined for fishmeal and fish oil production, and to reduce waste by improving fish processing and trading channels. However, the loss or wastage between harvesting and consumption still accounts for slightly over one-fourth of the harvested volume.

Table 7.1 Food product production and proportion traded in 2018.

Food type	Total production (mt)	Traded volume (mt)	Percentage
Aquatic food	175.9	65.3	37
Bovine meat	72.6	11.2	15
Poultry	134	13.9	10
Porcine meat	110	9.5	9
Ovine meat	16	1.0	6
Total meat	339	33.8	10
Rice	517	46.8	9
Wheat	1448	203.7	14
Total cereals	762	175.5	23

Source: FAO, 2020 - Food Outlook - Biannual Report on Global Food Markets (https://www.fao.org/3/ca9509en/CA9509EN.pdf).

Fishing and trade

Archaeological evidence shows that aquatic animals have been a human food source for over 40,000 years (Gartside and Kirkegaard, 2011). Whereas the oldest history of plant cultivation has been traced back to 23,000 years (Snir et al., 2015). The earliest domestication of land animals is believed to have been between 13,000 to 10,000 years ago (NGS, 2019). Egyptian history shows fishing existed 3500 BC i.e. 5,500 years ago. Among the earliest histories of catching wild fish using various methods and having trade to supply as food to the people were from Peru (2000–3000 years back), Vancouver, Canada (2,500 years ago), the Mediterranean Sea (2,000 years back) and Japan (1,300 years back). Some European records showed historical trade started catching fish for sale to the consumers living nearby during 10th Century AD (Hoffmann, 2005). Until the 11th century freshwater fishes were consumed fresh or lightly preserved. But only from the 12th century, when people started to processed items on a larger scale (drying, salting or brining) did real trade pick up. Aquatic food was transported to more distant markets when catches were abundant, during peak seasons and after processing, for examples, herring (*Clupea harengus*) and sardine (*Sardina pilchardus*). Transport was often arranged by fishmongers to pool regional catches in seasonal abundance. Some maintained live storage facilities and some helped finance fishers through advance purchase of their catches. Markets for aquatic food grew continuously from the 12th century with prices rising over time. From the 15th century, deep-sea fishing and the trade of fish expanded with the use of drift nets. Trawlers started appearing in the 17th century and the number and range of these boats expanded rapidly in the 19th century with the availability of steam engines. Bigger and stronger boats started operating to pull wide nets in deeper water. The design of trawlers changed with the shifts from sail to coal-fired steam by the First World War, and then to diesel turbine engines by the end of the Second World War. As a result, the seafood trade boomed. In England the small coastal town of Grimsby became a commercial fishing centre of Europe. That was connected to London's Billingsgate Fish Market (the world's biggest fish market at that time) by a direct railway line.

The large, powerful fishing vessels started to become the problem: their number increased, they harvested more efficiently and they adopted new technology, such as refrigeration equipment and on board processing. The first combined freezer and stern trawler appeared in 1947. 'Super trawlers' were developed that could carry 60 t. After the Second World War, competition for fishing ground and the catch became fierce among countries in Europe. The Cod Wars (1958–1967) were a confrontation between the UK and Iceland over white cod in the North Atlantic. Other wars over fishing ground were seen even between states within the USA. Competition and confrontations occurred in Asia as well between Japan and Korea, and also with China for sea territory. Similar competition and disputes appeared among the ASEAN countries of Indonesia, Malaysia, Thailand and Vietnam. As a result of overfishing and the fish wars, wild fish stock started to deplete; some of the fishing grounds were empty as early as the 1950s. For example, Monterey Bay in California used to catch, process and supply over 200,000 t of sardines annually (Parrish, 2000). It collapsed completely by the 1950s, and they had to convert it into a marine conservation park.

In 2018, there were 4.56 million fishing vessels in the world, 75% were in Asia catching the fish for trade (FAO, 2020). There are many cases of fishery collapses (Haw, 2013). Presently, 30% of species have been reduced to biologically unsustainable levels – overfished. Recognition of overfishing has prompted many to promote fish farming as the future food security.

Aquatic food production is not evenly distributed due mainly to geographical and climatic differences, e.g. salmonids can only be grown in the saline waters of temperate regions. However, salmon is desired globally. Therefore, its market is global. Similarly, most shrimp are grown in tropical regions in brackish to saline waters and their market too is global. Trade is indispensable. Aquatic food is the top food item, in terms of value, traded from the developing to the developed world. Trade is almost doubling each decade.

Trade plays an essential role in distributing aquatic foods, boosting their consumption and achieving global food and nutrition security by taking products from producers to distant markets where consumers access them. It also provides employment and generates income for millions of people working in a range of industries and activities around the world, particularly in developing countries. According to Wyatt et al. (2021) exports of aquatic products are important to many countries and numerous coastal, riverine, and lacustrine regions. They often exceed 40% of the total value of merchandise trade in Cabo Verde, Faroe Islands, Greenland, Iceland, Maldives, Seychelles and Vanuatu.

Present trade activity

Globally China is the top exporter of aquatic food, with a value of about US$19 billion annually in 2020 (Harkell, 2023). Norway is the leading aquatic food production country in Europe, of mainly salmon valued at around US$15 billion annually (Holland, 2023). Vietnam exports various aquatic food products with a total value of over US$11 billion (Thanh, 2023). Pangasius as the most prominent species, the value of its produce reached over US$2 billion in 2018 (FAO-Globefish, 2019). Over 90% of pangasius produced in Vietnam is exported. India and Thailand are more or less similar in terms of exports (US$8 billion per year). In many countries, the produce from fisheries and marine aquaculture are destined for the export markets. A typical example is Andhra Pradesh in India where about 60% of the population is vegetarian and where Indian major carp are produced. These carp are transported to other states or even exported to Nepal. The vast majority of carp production is consumed by the domestic market whereas over 80% of shrimp produced in India are destined for the export market, especially the EU and USA. Similarly, 90% of the shrimp produced in Bangladesh is exported. Some shrimp are even exported to Thailand as demand for shrimp is high in Thailand and production has fallen due to diseases. Other countries emerging as exporters, such as shrimp from Ecuador to China and tilapia from Indonesia, Colombia, Costa Rica, Mexico to the USA. Similarly, frozen fillets of tilapia are exported from Asia and Latin America to the USA and EU. Whereas for whole frozen tilapia large volumes are exported to Africa from China due to high demand in Kenya, Ivory Coast and Zambia. There is nascent aquaculture in some African countries, such as Ghana, Kenya, Nigeria, Uganda, Zambia and Zimbabwe, and local producers are urging their governments to ban imports from China in order to encourage their local production (Okai, 2019).

Among the importers, the USA, EU and Japan are top of the list importing US$20 billion, US$20 billion and US$15 billion, respectively. China is also emerging as a major main importer of aquatic foods, valued about US$8 billion annually. China is a good market for salmon and imports 70,000 t of salmon per year from Chile, Norway and Denmark. China is also importing shrimp and pangasius from Vietnam. There is concern in Europe and the USA that all aquatic food produced in Asia will be diverted to China. Recent export of shrimp from Vietnam and Ecuador are some of the examples and indicators of this trend.

Transportation

Live fish are transported from production sites by road and are kept in water-filled metal boxes. Collaboration between fishers and fish farmers with processors, distributors, retailers, restaurants and foodservice providers to resolve the environmental issues affecting live transport is expected to provide lucrative opportunities.

Transportation is an important part of the trade. For example, whole shrimps are normally transported using ice. They are de-iced, washed, weighed and graded. All discoloured, bruised pieces, soft shell and hanging meat are discarded. Only those of sufficient quality are selected and then treated with sodium-metabisulphite to avoid black spot before grading. The shrimps are arranged in trays and filled with water in case of plate freezing and are frozen at –40°C for 1 to 2 hours (Das and Kannuchamy, 2014). the trays are then de-panned and glazed. For blast freezing the shrimp are arranged in trays and are blast frozen at –40°C for 2 hours. The product is passed through a metal detector to check for contamination by small pieces of metal before packing according to specification and stored at –18°C. During the processing, there are opportunities for pathogens to enter if conditions are not stringently maintained. Therefore, certification and traceability matters.

Aquatic food marketing

Conventional marketing used to be done producing hard copy brochures, posters, billboards, advertisement in newspapers, magazines and on television. Similarly, TV interviews, documentaries and talk shows were common methods. With the advancement of technology, marketing methods have also evolved to more efficiently reach larger and specific targeted populations through websites, mail marketing, social media channels and blogs, often hoping to trigger a viral marketing response.

Producer presence at conferences, exhibitions or tradeshows can be an effective way of marketing to targeted groups. These are organized by various societies, such as World Aquaculture Society (WAS) and its regional chapters, the Asian Fisheries Society and the European Aquaculture Society (EAS). Private groups and organizations also organize events. Societies also publish trade magazines which accept advertisements, e.g. *Aquaculture Magazine* of the World Aquaculture Society, *Global Aquaculture Advocate*, *Aqua Culture* and so on.

As in other sectors, aquatic products are also sold online. One of the pioneer examples of the new-retail model is in China, called Hema Xiansheng and owned by Alibaba as mentioned above. Since the COVID-19 pandemic the number of online marketing and delivery systems has increased significantly.

The role of markets

Food and other essential items are produced remotely from their point of sale, they are transported to markets, cleaned, processed and stored to make them safely and easily available. It is not possible for every person to produce the food they need. Therefore, markets provide services to everyone and are meeting places for producers and consumers. A series of transactions occur in markets between producers, traders or consumers at various levels; markets are the power houses of the economy. Markets require gatherings of groups and individuals, they promote social cooperation and some level of democratic norms. Therefore, markets may contribute to social welfare and well-being beyond the monetized transactions (Dolfsma et al., 2005). Government or other stakeholders sometimes manage and promote markets. Markets help create employment: hundreds of people can be engaged at a small market and along the value chain supplying it.

Domestic and traditional markets provide support to a large number of small-scale farmers and fishers. International or export markets give opportunities to the well-equipped fishing groups and well-managed aquaculture farmers. They are normally well-off especially when export markets boom. For example, when shrimp (*Penaeus monodon*) farming boomed in Thailand during mid-1980s to mid-1990s, profits exceeded 100%. Similarly, international markets for pangasius fillets provide great opportunities to the farmers of Vietnam who can earn a good income. During its boom time, 40,000 farmers grew pangasius to sell to the exporting companies. The processing industry created over 200,000 jobs especially for women filleting manually.

Markets also serve as good sources of information about demand for each product type, sales volume, and prices. Producers and suppliers use this data to manage and plan their production and transportation. More importantly, market information, especially price and demand, may encourage producers, processors and suppliers to be more innovative. Therefore, it is true to say markets encourage innovation, leading to new products e.g. highly attractive and pink salmon fillets and red tilapia for Asian markets and white flesh pangasius fillets for the EU and US markets. Additional activities create more jobs.

As markets provide services to many sellers and buyers, competition occurs among them, which in turn results in lower prices making goods and services affordable to more and a wider range of people.

Markets reflect the views of participants; consumers can choose to buy or not in line with their ethical and environmental concerns. For example, shrimp from Thailand was once not accepted by European consumers in the assumption that its farming was destroying the mangroves and that slaved labour was used.

Markets can also supply feedback from buyers and consumers so that producers can determine whether they are using their resources wisely, efficiently and without destroying the environment. Tilapia produced in China is the cheapest, which may encourage farmers in other countries, such as Bangladesh, Thailand and Vietnam, to innovate and invest if they want to compete on price. Fair competition expressed through markets can result in the 'best' outcome for the limited resources available for the humankind.

Conclusion

Aquatic animals have been taken from the sea and other water bodies since the time before human civilization. Seafood has always contributed to human food security. Wild catch alone cannot fulfil market demand. At the same time, human induced pressure is threatening most natural stocks especially of anchovies, sardine, tuna, crabs, squid, shrimp and lobsters. The concept of aquaculture started in the 1930s, and took off in the 1980s. However, production is not evenly distributed and is related to climatic, geographic and other limitations. As a result, transfer of products from one place to another or even one country to another country or continent is a necessity. Demand in one (often developed) country results in higher production in another (often developing).

Market demand and consumer preferences have driven innovation and the adoption of more sustainable farming methods, such as the development of environmentally friendly methods of shrimp and salmon farming. Markets demand new and value-added products, in turn driving innovation in production, processing and packaging, transportation and storage. Continuing and increased investment, both financial and intellectual, in research and development to better the domestication techniques of the species currently farmed and at the same time to explore new species and strains will greatly benefit aquatic food security.

References

Asche, F. (2017) New markets, new technologies and new opportunities in aquaculture. *Aquaculture Economics and Management*, 21, 1–8. https://doi.org/10.1080/13657305.2016.1272649.

Cassady, R. (1967) *Auctions and Auctioneering*. University of California Press: Berkeley, CA.

CBI (2016) Fish and seafood market information. Centre for Market Promotion Centre for the Promotion of Imports from Developing Countries. https://www.cbi.eu/market-information/frozen-fish-seafood/channels-segments.

Chiber, A. (2013) Middle East hungry for more seafood. *Food Navigator News*. https://www.foodnavigator.com/Article/2013/12/04/Middle-East-hungry-for-more-seafood.

Das, O. & Kannuchamy, N. (2014) Processing of frozen seafood products. *INFOFISH International*, 4/2014, 30–34.

Dolfsma, W., Finch, J.H. & Mcmaster, R. (2005) Market and society: How do they relate, and how do they contribute to welfare? *Journal of Economic Issues*, 39, 347–356. https://doi.org/10.1080/00213624.2005.11506811.

FAO (2001) *Production, Accessibility, Marketing and Consumption Patterns of Freshwater Aquaculture Products in Asia: A Cross-Country Comparison. FAO Fisheries Circular No. 973*. Food and Agriculture Organization: Rome.

FAO (2019) GlobeFish. *Information and Analysis on World Fish Trade*. www.fao.org/in-action/globefish/fishery-information/world-fish-market/en/.

FAO (2020) *The State of World Fisheries and Aquaculture 2020 – Sustainability in Action*. Food and Agriculture Organization of the United Nations: Rome. https://doi.org/10.4060/ca9229en.

FAO (2022) *The State of World Fisheries and Aquaculture – Towards Blue Transformation*. FAO: Rome. https://www.fao.org/3/cc0461en/cc0461en.pdf.

Gabriel, A.S., Une Ninomiya, K. & Uneyama, H. (2018) The role of the Japanese traditional diet in healthy and sustainable dietary patterns around the world. *Nutrients*, 10, 173. https://doi.org/10.3390/nu10020173.

Gartside, D.F. & Kirkegaard, I.R. (2011) A history of fishing p. In: *The Role of Food, Agriculture, Forestry and Fisheries, in Human Nutrition* (edited by V. R. Squires). OLSS Publishers, Oxford, pp. 105–139.

Harkell, L. (2023) China's seafood imports surge to $19bn in 2022 despite COVID lockdowns. China Fisheries Seafood Expo, April 12, 2023: https://chinaseafoodexpo.com/chinas-seafood-imports-surge-to-19bn-in-2022-despite-covid-lockdowns/.

Haw, J. (2013). *The Historical Collapse of Southern California Fisheries and the Rocky Future of Seafood.* Scientific American.

Hoffmann, R.C. (2005) A brief history of aquatic resource use in medieval Europe. *Helgoland Marine Research*, 59, 22–30. https://doi.org/10.1007/s10152-004-0203-5.

Holland, J. (2023). Norway smashes seafood export record, earning USD 15 billion despite volume drop. Seafoodsource.com–online resource: https://www.seafoodsource.com/news/supply-trade/norway-smashes-seafood-export-record-earning-usd-15-billion-despite-volume-drop.

Matsuyama, Y. (2021) The fishing industry is in trouble due to the falling fish consumption and aging fishermen. The Canon Institute of Global Studies (CIGS). https://cigs.canon/en/article/20210629_5993.html.

NGS (National Geographic Society) (2019) The development of agriculture. https://www.nationalgeographic.org/article/development-agriculture/.

NOAA (2018) Fisheries of the United States, 2017 Fact sheet. https://www.fisheries.noaa.gov/resource/document/fisheries-united-states-2017-fact-sheet.

Okai, E.K. (2019) Africa's tilapia farmers rise to Chinese challenge. The FishSite. https://thefishsite.com/articles/africas-tilapia-farmers-rise-to-chinese-challenge.

Parrish, R. H. (2000) Monterey Sardine Story. Pacific Fisheries Environmental Group. https://swfsc-publications.fisheries.noaa.gov/publications/CR/2000/2000ParrR2.pdf.

Snir, A., Nadel, D., Groman-Yaroslavski, I., Melamed, Y., Sternberg, M., Bar-Yosef, O. & Weiss, E. (2015) The origin of cultivation and proto-weeds, long before Neolithic farming. *PLOS ONE*, 10, e0131422. https://doi.org/10.1371/journal.pone.0131422.

Srinivas, R. (2022) MPEDA Targets Rs. 1 lakh crore seafood exports by 2025. The Hindu. July 26, 2022. https://www.thehindu.com/news/national/andhra-pradesh/mpeda-targets-rs1-lakh-crore-seafood-exports-by-2025/article65684819.ece.

Thanh, V. (2023) Seafood sector striving to expand export market. Vietnam Investment Review. January 10, 2023. https://vir.com.vn/seafood-sector-striving-to-expand-export-market-99143.html.

Wyatt, T., Friedman, K., & Hutchinson, A. (2021) Are fish wild? *Liverpool Law Review*, *42*(3), 485–492. https://doi.org/10.1007/s10991-021-09285-0.

Zhu. C. & Mikhailova, V. (2020) Outlook of the seafood market in China after COVID-19. https://daxueconsulting.com/chinese-seafood-market/.

8

The role of processing and retail sectors in aquatic food security

Mala Nurilmala, Asadatun Abdullah, Roni Nugraha, Ruddy Suwandi, Nurjanah Nurjanah, Tati Nurhayati and Yoshihiro Ochiai

Introduction

Fish and shellfish inhabiting aquatic environments (in both sea- and freshwater) have acquired unique physiological systems through their evolution. The most notable difference between aquatic and terrestrial animals is the body temperature. The body temperatures of aquatic animals are generally low, depending on the water temperature of their respective habitats. Therefore, the substances which make up their bodies are much less stable compared with those of terrestrial animals, which makes fish and shellfish more quick to perish after death (or after harvesting). The tissues of aquatic animals are relatively soft allowing the invasion of microorganisms, and together with autolytic degradation caused by endogenous proteolytic and other digestive enzymes, the shelf-life of fish and shellfish is much shorter than that of terrestrial animals. Therefore, counteractions are essential to extend shelf-life.

Food can be classified by perishability, namely:

1. Non-perishable – foods that are not easily damaged. This group can be stored for a relatively long time at room temperature. The group includes rice and dried beans.
2. Semi-perishable – foods that are somewhat perishable. This group can be stored for a limited period of time as represented by onions and tubers.
3. Perishable – foods that are easily damaged. This group is quickly damaged when stored without any pretreatment (preservation), as represented by meat, fish, milk, ripe fruit and vegetables.

Freshness

Seafood deteriorates rapidly and thus requires proper handling. Initial handling will affect quality later on. Rotten seafood can never be brought back to a fresh state even with modern technology. Therefore, great effort to maintain the freshness of seafood should be made, starting from the point that the fish are harvested/caught right through to the point that they reach the consumer.

Margaret Crumlish and Rachel Norman (eds) **Aquatic Food Security**
DOI: 10.52517/9781789183368.008, © 5m Books Ltd, 2023

Tremendous care should be taken to prevent or minimize damage during storage, shipping, auctions and distribution as well as transportation.

Fresh fish have a bright appearance since there are little postmortem biochemical changes. However, shortly after death, they begin to deteriorate caused by autolysis and bacterial activity. Gradually the surface colour becomes dull, due to the onset of mucus as a result of further biochemical processes and the multiplication of microbes.

Various physical and chemical changes take place rapidly, leading to organoleptic unacceptability. The sequence of changes is classified into three states, namely, pre-rigor, rigor mortis, and post-rigor. The pre-rigor state is a phase starting in fish just after death, during which the chemical changes and bacterial growth proceed very slowly. In this phase, the released mucus from the gland below the surface of fish skin forms a thick layer around the fish body. This phase relaxes the muscles of the fish shortly after the fish died.

The second phase, rigor mortis, is recognized by hardening of the muscle. Fish muscles become stiff because of the irreversible reaction between muscle contractile proteins, actin and myosin, due to the absence of ATP. The last step, post-rigor, is characterized by re-softening of fish muscle. In this phase, fish begin to spoil through the actions of microbes.

There are various ways to evaluate the freshness of fish based on physical (organoleptic), chemical and microbiological measurements. In other words, fish freshness could be categorized into the subjective and objective parameters. Subjective methods are preferred to the latter, because they rely on simple sensory analysis of the appearance, texture, eyes, colour of gills, etc. High quality is characterized by brightness in skin colour, convex eyes, consistency in meat texture, etc., while fish of poor freshness can be recognized by dull colour, cloudy eyes, and so on.

Extending shelf-life

Many methods have been developed to extend the shelf-life of seafood. Drying or salting is the easiest way to delay the quality deterioration. Both methods reduce the water activity (ratio of free water) that is utilized by microorganisms to multiply. Freezing is also effective, and, additionally, reduces the rate of chemical reactions, such as oxidation and enzymatic reactions. However, there is a drawback to frozen storage, because at higher temperatures (roughly above –30°C), dehydration occurs, causing 'freezing burn' and lipid oxidation. It has now been established that many enzymatic reactions can proceed even under freezing (Li et al., 2019). Besides, drip loss, due to the rupture of cell membranes by the growth of ice crystals, takes place when thawing frozen food. Smoking can also prolong the shelf-life of seafood through the introduction of antibacterial and antioxidative substances contained in the smoke. However, recently, polycyclic aromatic hydrocarbons (PAHs), which can have health implications, have been identified in smoked food (Zelinkova and Wenzl, 2015).

The demand for seafood of high quality and freshness is increasing worldwide. To fulfil the demand, careful storage, processing, shipping and handling are essential to minimize wastage and to protect consumers from food poisoning caused by pathogenic bacteria and accumulation of biogenic amines (histamines and so on). Bacteria, as living organisms, develop, for example, and acquire resistance to human-made antibiotics (Roschanski et al., 2017). Intensive hygiene control at each stage at the farm, and during processing, shipping and storage is essential.

Hazards affecting food security are categorized into biological, chemical and physical catagories.

- Biological hazards mostly include pathogenic microorganisms that cause food poisoning, sometimes leading to fatal incidents (*Vibrio* sp. and *Escherichia* sp.), and viruses, such as norovirus.
- Chemical hazards encompass a range of toxic and harmful substances that can contaminate seafood. Owing to bioaccumulation, aquatic animals may contain toxic substances, such as the natural toxins tetrodotoxin, ciguatoxin and saxitoxin, as well as toxic elements, such as arsenic (Molin et al., 2015) and organic mercury (Costa et al., 2016). Endocrine disturbing substances in the environment are also of great concern when ingesting seafood (Ávalez-Muños et al., 2018). Climate changes due to global warming (including water temperature rise, acidification, sea level rise) will be threatening to the security of seafood (Kibria et al., 2017).
- Physical hazards include contamination by small pieces of metal, glass and wood. Asche et al. (2015) claimed that developing countries export high-quality seafood in exchange for lower quality seafood. Recently, pollution of aquatic organisms by microplastic is becoming another serious concern (Santillo et al., 2017).

Rationale

In order to provide high-quality seafood, it is essential to understand the uniqueness of fish and shellfish and gather and maintain information on quality maintenance. This chapter aims to cover up-to-date information on consumer-oriented strategies for the quality maintenance of seafood.

Traditional processing and products

Background

Traditional fish products still represent the main utilization of fish in Southeast Asia as well as major source of animal protein (Yeap and Tan, 2003). In some regions, for example Africa, most of the fish processing technology applied may be classified as traditional (Sefa-Dedeh, 1993). Traditional fishery processing technology plays an important role in many countries worldwide. The main reason is the seasonal time of fish harvesting, therefore, preservation becomes an important issue (Kurlansky, 2003; Skåra et al., 2015). Traditional fish processing technology might be correlated to reduced postharvest losses during high seasonal catch. The oldest method of fish preservation is drying often, in some combination with salting. Drying is practised worldwide and has a long history in, for example, China, Egypt and northern Europe (Kurlansky, 2003; Skåra et al., 2015). Marinating fish in spices, salt or sugar in certain combinations and concentrations may affect the nature of the protein components (aggregation and denaturation) and drive moisture content loss during processing. The changes in macromolecule formation would affect the texture, flavour and aroma (Ashton, 2002; Hall, 2010). Furthermore, fish smoking is another traditional fish processing method. The process of fish smoking reduces the moisture content and the compound from hot smoke will provide antimicrobials and antioxidants to preserve the highly perishable raw materials (Hall, 2010). Fermented fish products also have a long history, especially in tropical climates, producing distinct, locally popular tastes and aromas. Traditional processing techniques remain popular and are pragmatic and less costly as they do not require a cold-chain for transportation and storage (Saisithi, 1994; Kose and Hall, 2010).

Sun-drying and salting

Traditionally, to preserve excess fish harvests fishers used to sun-dry the wet products. Reduction of moisture content in food will delay the process of deterioration. In some regions, sun-drying is combined with the application of high concentrations of salt. Salt inhibits the growth of microorganisms and drives off water thereby accelerating the drying process. The salting process is mainly affected by the size of fish and the salt concentration. Small fish can be salted whole without gutting. Whereas, larger fish need to be gutted and split open (e.g. into butterfly form) to allow salt to penetrate into the meat. In order to improve the efficiency of sun-drying process the producers could: (1) increase the surface area of fish flesh (surface area) for water-content decrease; (2) cut the fish flesh thinner; and (3) carry out maximum exposure of fish flesh to sun and wind (Hall, 2010).

The quality of salted fish products depends on the preparation of the raw materials. Special care must be taken regarding product safety. Halotolerant or halophile microflora can grow and accelerate spoilage. In recognition of the role of dietary salt levels in increased incidences of high blood pressure, effort has been made to reduce the concentration of salt. A lighter salt concentration (<5%) is now applied to most seafood products. The process of salting mainly depends on local tradition, but the most popular technique is dry or wet salting.

Smoking

Traditionally, the smoking process has been applied to enhance flavour and preserve fish products. The smoking process can be divided into cold smoking, liquid smoking and hot smoking methods. The smoking method affects the sensory profiles of resultant seafood products with respect to texture, colour, and flavour. In Indonesia the product is called *sale ikan* or *cakalang fufu*. The products are processed from fresh fish flesh smoked with hot smoke from wood chips. Hot smoking using wood has been the most common traditional process in Southeast Asia regions. Today, the application of liquid smoke into food has replaced the traditional hot smoking process in urban cooking (Berhimpon et al., 2018; Simon et al., 2005). However, the shelf-life of such products is generally short compared with the hot-smoked ones, and thus they still need to be refrigerated.

Traditional fish processing using hot or cold smoke has been a common practice. The smoke obtained from burning wood contains chemicals that contribute to flavour and sensory profiles, e.g. phenols, aldehydes, acetic acid and a range of polycyclic aromatic hydrocarbons (PAHs) (Babić et al., 2018; Basak et al., 2010). The fishery industry must comply with international regulations, such as those issued by the European Union (EU), the USA and Canada. In the case of smoked fish, the regulations limit the content of PAHs. Some of PAH derivatives in the smoke are possibly carcinogenic (IARC, 1972). Recent studies have demonstrated the effectiveness of techniques to reduce PAHs content in smoked fish products using filters (Akpambang et al., 2009; Parker et al., 2017; Radojicic et al., 2012). Zeolite filtering was found to be the safest method for traditional smoked fish production. Recent scientific advances in safe and good practice for smoked fish must be disseminated into the traditional and semi-modern fishery industries (Babić et al., 2018).

Fermented products and fish paste

Fermentation of fish has been a preferred way to preserve seafood, enhancing nutritional values and sensory level, as well as reducing the energy required in the cooking process (Ohshima and Giri, 2014). Traditional fermentation processes often rely on naturally occurring enzymes or starter bacteria. Natural enzymes will change texture and impart specific flavours, meanwhile starter bacteria will develop aroma and flavour. Some fermented fish and shellfish products provide higher nutritional values with a longer shelf-life than those left unfermented. In the fermentation process, microorganisms will help convert complex macro-molecular compounds into simple forms, such as amino acids and fatty acids (Raksakulthai and Haard, 1992; Zeng et al., 2013).

There are several factors contributing to the success of the fermentation process: (1) salt concentration; (2) pH level; (3) the type of microorganism used as a starter; and (4) the characteristics of raw material used (Skåra et al., 2015). As can be seen in the Japanese product *nare-zushi*, keeping raw fish in fermented rice is effective in extending the shelf-life as well as offering favourable flavour. The point is that *Lactobacilii*, the dominant bacteria in the product, produces lactic acid, which reduces the pH level and thus prevents the growth of unfavourable bacteria. Marinating fresh fish with salt and vinegar provides a product of longer shelf-life and special texture and flavour. An example is marinated mackerel *shime-saba* in Japan. It is also noteworthy that the puffer toxin (tetrodotoxin) level in pufferfish ovary is dramatically reduced when kept in salted rice bran for years, known as *fugunoko-nukazuke* in Japan (Kobayashi et al., 1999).

The application of lactic acid bacteria (LAB) is considered a safe fermentation agent (generally recognised as safe [GRAS]). The popular fermentation technique in Southeast Asia is the combination of lightly salted fish products (2–7% sale content), a carbohydrate source, herbs or spices subjected to fermentation at ambient temperatures for 2 to 7 days (Anal, 2019; Paludan-Müller et al., 2002). In any food processing technology, one of the most important tasks is to assure the quality and safety of the food. Seafood safety becomes an important issue in fermentation process, particularly in ready-to-eat (RTE) products. Recent studies have investigated the survival of food-borne pathogenic bacteria as well as the incidence of biogenic amines (Özyurt et al., 2016; Verdos et al., 2019). The combination of high salt concentration and low pH levels will reduce the population of pathogenic bacteria during the fermentation process, this should be combined with proper handling of the raw materials and strictly followed the good hygiene practice (GHP) (Verdos et al., 2019).

In Southeast Asia fermented fish products are one of the region's staple food and are a significant source of nutrition. Two of the most ubiquitous fermented products across the Southeast Asian region are fish or prawn paste and fish sauce. Fermented food products are good sources of peptides and free amino acids (Hajeb and Jinap, 2012; Rajapakse et al., 2005; Sathivel et al., 2003). Fermented foods are widely consumed in Southeast Asia due to their savoury (umami) taste and rich flavour and are used in daily cooking (Hajeb and Jinap, 2012). In specific Southeast Asian countries, fish paste, shrimp paste and fermented shrimp have been consumed since the 15th century as condiments or seasonings, staples and daily food (Faithong et al., 2010; Hajeb and Jinap, 2012; Yoshida, 1998). Example of fish sauces are *nước mắm* in Vietnam, *nam pla* in Thailand and *shottsuru* and *ishiru* in Japan.

Fermented shrimp paste is known by many different names, for example, *terasi*, *belacan*, *kapi*, *ngapi*, *burong*, *tamayo* and *mam* (Hajeb and Jinap, 2012). Shrimp paste is produced by different methods depending on the local tradition using different salt concentrations and shrimp species. The genera of shrimps that are used as the main raw materials are *Acetes* sp., *Mesopodopsis* sp., *Schzopodes* and *Mytis* sp. (Hajeb and Jinap, 2012; Mizutani et al., 1987). Traditional methods to produce shrimp paste require a microbial starter culture and a salt concentration between 5% and 20%, and utilize pre-cooked heads and shells of shrimps.

Fermented shrimp paste is used as a condiment and a flavour enhancer. The nature of fermented shrimp paste is semi-solid with low moisture content, strong aroma and umami taste. The umami taste comes from the high content of glutamate and 5'-nucleotides (Jinap et al., 2010; Khairunnisak et al., 2009).

Exemplification of fermented products in Southeast Asia, particularly Indonesian regions, is represented by *terasi udang* and *petis udang*. Both are widely consumed as Indonesian ethnic foods. *Terasi udang* is a dark brown or reddish colour solid paste with strong aroma and a high level of glutamate (Mizutani et al., 1987). *Petis udang* is used to prepare the traditional salad *rujak petis* as well as a flavour enhancer. *Petis udang* contains high amounts of free amino acids such as glycine, alanine and glutamate (Yoshida, 1998).

Brine-cooked fish products

Brine-cooked fish or salted boiled fish is consumed in many Southeast Asian countries. The products have many different local names such as *pindang* (Indonesia), *sinaeng* (the Philippines) and *pla-tu-nung* (Thailand). The most widely consumed one in the Indonesian region is *pindang*. Indonesia is the biggest producer of marine fish catch in Southeast Asia. Therefore, to avoid postharvest losses, more than 15% of the fish are processed into *pindang*. The processing technique for brine-cooked fish requires only simple equipment. Initially, the fish is soaked in brine (<15% salt), then boiled, and air dried at room temperature. With such a simple process small enterprises and local fishers could produce *pindang* easily at the fish landing area (Fitriati et al., 1994).

Industrial seafood processing and products

In order to improve food security seafood needs to be processed. Habits and lifestyles change; consumption has shifted from buying or consuming whole fresh fish and shellfish to purchasing processed/ready-to-eat seafood. Seafood processing can be classified into traditional and modern techniques. In the section above, traditional seafood processing methods were discussed. This section will focus on modern seafood processing, including filleting, canning and surimi-based products. Preservation can be achieved by providing a treatment that can kill food deteriorating microbes through either high or low temperature treatments.

Nutritional content

Fish contains several elements needed by the human body being high in protein, polyunsaturated fatty acids (omega-3), vitamins and minerals.

Protein

It is well known that fish are a high-protein source. Protein comes from the Greek word *proteios*, literally meaning 'primary'. The functions of protein are to form body tissues, replace damaged cells and provide a source of calories. Enzymes belong to proteins. In addition, proteins can be neurotransmitters and oxygen carriers in the blood. The adult human needs 1 g of protein per 1 kg of body weight per day; 1 g of protein produces four calories.

Fish protein is characterized by complete amino acid balance, both essential and non-essential ones. Amino acids that are synthesized in the body are called non-essential amino acids, while essential amino acids are those which the human body cannot synthesize. Though amino acid content in fish muscle is very similar regardless of species, the amount of free histidine is higher in scombroid fish such as tuna and skipjack.

Lipids

Lipids are very important for the human body. The important constituent of lipids is fatty acid, which are categorized into saturated and unsaturated forms. Fish lipids contain polyunsaturated fatty acids (PUFA), especially omega-3 fatty acids (EPA and DHA) which can reduce blood cholesterol levels, improve intelligence and prevent degenerative diseases. Daily lipid requirements range from 0.5–1.0 g lipid per 1 kg body weight.

Carbohydrate

There are small amounts of carbohydrates in fish, namely glycogen. Fish is not a good source of carbohydrates. The glycogen is located mainly in muscles. Generally, the amount of carbohydrate is about 0.1–1.0%. Shellfish, however, can contain 1–7%. Flora such as seaweed is a good fibre source.

Vitamins and minerals

High levels of vitamins and minerals are also found in fish, so fish is beneficial both for growth and maintenance of a healthy body. Vitamins are organic nutrients that are generally not synthesized in the body, so they must be supplied from food. They are needed in small amounts and essential for various biochemical functions. Vitamins function as catalysts in the human body, allowing the transformation of macronutrients, called metabolism. Vitamins are classified into two groups based on their solubility, namely, the fat-soluble ones such as vitamins A, D, E and K, and the water-soluble ones such as C, B_6, B_{12}, biotin, and niacin.

Minerals are present in the body and play important roles in maintaining cells, tissues, and organs. The mineral balance in the body regulates the functions of enzymes, maintains the acid-base balance and regulates the sensitivity of muscles and nerves to stimulation. The minerals are classified into macro and micro groups. Macro minerals compose up to 1% of the total human body weight and the required amounts are about 1000 mg/day, while micro minerals are needed with much smaller amounts of less than 100 mg/day.

Filleting

Today's consumers prefer food products that are ready to be cooked or to be eaten. Seafood need to be processed for ease of consumption; particularly the removal of bones, scales and shells. One popular form is the fillet. The word 'fillet' from the French for a thread or strip. Fish fillets are strips or cuts made along the spine, starting from the back of the head toward the tail (CAC, 2003). Fish fillets are commonly found in supermarkets and retail shops. Almost all fish can be made into fillets, for example, seawater fish such as seabass/baramundi (*Lates calcalifer*) or freshwater fish such as pangasius (*Pangasius bocourti*) and tilapia (*Oreochromis niloticus*). Filleting usually requires a large-sized fish (>500 g). The raw materials used must be fresh. In general, the stages of filleting process are as follows.

1 *Acceptance of raw materials.* In this stage, the raw material will be received, selected according to size, and checked based on freshness and for biological contamination residues (must also be free from the bacteria such as *Salmonella* and *Escherichia coli*).
2 *Sizing and bleeding.* Fish will be sorted according to their sizes. Bleeding is also commonly done on some species, such as catfish, which are directly harvested from culture ponds.
3 *Separation of bones from meat.* This stage is the main process of filleting. The fish meat will be removed from the bone. Generally, this step is carried out manually using a special knife. Nowadays, the separation of meat from bones using a deboning machine has been developed. Figure 8.1 shows the separation process of fish meat from bones in a filleting factory.
4 Removing skin. In this step, the section of meat that has already been separated from the bone is skinned.

Figure 8.1 Separation of bones from fish meat.

Figure 8.2 Fish fillet after trimming and washing.

5 Trimming and washing. This step includes fat removal and clean up of the product. The remaining thorns and red meat (dark muscle) are removed (Figure 8.2). The fillet is washed to remove impurities and improve appearance before freezing.

6 Soaking. Soaking is carried out before the fillet is to be frozen (Figure 8.3). As a result, the product will be cleaner in addition to improvement of appearance and increased water holding capacity. Food additives are usually applied at this stage following the appropriate standards. The products are mixed using a tumbling machine containing water and dissolved food additives, and rotated for a given time (in the range of 30 to 60 minutes).

7 Freezing. Fillets are frozen before shipping. The freezing point of a liquid in food is on a temperature range where the liquid is in a state of balance with its solid material. Knowing the freezing point is very important to avoid unnecessary changes in the product including physical, chemical and biological ones during freezing and storage. The freezing point of fish meat is –0.6 to –2.0°C (Desrosier and Desrosier, 1977; Fellows, 1992).

There are two types of freezing, i.e., quick and slow freezing. Quick freezing is a process by which the temperature of materials can pass through the maximum ice crystal formation zone within 30 minutes or less (Figure 8.4). In contrast, slow freezing takes much longer time, i.e. 3–72 hours. Freezing speed has a significant impact on the formation of ice crystals in muscle tissues. The slower the freezing rate, the larger the ice crystals become. The freezing above –18°C is categorized as slow freezing, and large ice crystals tend to be formed. On the other hand, quick freezing has several advantages over slow freezing as follows.

Figure 8.3 Fish fillet after soaking process.

a The ice crystals formed are smaller and cell damage can be reduced.
b The shorter freezing time reduces the diffusion of salt and water to form ice crystals.
c The product is quickly brought below the temperature range for the growth of bacteria, moulds, and yeast, thereby preventing decomposition due to microorganisms.

8 Glazing. Glazing is carried out after the fillet is frozen by dipping the frozen product in cold water. The function of glazing is to prevent dehydration and lipid oxidation, and also to slow down denaturation of the proteins. In addition, glazing can increase the weight of fish fillets. Adding gelatine and vitamin C to the glazing water reinforces the glaze and prevents oxidation, respectively.
9 Packaging. Frozen fish fillets are generally vacuum packaged: removing the air from the package before sealing.

Figure 8.4 Fish fillet processed using quick freezing.

Freezing and frozen storage

There are several biological and chemical changes caused by the freezing process, namely, the formation of ice crystals, increase in fish volume and dehydration. Ice crystal formation can generally result in deterioration. Slow freezing or fluctuating storage temperatures cause the growth of ice crystals which damages the cell membrane, leading to drip loss after thawing.

Dehydration during frozen storage is characterized by the transfer of water vapor from the surface of fillet to cooling pipes in a freezing unit. Dehydration can easily occur if the fillets are loosely packed or the frozen storage conditions are unsatisfactory. Frozen fillets that are excessively dehydrated will become dry with the occurrence of wrinkles, discoloration to a pale color and pores on the meat. These characteristics indicate the damage known as 'freezer burn' (Londahl, 1981). Protein denaturation in frozen fillets can also occur as a result of drastic changes in solubility (aggregation, polymerization) and lipid oxidation. The rate of protein denaturation depends on storage temperature.

Biological changes can also occur because of bacteria. Psychotropic bacteria (*Pseudomonas* spp., *Alcaligenes* spp. and *Flayobacterium* spp.) can survive in the fillet during frozen storage. These microorganisms will start to grow during thawing process.

Although it is difficult to strictly control changes in the quality of fillet during the freezing and storage process, the intensity of these changes can be reduced in several ways as follows.

1 The use of cryoprotectant chemicals can protect proteins in the fillets and prevent loss of function during freezing. Cryoprotectants can improve the quality and extend the shelf-life of frozen fillets by preventing irreversible changes in the myofibrillar proteins caused by freezing, frozen storage

and thawing. Cryoprotectant ingredients that are added to protect the products during the freezing and storage process include sucrose, sorbitol, sodium lactate and sodium phosphate.

2 Glazing by coating of frozen fillets with ice films can prevent adhesion between raw materials, protect products from dehydration during storage, prevent lipid oxidation (rancidity) and improve surface appearance.
3 Packing frozen fillets in water and oxygen proof materials or vacuum packaging.

Canning

Canning is a processing method to preserve food in a hermetically sealed container (can) by heat sterilization. Canning is a way of preserving food by high-pressure heating to kill microorganisms. The purpose of food canning is to avoid spoilage during long-term storage at room temperature. Canning is an effective processing method for seafood, since fish and shellfish are very perishable. This is the most common method for maintaining nutritional values and taste. The shelf-life of canned products can exceed 2 years.

In general, seafood canning consists of the preparation, filling, exhausting, sterilization, cooling and storage. The preparation consists of selecting material fish, washing and cutting for further processing. Washing removes contaminated materials and is expected to reduce the number of initial microbes in correlation with the effectiveness of the sterilization process.

In the filling step, cuts of fish are put into cans, leaving a space at the top of the can, called the 'head space'. The space is a gap between the surface of the contents and the lid that serves as a reserve space during sterilization to avoid the swelling of cans. Basically, the head space varies depending on the type of products and containers. When the canning medium is filled, the head space height must not be less than 0.25 inches or 10% of the can capacity. The filling of materials must be uniform in order to maintain the air cavity, to obtain consistent products, and to maintain the constant weight of the material. The can will be convex and easily damaged when the contents are full.

Salts and acids are added to maintain shape, color, texture and taste. Salt will also help distribute heat in the can. The next process is called exhausting, which is the removal of air contained in the head space. The purpose of exhausting is to reduce pressure in the can during the heating process. Air, especially oxygen, is eliminated to prevent corrosion in the can and oxidation of the contents.

Sealing is carried out to cover the can before sterilization. The purpose of sealing is to install the lid of the container in such a way that contaminants cannot enter the can after sterilization. Sealing is done with a sealing machine. The can closure must be perfect, because any leakage can damage the product.

Commercially sterile food means that all microorganisms causing poisoning and spoilage have been inactivated. Microbial spores may still remain that are heat-resistant. However, they will not germinate under anaerobic conditions. The sterilization step is usually done at 121°C for 20–40 minutes, depending on the type of food. The sterilization process is followed by immediate cooling to prevent over cooking and the growth of thermophilic bacteria. Generally, cooling can prevent the growth of spores from food spoiling bacteria that have survived. Cooling is done through soaking the can in cold water until the temperature drops to 35–40°C. Cooling is performed slightly above the room temperature, so that the water attached to the walls of the container quickly evaporates and thus rusting is prevented.

Surimi-based products

It has been established that fish muscle proteins consist of myofibrillar, sarcoplasmic and stroma. Surimi is a type of fishery product mainly containing myofibrillar protein. The addition of cryoprotectants is necessary to avoid protein denaturation during frozen storage. Originally, fish surimi was developed in Japan. Usually, surimi is used as a raw material for processing kamaboko, fish sausages and ham (Suzuki, 1981). Nowadays, surimi is known all over the world and shows much potential for further development. There are two types of surimi commonly used in the industry, namely, *mu-en* surimi in Japanese (made without the addition of salt) and *ka-en* surimi (made with the addition of salt). For surimi Alaska pollock (*Gadus chalcogrammus*) and Pacific whiting (*Merluccius productus*) are used as the main common raw materials. Recently, other types of fish, from both fresh- and seawater, are being used to prepare surimi.

Generally, there are four basic stages in the processing of surimi, namely, washing, milling, packaging and freezing. Washing (or leaching) of fish meat is carried out two to five times. By washing, water-soluble proteins and other low molecular substances (free amino acids, odor-forming molecules such as trimethylamine) are removed. Usually, the last washing water contains 0.01% to 0.3% sodium chloride to facilitate water disposal. The water used for washing has a temperature in the range of 5–10°C (Suzuki, 1981). Washing with tap water (room temperature water) can damage the texture of products and accelerate fat oxidation, while washing with seawater can increase protein loss. However, the addition of sucrose and sorbitol can prevent protein denaturation. Addition of polyphosphates can reduce the amount of drip and shrinkage during cooking, and stabilize the emulsion.

One example of surimi products are sausages. Sausage comes from the Latin word, *salsus*, meaning literally 'meat prepared by salting'. Sausage is made by mixing ground meat and seasoning, then injection into a sausage casing. Recently, fish meat is becoming popular for making sausage products.

Retailing to optimize accessibility

The global seafood market has been expanding for a number of years and is estimated to grow at a CAGR (compound average growth rate) of 4.97% from 2018 to 2025 (Coherent Market Insight, 2018). Fish and seafood export values have increased from US$15 billion in 1980 to US$151 billion in 2020 due to high demand for seafood in the USA, EU and Japan (FAO, 2022). To date, the growth of seafood demand is faster than population growth (Merino et al., 2012) owing to the emergence of a larger proportion of middle-class people who have greater spending power and typically consume more animal proteins than those on lower incomes (Kharas, 2010). Asia-Pacific countries are the biggest seafood producers, accounting for nearly 50% of the world fishery capture and nearly 90% of the aquaculture production.

In terms of food security, seafood contributes approximately 17% of global animal proteins, omega-3 fatty acids and essential micronutrients (vitamins A, B and D, zinc, iodine, selenium, calcium, and iron). The relative contribution can be as high as 40% in regions where seafood is the main animal protein source, such as low income food-deficit countries (LIFDCs) and small island developing states (Belton and Thilsted, 2014). However, seafood can only contribute to food security provided that it is adequate, safe, sustainable and resilient to shocks and also meets the legal standards during production (Jennings et al., 2016). These critical elements can be achieved when proper management of the supply chain is realized. The seafood supply chain can be defined as a set of interdependent actors (fishers, agents, processors, distributors, wholesalers, retailers and food services) who work together to deliver seafood-derived products to the consumer (Thorpe and Bennett, 2004). Supply chains differ among countries and regions due to their socio-economic, environmental and cultural differences, and their native or introduced species, harvesting techniques and the products they produced.

Globalization of seafood trade has positively impacted the food security programme by providing sustainable seafood supply, improving resilience against supply shocks and generating positive net income from the export of high-value seafood (Gephart and Pace, 2015). For example, consumers from many developed countries still have access to seafood products from developing countries, despite seafood production in their countries being at its lowest level since the Second World War due to overexploitation. Globalization has removed the barriers for seafood trade; seafood is traded far from catching areas, and about 38% of the seafood produced is being traded internationally in different shapes and forms. Out of the internationally traded seafood, about 53% comes from developing countries, contributing US$35 billion to their net income in 2012. Improvement in transportation and logistics, as well as standardization, allows seafood to access global markets through retail sectors, which replaced fishmongers and local markets as the main outlets. For example, the supermarket share in UK increased from about 16% in 1988 to about 66% in 2001, while fishmongers' share fell from about 49% to about 18%. The increase of retail sector share in the seafood market could also be attributed, in part, to consumer convenience and economies of scale (Murray and Fofana, 2002). Improvements in storage and preservation technologies have continued, allowing a wider range of seafood products to be traded. Most importantly, freezing technology has improved to such an extent that many product forms can be frozen twice, allowing products to be processed in locations with competitive advantages in processing fish rather than in locations close to where the fish was caught. This has created the seafood processing clusters, commoditization with the global sourcing to these clusters and increase in the trade to and from them.

Farm to consumers (fresh or live)

The global demand for seafood is still dependent on fresh and live fish. It was estimated that up to 44% of fresh or live seafood is distributed globally (FAO, 2022). This number could go up to over 80% in some rural markets of the developing countries. In these countries, fresh or live fish is usually marketed for domestic consumption. For example, in Latin America over 70% of fish produced by small-scale fisheries is marketed fresh. Meanwhile in India, the proportion of fresh fish consumed domestically reaches over 95% (FAO, 2018).

Farm to market then to consumers

In artisanal fisheries, seafood distribution from the fishers often involves large numbers of intermediaries. These intermediaries include retail outlets (from locally owned fish markets to national supermarket chains), restaurants and foodservice establishments, such as hotels, hospitals and schools, to sell the seafood to a consumer. Although a fisher can bypass the supply chain and sell the seafood directly to a consumer, these intermediaries exist in the more formal markets. For example, the fish supply chain from aquaculture and sea catchment areas in Java Island, Indonesia consists of six stages (Figure 8.5); each of the stages has a different role in the supply chain (Guritno and Tanuputri, 2017). However, in a modern market where seafood is continuously traded, these supply chain

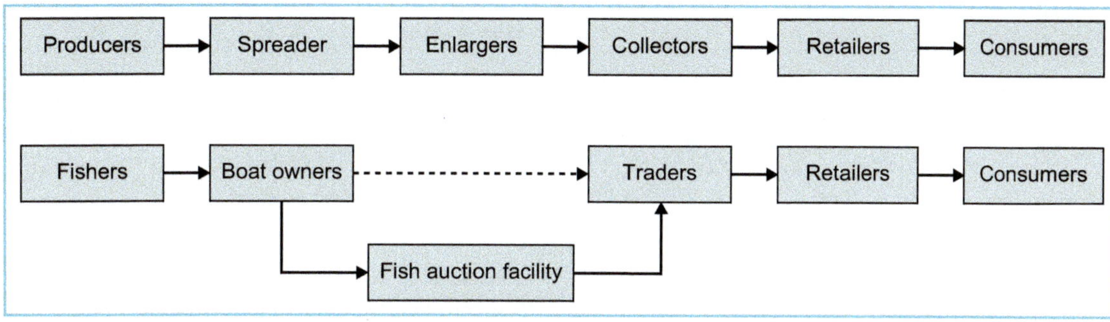

Figure 8.5 Supply chain of aquaculture (top) and capture fisheries (bottom) in Java, Indonesia (Guritno and Tanuputri, 2017).

organizations are too complicated and could delay flows. Large supermarket chains simplified the chain by removing some intermediaries, establishing efficient logistic and distribution.

Farm to processing and then to consumers

The perishability of seafood still limits distribution and trading. As mentioned above, processing preserves and improves the quality of seafood. Until the early 20th century, fish were processed in traditional ways: dried, dried salted and heavily salted fish. Nowadays, processing plants use advanced and automatic technologies, although smaller companies still rely on manual handling. More than one-half of fish production for human consumption is being processed by curing, preparing or freezing. Frozen fish form the largest portion of processed fish, accounting for more than 35% of fish, followed by preserved (11%) and cured (10%) (FAO, 2022). The growth in seafood processing for value addition has, in turn, led to more residual by-products. Fish by-products are utilized for a range of purposes including fish sausages, cakes, gelatin, sauces, pharmaceuticals, cosmetics, biodiesel, fertilizer and fish meal (animal feed).

Seafood product development

Consumer recognition of the health benefits of fresh and live seafood is driving high demand. However, in some countries, such as Japan and Nordic countries, seafood consumption among young generations is falling. Numerous studies have pointed out the main driving forces to seafood consumption are health promotion, taste, convenience and a desire for diet variety (Altintzoglou et al., 2010.; Birch et al., 2012; Christenson et al., 2017). These studies suggest that the seafood industry should prioritize providing greater convenience, improvement in product quality and maintaining the nutritional content. Thus, to improve the accessibility of seafood products to consumers, the seafood processing industry must consider product characteristics, marketing strategy and some product concepts as illustrated in Figure 8.6.

Convenience is preferred by today's consumers because of reduced leisure and cooking time. There are two products that are increasingly demanded: the first is *fresh and chilled* products that are conveniently packaged, processed and ready-to-cook, and the second is the *processed, chilled and ready-to-eat* seafood (Venugopal, 2005). According to Altintzoglou et al. (2010), consumers in Nordic countries incline towards the ready-to-cook fresh seafood, which allow them some involvement in preparation. Meanwhile, their Australian counterparts, who prefer more convenient and easy-to-prepare food,

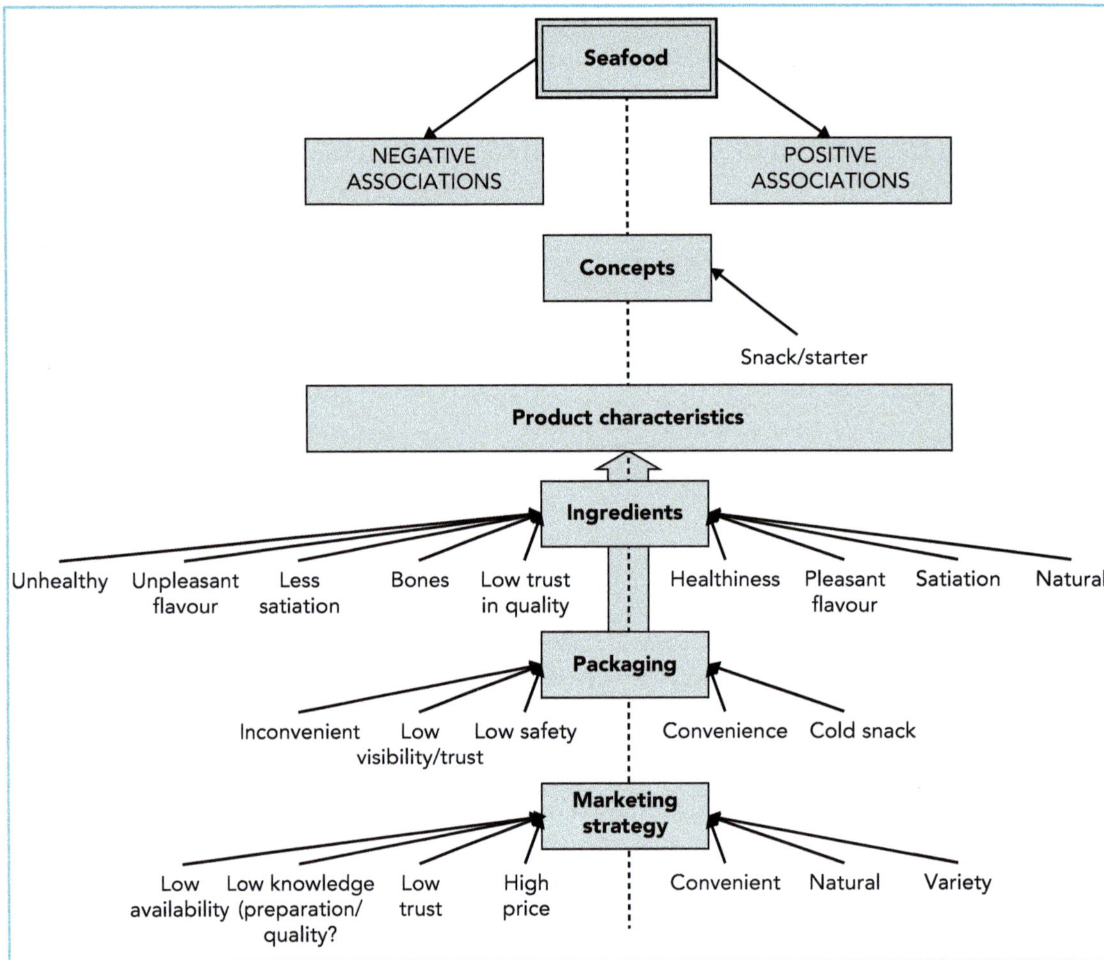

Figure 8.6 Important key characteristics of seafood product development (modified from Altintzoglou et al., 2010).

choose pre-packaged ready-to-eat seafood products. Processed products also include packaging dates and use by or best before dates informing consumers about the freshness of the products and removing the uncertainty of judging the freshness of whole fish products themselves (Birch et al., 2012).

Value-added seafood products are becoming increasingly important in satisfying consumers' demand for convenient, safe and healthy products. Value-added seafood products have been developed to break the stagnancy of market growth. These products provide an opportunity to open new markets, maintaining safety and quality at the same time. By adding value throughout processing, problems of over-supply could be solved. The breakthrough in the development of value-added products was driven by the emerging new technologies such as high-pressure processing, radio-frequency heating and flexible retort packaging. Although some demographic sectors, such as 'baby boomers' (aged 65–84) prefer still to purchase traditional seafood, younger generations, 'Gen Y'ers' (aged 17–34), are more flexible about what they consume, more adventurous and quicker to accept new ideas (Morrissey and DeWitt, 2013).

Conclusion and recommendations

Declining global fish resources and changing consumption patterns due to the changes in life style are likely to increase food insecurity in the future (McClanahan et al., 2015). Education of not only fishers but also those who are involved in processing would mitigate food insecurity. Rigorous implementation of HACCP systems would greatly help in both developed and developing countries (Young et al., 2018). HACCP systems are designed to prevent the introduction of hazards to a given food, instead of relying on the inspection of the final products when it is too late to take remedial action, resulting in wastage and hazardous produce. In the case of seafood, temperature control is the most important point through farming, processing and distribution. To achieve a higher level of food security the International Organization for Standardization (ISO) 22000 Food Safety Management System, consisting of ISO 9001 and HACCP, is being widely introduced within the seafood industry (Soares et al., 2015). The introduction of Halal requirements to seafood processing has clearly improved the quality of seafood (Riaz et al., 2018). Such a concept is also important for the utilization of by-products from fish and shellfish (Nurilmala et al., 2019). Meanwhile, the eco-labels and sustainability certifications of seafood (such as those of the Marine Stewardship Council, MSC) are now gaining traction in advanced countries (Swartz et al., 2017) with notable success in enhancing consumer awareness of the need to conserve our fisheries resources. The same concept has been introduced to farmed fish (Aquaculture Stewardship Council, ASC). However, it should be noted that these accreditation systems do not guarantee the safety of seafood.

Food allergies are becoming of greater concern as the number of people allergic to seafood is increasing. Fish and shellfish can be hazardous when the parasite *Anisakis* sp. is present (Sharp and Lopata, 2014), because the epitopes of the allergenic proteins still remain, even in processed forms. Since it is almost impossible to remove the allergenicity of food, what can be done to reduce the incidents of poisoning is to avoid the foods that contain allergic substances. The development of simple detection systems for parasites is recommended to reduce the risk of contamination. Some attempts have been made to eliminate the allergen (tropomyosin) from squid and shrimp muscles by boiling (Ozawa et al., 2017).

To conclude, action to guarantee and improve the security of seafood should be continuous. There is great scope for scientific and academic study to further knowledge and address the challenges to aquatic food security.

References

Akpambang, V.O.E., Purcaro, G., Lajide, L., Amoo, I.A., Conte, L.S. & Moret, S. (2009) Determination of polycyclic aromatic hydrocarbons (PAHs) in commonly consumed Nigerian smoked/grilled fish and meat. *Food Additives and Contaminants. Part A, Chemistry, Analysis, Control, Exposure and Risk Assessment*, 26, 1096–1103. https://doi.org/10.1080/02652030902855406.

Altintzoglou, T., Birch Hansen, K., Valsdottir, T., Øyvind Odland, J., Martinsdóttir, E., Brunsø, K. & Luten, J. (2010) Translating barriers into potential improvements: The case of new healthy seafood product development. *Journal of Consumer Marketing*, 27, 224–235. https://doi.org/10.1108/07363761011038293.

Álvarez-Muñoz, D., Rodríguez-Mozaz, S., Jacobs, S., Serra-Compte, A., Cáceres, N., Sioen, I., Verbeke, W., Barbosa, V., Ferrari, F., Fernández-Tejedor, M., Cunha, S., Granby, K., Robbens, J., Kotterman, M.,

Marques, A. & Barceló, D. (2018) Pharmaceuticals and endocrine disruptors in raw and cooked seafood from European market: Concentrations and human exposure levels. *Environment International*, 119, 570–581. https://doi.org/10.1016/j.envint.2018.07.006.

Anal, A.K. (2019) Quality ingredients and safety concerns for traditional fermented foods and beverages from Asia: A review. *Fermentation*, 5, 8. https://doi.org/10.3390/fermentation5010008.

Asche, F., Bellemare, M.F., Roheim, C., Smith, M.D. & Tveteras, S. (2015) Fair enough? Food security and the international trade of seafood. *World Development*, 67, 151–160. https://doi.org/10.1016/j.worlddev.2014.10.013.

Ashton, I.P. (2002) Understanding lipid oxidation in fish. In: *Safety and Quality Issues in Fish Processing* (edited by H. A. Bremner). Woodhead Publishing: Cambridge, pp. 254–285.

Babić, J.M., Kartalović, B.D., Škaljac, S., Vidaković, S., Ljubojević, D., Petrović, J.M., Ćirković, M.A. & Teodorović, V. (2018) Reduction of polycyclic aromatic hydrocarbons in common carp meat smoked in traditional conditions. *Food Additives and Contaminants. Part B, Surveillance*, 11, 208–213. https://doi.org/10.1080/19393210.2018.1484821.

Basak, S., Şengör, G.F. & Karakoç, F.T. (2010) The detection of potential carcinogenic PAH using HPLC procedure in two different smoked fish, case study: Istanbul/Turkey. *Turkish Journal of Fisheries and Aquatic Sciences*, 10, 351–355.

Belton, B. & Thilsted, S.H. (2014) Fisheries in transition: Food and nutrition security implications for the global South. *Global Food Security*, 3, 59–66. https://doi.org/10.1016/j.gfs.2013.10.001.

Berhimpon, S., Montolalu, R.I., Dien, H.A., Mentang, F. & Meko, A.U.I. (2018) Concentration and application methods of liquid smoke for exotic smoked skipjack (*Katsuwonus pelamis* L.). *International Food Research Journal*, 25, 1864–1869.

Birch, D., Lawley, M. & Hamblin, D. (2012) Drivers and barriers to seafood consumption in Australia. *Journal of Consumer Marketing*, 29, 64–73. https://doi.org/10.1108/07363761211193055.

CAC (Codex Alimentarius Commission) (2003) *Code of Practice for Fish and Fishery Products*. Codex Alimentarius Commission: Rome.

Christenson, J.K., O'Kane, G.M., Farmery, A.K. & McManus, A. (2017) The barriers and drivers of seafood consumption in Australia: A narrative literature review. *International Journal of Consumer Studies*, 41, 299–311. https://doi.org/10.1111/ijcs.12342.

Coherent Market Insight (2018) Global sustainable seafood market to surpass US$18.63 billion by 2025. https://www.coherentmarketinsights.com/.

Costa, Fdo N., Korn, M.G.A., Brito, G.B., Ferlin, S. & Fostier, A.H. (2016) Preliminary results of mercury levels in raw and cooked seafood and their public health impact. *Food Chemistry*, 192, 837–841. https://doi.org/10.1016/j.foodchem.2015.07.081.

Desrosier, N.W & Desrosier, J.N. (1977) *The Technology of Food Preservation*. The AVI Publishing Co. Inc.: Westport, CT.

Faithong, N., Benjakul, S., Phatcharat, S. & Binsan, W. (2010) Chemical composition and antioxidative activity of Thai traditional fermented shrimp and krill products. *Food Chemistry*, 119, 133–140. https://doi.org/10.1016/j.foodchem.2009.06.056.

FAO (2018) *The State of World Fisheries and Aquaculture 2018 – Meeting the Sustainable Development Goals*. FAO: Rome.

FAO (2022) *The State of World Fisheries and Aquaculture 2022 –Towards blue transformation*. FAO: Rome.

Fellows, P.J. (1992) *Food Processing Technology (Principles and Practice)*. Ellis Horwiid Limited: Chichester.

Fitriati, M., Nurzain, M. & Sunarya (1994). *Monitoring of Nutritional Composition of Raw Material and End Product of Pindang. A Report of NCQC Research Project*. NCQC: Jakarta.

Gephart, J.A. & Pace, M.L. (2015) Structure and evolution of the global seafood trade network. *Environmental Research Letters*, 10, 125014. https://doi.org/10.1088/1748-9326/10/12/125014.

Guritno, A.D. & Ryanjani Tanuputri, M. (2017) Comparison of fish supply chain from aquaculture and sea catchment areas. *Journal of Service Science and Management*, 10, 353–359. https://doi.org/10.4236/jssm.2017.104028.

Hajeb, P. & Jinap, S. (2012) Fermented shrimp products as source of umami in Southeast Asia. *Journal of Nutrition and Food Sciences*, S10. https://doi.org/10.4172/2155-9600.S10-006.

Hall, G.M. (2010) Introduction: Challenges to the fish-processing industry in a resource-starved world. In: *Fish Processing: Sustainability and New Opportunities* (edited by G. M. Hall). Wiley-Blackwell: Hoboken, NJ, USA, pp. 1–29.

International Agency for Research on Cancer (IARC) (1972) *IARC Monographs on the Evaluation of Carcinogenic Risks to Humans*, 1, 184.

Jennings, S., Stentiford, G.D., Leocadio, A.M., Jeffery, K.R., Metcalfe, J.D., Katsiadaki, I., Auchterlonie, N.A., Mangi, S.C., Pinnegar, J.K., Ellis, T., Peeler, E.J., Luisetti, T., Baker-Austin, C., Brown, M., Catchpole, T.L., Clyne, F.L., Dye, S.R., Edmonds, N.J., Hyder, K., Lee, J., Lees, D.N., Morgan, O.C., Oidtmann, B., Posen, P.E., Santos, A.R., Taylor, N.G.H., Turner, A.D., Townhill, B.L. and Verner-Jeffreys, D.W. (2016) Aquatic food security: insights into challenges and solutions from an analysis of interactions between fisheries, aquaculture, food safety, human health, fish and human welfare, economy and environment. *Fish and Fisheries*, 17, 893–938.

Jinap, S., Ilya-Nur, A.R., Tang, S.C., Hajeb, P., Shahrim, K. & Khairunnisak, M. (2010) Sensory attributes of dishes containing shrimp paste with different concentrations of glutamate and 5'-nucleotides. *Appetite*, 55, 238–244. https://doi.org/10.1016/j.appet.2010.06.007.

Kharas, H. (2010) The emerging middle class in developing countries. OECD Development Centre. Working Paper, no. 285.

Kibria, G., Haroon, A.K.Y. & Nugegoda, D. (2017) Climate change impacts on tropical and temperate fisheries, aquaculture, and seafood security and implications – A review. *Livestock Research for Rural Development*, 29, 1–29, article no.022.

Khairunnisak, M., Azizah, A.H., Jinap, S. & Nurul Izzah, A. (2009) Monitoring of free glutamic acid in Malaysian processed foods, dishes and condiments. *Food Additives and Contaminants. Part A, Chemistry, Analysis, Control, Exposure and Risk Assessment*, 26, 419–426. https://doi.org/10.1080/02652030802596860.

Kobayashi, T., Kimura, B. & Fujii, T. (1999) Decrease of toxicity during storage of fermented puffer fish ovaries in vitro. *Food Hygiene and Safety Science*, 40, 178–182. https://doi.org/10.3358/shokueishi.40.178.

Kose, S. & Hall, G.M. (2010) Sustainability of fermented fish-products. In: *Fish Processing: Sustainability and New Opportunities* (edited by G. M. Hall). Wiley-Blackwell: Hoboken, NJ, USA, pp. 138–166.

Kurlansky, M. (2003) *Salt: A World History*. Vintage Books: London.

Li, D., Qin, N., Zhang, L., Li, Q., Prinyawiwatkul, W. & Luo, Y. (2019) Degradation of adenosine triphosphate, water loss and textural changes in frozen common carp (*Cyprinus carpio*) fillets during storage at different temperatures. *International Journal of Refrigeration*, 98, 294–301. https://doi.org/10.1016/j.ijrefrig.2018.11.014.

Londahl, G. (1981) Refrigerated storage in fisheries. In: *FAO Fisheries Technical Paper*. Food and Agriculture Organization of the United Nations: Rome, 214, 1–5.

McClanahan, T., Allison, E.H. & Cinner, J.E. (2015) Managing fisheries for human and food security. *Fish and Fisheries*, 16, 78–103. https://doi.org/10.1111/faf.12045.

Merino, G., Barange, M., Blanchard, J.L., Harle, J., Holmes, R., Allen, I., Allison, E.H., Badjeck, M.C., Dulvy, N.K., Holt, J., Jennings, S., Mullon, C. & Rodwell, L.D. (2012) Can marine fisheries and aquaculture meet fish demand from a growing human population in a changing climate? *Global Environmental Change*, 22, 795–806. https://doi.org/10.1016/j.gloenvcha.2012.03.003.

Mizutani, T., Kimizuka, A., Ruddle, K. & Ishige, N. (1987) A chemical analysis of fermented fish products and discussion of fermented flavors in Asian cuisines. *Bulletin of the National Museum of Ethnology*, 12, 801–864.

Molin, M., Ulven, S.M., Meltzer, H.M. & Alexander, J. (2015) Arsenic in the human food chain, biotransformation and toxicology – Review focusing on seafood arsenic. *Journal of Trace Elements in Medicine and Biology: Organ of the Society for Minerals and Trace Elements*, 31, 249–259. https://doi.org/10.1016/j.jtemb.2015.01.010.

Morrissey, M. & DeWitt, C. (2013) Value-added seafood. In: *Seafood Processing: Technology, Quality and Safety* (edited by I. S. Boziaris). John Wiley & Sons, Ltd: Hoboken, NJ, USA, pp. 343–358.

Murray, A.D. & Fofana, A. (2002) The changing nature of UK fish retailing. *Marine Resource Economics*, 17, 335–339. https://doi.org/10.1086/mre.17.4.42629373.

Nurilmala, M., Pertiwi, R.M., Nurhayati, T., Fauzi, S., Batubara, I. & Ochiai, Y. (2019) Characterization of collagen and its hydrolysate from yellowfin tuna Thunnus albacares skin and their potencies as antioxidant and antiglycation agents. *Fisheries Science*, 85, 591–599. https://doi.org/10.1007/s12562-019-01303-5.

Ohshima, T. & Giri, A. (2014) Traditional fish fermentation technology and recent development. In: *Encyclopedia of Food Microbiolog.* Elsevier: Amsterdam, pp. 753–759.

Ozawa, H., Koyama, M., Ishizaki, S. & Ochiai, Y. (2017) Effect of boiling on the elution of the major allergen tropomyosin from shrimp and squid muscles. *Food Science and Technology Research*, 23, 41–43. https://doi.org/10.3136/fstr.23.41.

Özyurt, G., Gökdoğan, S., Şimşek, A., Yuvka, I., Ergüven, M. & Kuley Boga, E. (2016) Fatty acid composition and biogenic amines in acidified and fermented fish silage: A comparison study. *Archives of Animal Nutrition*, 70, 72–86. https://doi.org/10.1080/1745039X.2015.1117696.

Paludan-Müller, C., Valyasevi, R., Huss, H.H. & Gram, L. (2002) Genotypic and phenotypic characterization of garlic-fermenting lactic acid bacteria isolated from som-fak, a Thai low-salt fermented fish product. *Journal of Applied Microbiology*, 92, 307–314. https://doi.org/10.1046/j.1365-2672.2002.01544.x.

Parker, J.K., Lignou, S., Shankland, K., Kurwie, P., Griffiths, H.D. & Baines, D.A. (2018) Development of a zeolite filter for removing polycyclic aromatic hydrocarbons (PAHs) from smoke and smoked ingredients while retaining the smoky flavor. *Journal of Agricultural and Food Chemistry*, 66, 2449–2458. https://doi.org/10.1021/acs.jafc.6b05399.

Radojicic, V., Alagic, S., Adnadjevic, B. & Maktouf, A.M. (2012) Effect of varied quantities of zeolite on the reduction of polycyclic aromatic hydrocarbons in tobacco smoke. *African Journal of Biotechnology*, 11, 10041–10047.

Rajapakse, N., Mendis, E., Jung, W.K., Je, J.Y. & Kim, S.K. (2005) Purification of a radical scavenging peptide from fermented mussel sauce and its antioxidant properties. *Food Research International*, 38, 175–182. https://doi.org/10.1016/j.foodres.2004.10.002.

Raksakulthai, N. & Haard, N. (1992) Correlation between the concentration of peptides and amino acids and the flavour of fish sauce. *Asian Food Journal*, 7, 286–290.

Riaz, M.N., Shaik, R. & Chaudry, M.M. (2018) Halal production requirements for fish and seafood. In: *Handbook of Halal Food Production* (edited by M. N. Riaz & M. M. Chaudry). Taylor & Francis Group:, Abingdon.

Roschanski, N., Guenther, S., Vu, T.T.T., Fischer, J., Semmler, T., Huehn, S., Alter, T. & Roesler, U. (2017) VIM-1 carbapenemase-producing Escherichia coli isolated from retail seafood, Germany 2016. *Eurosurveillance*, 22, 17–00032. https://doi.org/10.2807/1560-7917.ES.2017.22.43.17-00032.

Saisithi, P. (1994) Traditional fermented fish: Fish sauce production. In: *Fisheries Processing* (edited by A. M. Martin). Springer: Boston, USA, pp. 111–131.

Santillo, D., Miller, K. & Johnston, P. (2017) Microplastics as contaminants in commercially important seafood species. *Integrated Environmental Assessment and Management*, 13, 516–521. https://doi.org/10.1002/ieam.1909.

Sathivel, S., Bechtel, P.J., Babbitt, J., Smiley, S., Crapo, C., Reppond, K.D. & Prinyawiwatkul, W. (2003) Biochemical and functional properties of herring (*Clupea harengus*) byproduct hydrolysates. *Journal of Food Science*, 68, 2196–2200. https://doi.org/10.1111/j.1365-2621.2003.tb05746.x.

Sefa-Dedeh, S. (1993) Traditional food technology. In: *Encyclopaedia of Foods, Food Technology and Nutrition* (edited by R. Macrae, F. Robinson & M. Sadler). Academic Press: New York, USA, pp. 460–466.

Sharp, M.F. & Lopata, A.L. (2014) Fish allergy: In review. *Clinical Reviews in Allergy and Immunology*, 46, 258–271. https://doi.org/10.1007/s12016-013-8363-1.

Simon, R., de la Calle, B., Palme, S., Meier, D. & Anklam, E. (2005) Composition and analysis of liquid smoke flavouring primary products. *Journal of Separation Science*, 28, 871–882. https://doi.org/10.1002/jssc.200500009.

Skåra, T., Axelsson, L., Stefánsson, G., Ekstrand, B. & Hagen, H. (2015) Fermented and ripened fish products in the northern European countries. *Journal of Ethnic Foods*, 2, 18–24. https://doi.org/10.1016/j.jef.2015.02.004.

Soares, N.F., Vicente, A.A. & Martins, C.M.A. (2015) *Food Safety in the Seafood Industry: A Practical Guide for ISO 22000 and FSSC 22000 Implementation*. John Wiley & Sons, Ltd: Hoboken, NJ, USA.

Suzuki, T. (1981) *Fish and Krill Protein: Processing Technology*. Applied Science: London.

Swartz, W., Schiller, L., Rashid Sumaila, U.R. & Ota, Y. (2017) Searching for market-based sustainability pathways: Challenges and opportunities for seafood certification programs in Japan. *Marine Policy*, 76, 185–191. https://doi.org/10.1016/j.marpol.2016.11.009.

Thorpe, A. & Bennett, E. (2004) Market-driven international fish supply chains: The case of Nile perch from Africa's Lake Victoria. *International Food and Agribusiness Management Review*, 7, 40–57.

Venugopal, V. (2005). *Seafood Processing: Adding Value Through Quick Freezing, Retortable Packaging, Cook-Chilling, and Other Methods*. Taylor & Francis: Abingdon, UK.

Verdos, G.I., Makrigiannis, A., Tsigaras, E. & Boziaris, I.S. (2019) Survival of food-borne bacterial pathogens in traditional Mediterranean anchovy products. *Journal of Food Safety*, 39, e12576. https://doi.org/10.1111/jfs.12576.

Yeap, S.E. & Tan, S.M. (2003) Issues facing the traditional fish products industry in Southeast Asia. *JIRCAS International Symposium Series. Japan*, 11, 115–121.

Yoshida, Y. (1998) Umami taste and traditional seasonings. *Food Reviews International*, 14, 213–246. https://doi.org/10.1080/87559129809541158.

Young, K.R., Kamat, M.R. & Hoskin, G.P. (2018) HACCP, seafood, and the US. In: *Seafood and Environmental Toxins Foodborne Disease Handbook* (edited by Y. H. Hui). Food and Drug Administration. Taylor & Francis: Abingdon, UK.

Zelinkova, Z. & Wenzl, T. (2015) The occurrence of 16 EPA PAHs in food – A review. *Polycyclic Aromatic Compounds*, 35, 248–284. https://doi.org/10.1080/10406638.2014.918550.

Zeng, X., Xia, W., Jiang, Q. & Yang, F. (2013) Chemical and microbial properties of Chinese traditional low-salt fermented whole fish product Suan yu. *Food Control*, 30, 590–595. https://doi.org/10.1016/j.foodcont.2012.07.037.

9

Raw seafood practices, risk and public health in Southeast Asia

Fiona Harris

Introduction

Seafood is a critical and integral aspect of diet and nutrition across Southeast Asia, where both farmed and captured seafood are an important contributor to food security. However, food practices associated with seafood consumption carry known risks to human health. 'Food practices' is the term used by social scientists to encapsulate the social, economic and political contexts of how food is used, prepared and consumed by humans. Rather than simply focusing on food safety or the more encompassing term 'food security', social and behavioural scientists take account of the social and cultural contexts of food production and consumption, going beyond issues of nutrition and diet to explore what different foods mean to those who eat them. As public health scientists are all too aware, in order to implement successful behaviour change interventions targeting unhealthy behaviours, it is important to first understand the drivers for the behaviour and what might be amenable to change. This is also the case for behaviour linked to seafood consumption: referred to in this chapter as 'seafood practices'.

Diet and nutrition have long been a focus of public health initiatives, but even where a behaviour is known to carry risk, this does not necessarily lead to a change in behaviour as decades of smoking cessation interventions clearly illustrate. It is important therefore to contextualize food practices and understand what particular cuisines mean to those who eat them, before addressing any known health risk.

The anthropology of food

Growing, hunting, gathering or catching food and then preparing it are bound up with processes of engagement with the natural environment; in anthropological terms, this is understood as domesticating 'nature'. Indeed, food preparation (whether this involves heat or other processes) is both a way of taking food from the source to the plate and also a means of applying 'culture' to 'nature' (Lévi-Strauss, 1997). The act of sharing food creates and strengthens familial and community relationships, establishing micro economies of reciprocal obligations (Sahlins, 2004). Food provides a language to

Margaret Crumlish and Rachel Norman (eds) **Aquatic Food Security**
DOI: 10.52517/9781789183368.009, © 5m Books Ltd, 2023

be read: certain foods may indicate social position/wealth, eating together confirms relationships and so on. Thus, by exploring food practices, we can learn a lot about what it means to be a person in different parts of the world. Indeed, as Counihan (1999: 17) puts it: 'Food offerings connect the living and the dead, humans and their gods, neighbors and kin, and family members.'

Food practices are also embedded in history. If we were to explore the genealogy of a family recipe, it could tell a story of relatedness in which the knowledge of how to prepare the dish is passed lineally between generations and horizontally across both family and friendship networks. Food can thus become a metaphor for kinship and relatedness, travelling through both history and geographical spaces. Furthermore, food preparation (which includes the notion of both cooking with the application of heat as well as other forms of preparation) may become strongly associated with geographic regions, language groups or ethnicities, thus representing 'heritage cuisines' or what have been called 'edible identities' (Brulotte and Di Giovine, 2016). Indeed, the production and consumption of food is often bound up with social change, identity, ritual and politics as well as food security (Mintz and Du Bois, 2002), as I shall explain further below. As public health scientists would readily recognize, particular foodstuffs can be linked to morality: to notions of 'good' food and 'bad' food (Schwarz, 2018) that may bear no relation to nutritional properties. Furthermore, Schwarz explains that the sharing of 'good' food expresses emotion and 'caring' for those that it is shared with.

As contributors to this volume have noted, some seafood practices carry a risk to human health, particularly raw or undercooked seafood, which for the sake of brevity I shall refer to as raw seafood. Various food practices associated with raw seafood are found across Southeast Asia. For instance, *yu sheng* is a raw seafood dish generally associated with Chinese New Year and enjoyed in Singapore, Malaysia and Indonesia, if not more widely, with recipes readily available on the internet (https://www.nyonyacooking.com/recipes/yu-sheng-prosperity-raw-fish-salad~rk-AwvoDf5ZQ). This is a salad dish consisting of raw slices of fish placed among colourful strips of vegetables: an example of seafood prepared without the application of heat and considered to be a heritage cuisine of those who identify themselves as ethnically Chinese. Other examples include the raw fish dishes eaten in Thailand, such as *koi pla*, a raw fish dish served with condiments and rice, and fermented fish dishes, such as *pla som* and *pla ra* (Kaewpitoon et al., 2008). My own doctoral research in Sarawak, Malaysia, confirms that variations on raw seafood practices including fresh fish dishes marinaded in lime juice and chillies (*umai*), or preserved fish (*kasom ikien*) are enjoyed across this part of Borneo, if not more widely in island Southeast Asia (Harris, 2002).

The consumption of raw seafood is widespread in Southeast Asia and although there have been decades of research and public health interventions aimed at reducing or preventing human infections related to this, these food practices appear not to be amenable to change (Sripa and Echaubard, 2017). The recently detected threat to public health posed by a hypervirulent strain of Group B Streptococcus (ST283) was also linked to the consumption of raw seafood (Barkham et al., 2019; Chau et al., 2017), suggesting a need to learn more about the social and cultural contexts behind these food practices. Bringing an anthropology of food perspective to bear provides a different disciplinary lens to these public health issues.

Raw seafood and public health

A joint workshop conducted by the World Health Organization in collaboration with the Food and Agriculture Office of the United Nations (WHO, 2004) noted in their report that raw fish

Figure 9.1 Sharing food during the harvest festival.

consumption in Asia is common and is associated with a high risk of human infection with fish-borne zoonotic trematodes that cause liver and intestinal diseases. Food safety is a critical component of food security, and the mechanisms of human health risks related to the consumption of raw seafood are poorly understood in Southeast Asia. The main studies to date focus on liver fluke infections (Sripa and Echaubard, 2017) that are found in Northeast Thailand, Lao PDR, Vietnam and Cambodia (Kaewpitoon et al., 2008). Liver fluke infection has been linked to the consumption of undercooked fish products and there is a high prevalence of infection in Southeast Asia. Epidemiological studies of the common Southeast Asian liver fluke, *Opisthorchis viverrine*, a main cause of liver fluke infections in these countries, found prevalences of infection of 85% in parts of Cambodia, 78% in a survey of villages in northeast Thailand with an estimated 6 million people infected in Thailand alone. Liver fluke infection is associated with hepatobiliary diseases including cholangiocarcinoma (CCA), a cancer of the liver (Sripa, 2008). Indeed, CCA has been identified as being responsible for 86.5% of all cancers in one of the regions of northeast Thailand; the highest incidence in the world (Sripa, 2008). In Khon Khaen province, northeast Thailand, there are more than 30,000 annual cases of CCA, which are usually detected at a point when end-of-life care is the only option (Asavarut et al., 2016).

Despite the risks associated with liver fluke infection being common knowledge in these communities, raw seafood practices continue. It is therefore necessary to understand the socio-cultural contexts for this health behaviour in order to determine the barriers to change. Some of these reasons include the politics of identity, poverty and food scarcity, and attitudes towards risk. I shall explore these points with reference to the literature from mainland Southeast Asia, since the majority of social science-oriented research on liver fluke control is centred on northeast Thailand.

The politics of identity in northeast Thailand

Those who live in northeast Thailand are referred to as the Isan or *Khon Isan*, which literally means people of the northeast. Although constituting around one-third of the population of Thailand, the Isan region is the poorest and has an ethnic minority status along with a self-perceived marginality and disadvantage compared with the people living in central Thailand (Keyes, 1967). Historically, the Isan region belongs to those 'upland' areas in Southeast Asia that have been described as being on the margins of lowland, centres of power which today continue to generate complex ethno-regional differences (Keyes, 1995). The contemporary situation is significantly more nuanced than a polarized us–them identity, but regional differences are nevertheless clear in terms of poverty compared with central Thailand (particularly Bangkok) and clear language differences since the majority of the Isan population speak a version of Lao, among other languages. However, ethnic identity is complex in this region and as in other parts of the world, there is no contradiction between claiming both 'Isan-ness' and being Thai at the same time or for different purposes. What is salient here is the perceived difference, including an awareness that they are regarded by the Thais of Bangkok, for instance, to be provincial and subordinate (McCargo and Hongladarom, 2004).

In the early 1900s, there were rebellions and resistance to the power of the Thai state that continued sporadically in various forms over the ensuing decades. There is not space here to do justice to the complex history and relationship between Isan and the Thai state, save to emphasize that there remains a politics of identity in which Isan people themselves switch between Thai and Isan identities and languages, according to context, where a Thai identity is regarded as potentially conferring higher social status (McCargo and Hongladarom, 2004).

This complex historical and political landscape underpins the public health interventions to date in the region. Public health interventions have thus far met with little success and have failed to reduce the prevalence of liver fluke infection in northeast Thailand (Sripa and Echaubard, 2017). It has been argued that one of the main barriers to success has been the fact that the public health campaigns were regarded by their target audience as paternalistic, drawing on a 'condescending and top-down' approach (Asavarut et al., 2016). Asavarut and colleagues suggest the need to provide messages in local dialects rather than the Thai language as one means of engaging local communities. There is no doubt that these were well-intended campaigns that represented important efforts to address this risky behaviour, however, it seems that these public health campaigns may not have been informed by what these raw seafood dishes mean to those who consume them. As a regional, traditional cuisine they are examples of 'edible identities' (Brulotte and Di Giovine, 2016) and continuing to eat these foods may constitute subtle acts of resistance to health messages emanating from those perceived as more powerful neighbours.

Domestic economies and food poverty

One of the issues that public health campaigns may fail to take account of is the importance in the Isan region of fish and the popular, raw seafood dishes that are made from them. While large fish are consumed or sold soon after capture, small fish such as *cyprinids* (fish of the carp family) are preserved and stored and can provide a food resource during times of food scarcity. However, these are the species most likely to be infected with zoonotic liver fluke when prepared without the application of heat.

As noted above, food sharing is important to domestic economies as well as for strengthening family, neighbourhood and community networks (Counihan, 1999; Delormier et al., 2009; Errington and Gewertz, 2008). The sharing of cyprinids between households is common in northeast Thailand (Phimpraphai et al., 2018; Sripa and Echaubard, 2017) if not more widely in the region, where food sharing is a valued activity. A social network analysis of cyprinid fish sharing in northeast Thailand found this to be a common practice, where both fish and prepared dishes such as *pla som* (a fermented fish dish) were shared between households in areas with high rates of infection with the liver fluke, *O. viverrini*, causing liver fluke infection (Phimpraphai et al., 2018). However, this food sharing was important to members of the community who were disabled or elderly and unable to reciprocate. Therefore, a potentially risky food source is also associated with issues of reciprocity, community resilience and a moral economy of food practices. Indeed, in this region of Thailand fish is an important source of protein that is cheap and readily available in the lakes and rivers. The popularity of these raw fish dishes is illustrated by the fact that they are also found for sale pre-prepared in local markets (Phimpraphai et al., 2018). For many of the people in the Mekong Delta region where these fish are eaten, food security is at times a real and pressing issue, therefore simply advocating avoidance of these food practices (that include preserved fish) has met with little success.

Attitudes towards risk and beliefs

The attribution of meaning to risk, its perception and management is complex and multifaceted (Muñoz et al., 2019; Pumarega et al., 2017) and it is important to gain an understanding of Isan risk perception within their local communities. Risk perceptions change over time, are context dependent (Stevenson et al., 2019) and it is the context-dependent character of risk assessment that is the determining aspect (Renn, 2008). Knowing about a risk does not necessarily translate into acceptance of that risk as salient to a particular individual, since acceptance of risk is based on values, beliefs and social interactions. The identification of consumption behaviours and attitudes toward raw seafood and its associated risks is a crucial precursor to future public health interventions. The beliefs and practices surrounding eating raw seafood are as important as the questions of how it is consumed, who consumes it and for what purpose (Holtzman, 2010). In northeast Thailand, the perceived risk of contracting cancer is regarded as neither an immediate nor certain threat, since the time between consumption of raw seafood and the diagnosis of CCA may be separated by many years and community members are aware that all of those who eat raw fish do not necessarily become ill (Phimpraphai et al., 2018).

The anthropology of food also reveals that *belief* is an important aspect of attitudes towards food and food practices. For instance, anthropologists have explored the prohibitions or taboos on eating particular foods that are often cultural markers of particular religions or ethnicities, such as the prohibition on eating pork among Jews and Muslims (Counihan, 1999; Douglas, 2003), or in not eating certain foods during pregnancy or in the postpartum period (Laderman, 1987). However, beliefs might also affect attitudes towards risk as a recent qualitative study of raw seafood practices in northeast Thailand found. Here 'kamma' (sic) (i.e. karma or a belief in a pre-ordained destiny/fate) was implicated in a lack of interest in health risk and was identified as a key factor in the reluctance to change raw seafood practices (Samiphak and Syme, 2017). This suggests the complexity of risk perception may be bound up with religious beliefs in a pre-ordained fate or destiny as well as the passage of time.

Raw seafood practices and global public health concerns

While there is a growing body of literature associated with raw seafood consumption and cancer risk, other risks linked to these food practices have been identified, including a new and hypervirulent form of a bacterial foodborne disease (Group B Streptococcus) as well as other parasitic infections such as gnathostomiasis, which is a food-borne parasitic infection. Given the globalization of the food industry, live fish exports and a growing appreciation of Southeast Asian cuisines (including raw seafood dishes), suggest that these issues may become a more global concern.

In 2015, a public health risk was identified in Singapore related to the consumption of raw fish contaminated by a unique hyper-virulent strain of a bacterium called Group B Streptococcus ST283 (Tan et al., 2016). This particular strain of GBS has been identified as causing bacterial sepsis, including: meningitis, septic arthritis and other conditions associated with bacteraemia infections in people who consumed contaminated seafood products (Barkham et al., 2019; Chau et al., 2017; Kalimuddin et al., 2017). These conditions can be fatal, or have life-altering physical sequalae, such as loss of limbs or sensory function (e.g. hearing). Furthermore, it has been established that GBS ST283 is able to cause disease in fish as well as people across a more diverse geographical landscape in Southeast Asia than previously thought (Barkham et al., 2019; Chau et al., 2017).

The scale of this emerging foodborne risk to human health is unknown, since routine healthcare data commonly collected in high-income countries is not available to support epidemiological research in low- and middle-income countries (LMICs). It is worth noting that the practice of raw seafood consumption is not confined to Southeast Asia as illustrated by the wider popularity of, for instance, *sushi*, *sashimi* (raw fish dishes) and *ceviche* (marinaded fish). It is not known whether

GBS ST283 poses a threat to those eating these other cuisines as it has not yet been established what sources or types of fish are most likely to be infected or become contaminated.

It is also becoming increasingly apparent that with globalization and the rapid increase in foreign travel, food practices that were hitherto unique to certain regions and ethnicities are now travelling far beyond regional borders. This has led to the global import of live fish from Southeast Asia to supply the emerging demand for raw seafood dishes (Diaz, 2015). A review of gnathostomiasis (Diaz, 2015) found evidence that these fish imports carried with them the nematodes or worms that caused infection in humans, which raises the question if GBS ST283 might also be exported along with live fish hosts to other parts of the world.

Transdisciplinarity, citizen science and new directions for food safety

The evidence presented here suggests a need to situate an understanding of any public health risk within wider determinants of health, in order to develop interventions that might be both feasible and acceptable to target populations. Rather than a single disciplinary perspective, a multidisciplinary socio-ecological approach is required, that takes into account the needs of rural (often poor) communities (Sripa and Echaubard, 2017).

The practice of preparing and eating raw fish is linked to ethno-regional identities and is likely to be a food practice that will be resistant to change, given the politics of identity discussed above. The potential risk to human health posed by the consumption of raw fish is bound up with both identity, community resilience and food security, which poses a considerable challenge to public health intervention.

There is a need to better understand how individuals perceive and act on risk. For instance, common understandings of food safety across the Southeast Asian region include the notion that food can be made 'safe' by the addition of health-promoting/protecting additives such as ginger, lime juice, garlic, salt and chilli. While there is some scientific evidence that these additives have antimicrobial properties (Ankri and Mirelman, 1999; Islam et al., 2014; Wijnker et al., 2006), it remains uncertain whether they would kill harmful bacteria or the liver fluke parasites, thus making it 'safe or safer'. It is also unknown what ratio of additive to the raw fish preparation might be required to achieve the perceived protective effect. There is a need to explore these practices, indeed, to acknowledge folk models of biology and bacteriology to determine whether food additives, such as ginger and lime juice, might protect against infection and provide an acceptable way to address the potentially lethal or life limiting consequences of raw seafood consumption. Thus, a form of participatory citizen science might be one route towards exploring, understanding and finding ways of mitigating or eradicating the risk associated with raw food practices.

Questions remain as to whether the GBS risk might also be associated with the preparation and handling of infected fish, as in the case of gnathostomiasis (Daengsvang, 1981) or whether infection is via ingestion as part of food preparation. Therefore prior to the development of public health interventions, further research is required to identify all sources of risk and how to mitigate these.

Work across disciplines is required to address the emerging public health risk associated with raw fish consumption, with a need to draw on the natural, clinical and social sciences. A horizontal interdisciplinarity, which includes multiple disciplines, would be enhanced by a vertical approach to include engagement with community and policy-relevant stakeholders (Frodeman, 2013). Indeed, it has been argued that a transdisciplinary approach goes a step further and may support a more holistic view of a complex problem (Klein, 2004). A transdisciplinary approach draws together multiple disciplines along with the local knowledge and expertise of community members from the affected populations. Rather than simply viewing community members as research participants, instead they should be involved as partners in the research. Any public health initiative responding to the threat to human health presented by GBS ST283 must be developed in partnership with local communities, indigenous scientists and policy makers in order to achieve any likelihood of success.

References

Ankri, S. & Mirelman, D. (1999) Antimicrobial properties of allicin from garlic. *Microbes and Infection*, 1, 125–129. https://doi.org/10.1016/s1286-4579(99)80003-3.

Asavarut, P., Norsworthy, P.J., Cook, J., Taylor-Robinson, S.D. & Harrison, R.V. (2016) Diet and disease: Transgressing boundaries between science and society—Understanding neglected diseases through the lens of cultural studies and anthropology. *Medical Humanities*, 42, 181–183. https://doi.org/10.1136/medhum-2016-010893.

Barkham, T., Zadoks, R.N., Azmai, M.N.A., Baker, S., Bich, V.T.N., Chalker, V., Chau, M.L., Dance, D., Deepak, R.N., van Doorn, H.R., Gutierrez, R.A., Holmes, M.A., Huong, L.N.P., Koh, T.H., Martins, E., Mehershahi, K., Newton, P., Ng, L.C., Phuoc, N.N., Sangwichian, O., Sawatwong, P., Surin, U., Tan, T.Y., Tang, W.Y., Thuy, N.V., Turner, P., Vongsouvath, M., Zhang, D., Whistler, T. & Chen, S.L. (2019) One hypervirulent clone, sequence type 283, accounts for a large proportion of invasive Streptococcus agalactiae isolated from humans and diseased tilapia in Southeast Asia. *PLOS Neglected Tropical Diseases*, 13, e0007421. https://doi.org/10.1371/journal.pntd.0007421.

Brulotte, R.L. & Di Giovine, M.A. (2016). *Edible Identities: Food as Cultural Heritage*. Routledge: Abingdon, UK.

Chau, M.L., Chen, S.L., Yap, M., Hartantyo, S.H.P., Chiew, P.K.T., Fernandez, C.J., Wong, W.K., Fong, R.K., Tan, W.L., Tan, B.Z.Y., Ng, Y., Aung, K.T., Mehershahi, K.S., Goh, C., Kang, J.S.L., Barkham, T., Leong, A.O.K., Gutiérrez, R.A. & Ng, L.C. (2017) Group B Streptococcus infections caused by improper sourcing and handling of fish for raw consumption, Singapore, 2015–2016. *Emerging Infectious Diseases*, 23, 2002–2010. https://doi.org/10.3201/eid2312.170596.

Counihan, C. (1999) *The Anthropology of Food and Body: Gender, Meaning, and Power*. Psychology Press: Abingdon, UK.

Daengsvang, S. (1981) Gnathostomiasis in Southeast Asia. *Southeast Asian Journal of Tropical Medicine and Public Health*, 12, 319–332.

Delormier, T., Frohlich, K.L. & Potvin, L. (2009) Food and eating as social practice–understanding eating patterns as social phenomena and implications for public health. *Sociology of Health and Illness*, 31, 215–228. https://doi.org/10.1111/j.1467-9566.2008.01128.x.

Diaz, J.H. (2015) Gnathostomiasis: An emerging infection of raw fish consumers in Gnathostoma nematode-endemic and nonendemic countries. *Journal of Travel Medicine*, 22, 318–324. https://doi.org/10.1111/jtm.12212.

Douglas, M. (2003). *Purity and Danger: An Analysis of Concepts of Pollution and Taboo*. Routledge: Abingdon, UK.

Errington, F. & Gewertz, D. (2008) Pacific Island gastrologies: Following the flaps. *Journal of the Royal Anthropological Institute*, 14, 590–608. https://doi.org/10.1111/j.1467-9655.2008.00519.x.

Frodeman, R. (2013) *Sustainable Knowledge: A Theory of Interdisciplinarity*. Springer: Dordtrecht.

Harris, F.M. (2002) Growing gods: Bidayuh processes of religious change in Sarawak, Malaysia. University of Edinburgh. PhD thesis.

Holtzman, J. (2010) Remembering bad cooks: Sensuality, memory, personhood. *Senses and Society*, 5, 235–243. https://doi.org/10.2752/174589210X12668381452881.

Islam, K. et al. (2014) Antimicrobial activity of ginger (Zingiber officinale) extracts against food-borne pathogenic bacteria. *International Journal of Science, Environment and Technology*, 3, 867–871.

Kaewpitoon, N., Kaewpitoon, S.J., Pengsaa, P. & Sripa, B. (2008) Opisthorchis viverrini: The carcinogenic human liver fluke. *World Journal of Gastroenterology*, 14, 666–674. https://doi.org/10.3748/wjg.14.666.

Kalimuddin, S., Chen, S.L., Lim, C.T.K., Koh, T.H., Tan, T.Y., Kam, M., Wong, C.W., Mehershahi, K.S., Chau, M.L., Ng, L.C., Tang, W.Y., Badaruddin, H., Teo, J., Apisarnthanarak, A., Suwantarat, N., Ip, M., Holden, M.T.G., Hsu, L.Y. & Barkham, T. (2017) 2015 Epidemic of Severe *Streptococcus agalactiae* Sequence Type 283 Infections in Singapore Associated With the Consumption of Raw Freshwater Fish: A Detailed Analysis of Clinical, Epidemiological, and Bacterial Sequencing Data, *Clinical Infectious Diseases*, 64, S145–S152, https://doi.org/10.1093/cid/cix021.

Keyes, C.F. (1967) ISAN: Regionalism in northeastern Thailand. Cornell Thailand Project. Interim Reports Series. Number Ten.

Keyes, C.F. (1995) *The Golden Peninsula: Culture and Adaptation in Mainland Southeast Asia*. University of Hawaii Press: Honolulu.

Thompson Klein, J.T. (2004) Prospects for transdisciplinarity. *Futures*, 36, 515–526. https://doi.org/10.1016/j.futures.2003.10.007.

Laderman, C. (1987). *Wives and Midwives: Childbirth and Nutrition in Rural*. University of California Press: Malaysia.

Lévi-Strauss, C. (1997) The culinary triangle. In: *Food and Culture: A Reader*. Psychology Press: Abingdon, UK, pp. 28–35.

McCargo, D. & Hongladarom, K. (2004) Contesting Isan-ness: Discourses of politics and identity in Northeast Thailand. *Asian Ethnicity*, 5, 219–234. https://doi.org/10.1080/1463136042000221898.

Mintz, S.W. & Du Bois, C.M. (2002) The anthropology of food and eating. *Annual Review of Anthropology*, 31, 99–119. https://doi.org/10.1146/annurev.anthro.32.032702.131011.

Muñoz, A., Fontalba-Navas, A., Arrebola, J.P. & Larrea-Killinger, C. (2019) Trust and distrust in relation to food risks in Spain: An approach to the socio-cultural representations of pregnant and breastfeeding women through the technique of free listing. *Appetite*, 142, 104365. https://doi.org/10.1016/j.appet.2019.104365.

Phimpraphai, W., Tangkawattana, S., Kasemsuwan, S. & Sripa, B. (2018) Social influence in liver fluke transmission: Application of social network analysis of food sharing in Thai Isaan culture. *Advances in Parasitology*, 101, 97–124. https://doi.org/10.1016/bs.apar.2018.05.004.

Pumarega, J., Larrea, C., Muñoz, A., Pallarès, N., Gasull, M., Rodríguez, G., Jariod, M. & Porta, M. (2017) Citizens' perceptions of the presence and health risks of synthetic chemicals in food: Results of an online survey in Spain. *Gaceta Sanitaria*, 31, 371–381. https://doi.org/10.1016/j.gaceta.2017.03.012.

Renn, O. (2008). *Risk Governance: Coping with Uncertainty in a Complex World.* EarthScan: Abingdon, UK.

Sahlins, M. (2004) *Stone Age Economics*. Routledge: London.

Samiphak, S. & Syme, S.L. (2017) The battle of worldviews: A case study of liver fluke infection in Khon Kaen, Thailand. *Journal of Evidence-Based Complementary and Alternative Medicine*, 22, 902–908. https://doi.org/10.1177/2156587217723497.

Schwarz, C. (2018) Eating morality: Food and the goodness of care in northern aboriginal Australia. *Asia Pacific Journal of Anthropology*, 19, 19–34. https://doi.org/10.1080/14442213.2017.1394363.

Sripa, B. (2008) Concerted action is needed to tackle liver fluke infections in Asia. *PLOS Neglected Tropical Diseases*, 2, e232. https://doi.org/10.1371/journal.pntd.0000232.

Sripa, B. & Echaubard, P. (2017) Prospects and challenges towards sustainable liver fluke control. *Trends in Parasitology*, 33, 799–812. https://doi.org/10.1016/j.pt.2017.06.002.

Stevenson, M., Savage, B. & Taylor, B.J. (2019) Perception and communication of risk in decision making by persons with dementia. *Dementia*, 18, 1108–1127. https://doi.org/10.1177/1471301217704119.

Tan, H.S.G., van den Berg, E. & Stieger, M. (2016) The influence of product preparation, familiarity and individual traits on the consumer acceptance of insects as food. *Food Quality and Preference*, 52, 222-231, https://doi.org/10.1016/j.foodqual.2016.05.003.

WHO (2004) Joint WHO/FAO Workshop on Foodborne Trematode Infections in Asia.

Wijnker, J.J., Koop, G. & Lipman, L.J. (2006) Antimicrobial properties of salt (NaCl) used for the preservation of natural casings. *Food Microbiology*, 23, 657–662. https://doi.org/10.1016/j.fm.2005.11.004.

10

Aquatic food safety

Marine Furhmann, Joy Becker and Ruth N. Zadoks

Introduction

Aquatic animals play a vital role in global food security by providing nearly 17% of animal protein consumed by people. In 2016, the food supply from aquatic animals reached 171 mt valued at US$362 billion (FAO, 2018). Aquaculture contributes about one-half of this tonnage but represents nearly two-thirds of the commercial value. China is by far the world's largest aquatic animal consuming country with per capita consumption of 41 kg compared with the rest of the world at 15.5 kg per year (FAO, 2018). Nearly 40% of the world's supply of aquatic animals is consumed in China (FAO, 2018).

Aquatic food includes products of animal and non-animal origins (e.g. macroalgae or seaweed), where finfish, crustaceans and molluscs contribute the larger proportion of aquatic animals. Others, for example, turtles, sea cucumbers, sea urchins, frogs and jellyfish, do contribute, but to a lesser extent with an approximate 938,500 t of food annually (FAO, 2018). The jellyfish market resides mainly in Asia, but consumer interest is emerging in Australia, India, the USA and Europe (Hsieh et al., 2000) and the demand for reptile meat is increasing (EFSA, 2007). There is also a growing global demand for macro- and microalgae, which are mainly produced for human consumption, but also for use in pharmaceuticals, cosmetics and animal feeds, with the potential to reduce greenhouse gas emissions by ruminants (Abbott et al., 2020; Wells et al., 2017). In this chapter, we mostly focus on finfish, molluscs, echinoderms, crustaceans, cephalopods and aquatic plants. Consideration is also given where appropriate to jellyfish, amphibians and reptiles, including species from freshwater, estuarine and marine environments.

The consumption of aquatic food is associated with human health benefits including reducing the risk of heart and inflammatory diseases, depression and anxiety (Tacon and Metian, 2013). In addition to protein, aquatic species provide essential amino acids, polyunsaturated fatty acids (PUFA), vitamins, minerals, trace elements and other compounds that could be beneficial for human health (Bushkin-Bedient and Carpenter, 2010; Tacon and Metian, 2013). PUFAs are components of cell membranes and play a crucial role in maintaining cellular functions, which are essential to humans, and they are only available from our diet (Bushkin-Bedient and Carpenter, 2010). To derive maximum nutritional and health benefits from aquatic food, products must be safe for consumption and not cause food-borne diseases.

Margaret Crumlish and Rachel Norman (eds) **Aquatic Food Security**
DOI: 10.52517/9781789183368.010, © 5m Books Ltd, 2023

In the Anthropocene, aquatic food safety is facing new challenges as pressure on waterways and oceans is increasing. Aquatic ecosystems and their inhabitants are subjected to changes, such as increased water temperature, ocean acidification and pollution by anthropogenic chemical and biological waste, including sewage and agricultural run-off (Landrigan et al., 2020 Lloret et al., 2016). It is paramount that both the producer and consumer stay informed of the potential risks to ensure safe and nutritious aquatic food supply and consumption. A further challenge is the risk of spreading food-borne diseases with the expansion of the international food trade and changes in consumer preferences. For the context of this chapter, the definition of food safety was adapted from Jennings et al. (2016), that is, a food supply is considered safe when contamination with biological (e.g. human pathogens, biotoxins), chemical (e.g. heavy metals, pesticides, veterinary drug residues) or physical hazards (e.g. metals, plastic) is within safe limits based on scientific evidence, and when aquatic products are harvested, processed and stored in ways that avoid contamination and growth of pathogens. Generally, consumers of raw molluscs have the greatest risk to contract food-borne diseases due to the uncooked nature of the product (i.e. no heat treatment) and environmental contamination with human pathogens. However, there are food safety risks associated with aquatic food from all sectors (e.g. recreational fishing, commercial wild-caught fish and aquaculture) and with consuming both fresh and frozen products.

This chapter will focus on biological, chemical and physical hazards from water to consumer. To minimize the risks associated with these hazards and to facilitate the international trade of aquatic food, there is a need for harmonization of procedures and regulations including monitoring, surveillance and traceability. Examples of control measures will be given from the pre-harvest to post-harvest stages. Tools to educate and inform consumers will also be discussed, as well as socio-economic facilitators or barriers to uptake of practices to reduce aquatic food safety risks.

From the water to the plate: where is food safety at risk?

Aquatic food production

A multitude of steps separate the growth of aquatic organisms in the water from the plate or bowl of the consumers. Wild fish and shellfish are harvested by fisheries or recreational fishers in lakes, rivers, near shore or in the open ocean, while aquaculture produces a variety of aquatic organisms on land in ponds, cages, happas and recirculating tanks, in rivers and lakes, or near shore in coastal waters (Hicks, 2016) (Figure 10.1a). Aquatic organisms are inherently linked to their environment (Bosch et al., 2016) and whether they are caught or farmed, they may have been exposed to pathogens, biotoxins and contaminants before being harvested. These hazards are difficult to prevent and control (Huss et al., 2000). In addition, there are risks associated with farm inputs, such as the water supply, the food provided to the animals and the drugs administrated during the production cycle (Stringer and Hall, 2007).

Downstream, or post-harvest, aquatic organisms can be consumed directly when caught for recreational purpose or enter the supply chain to be sold on fresh markets, or in supermarkets, hypermarkets, online and in restaurants (Figure 10.1b). Throughout the supply chain, products may be cut, chilled, frozen, cooked, chemically treated to improve preservation or for smoking, dried, crushed, mixed, and packed or canned (Hicks, 2016; Stringer and Hall, 2007). The aquatic products are finally sold to consumers as whole, eviscerated or shelled, in fillets, raw, smoked, cooked, frozen or

Figure 10.1 (a) Oyster farm in Port Stephen, NSW, Australia. (b) Marine invertebrates, bivalves, sea urchins, shrimps and cephalopods sold at a fresh market in Val Paraiso, Chile.

as part of a meal (Hicks, 2016). Throughout these steps, many products are stored and transported several times (Stringer and Hall, 2007).

One of the major challenges in the food supply chain is that the product is sensitive to handling and storage conditions, especially for fish, which is highly perishable, and starts to deteriorate as soon as it is harvested. There is a need for all actors in the supply chain to work synergistically to ensure proper handling of aquatic products because inadequate heating or cooling, washing, and mishandling can promote the proliferation of pathogens or contamination by chemicals or physical items. Finally, the care for food safety should not end with the retailer as consumers also have a responsibility for the way they handle, store and consume seafood (Hicks, 2016).

Biological hazards

Pathogens posing a risk for human health

The major biological food safety concerns are hazards originating from food-borne infection with viable organisms and those posed by biotoxins. Other mechanisms of disease causation exist, such as allergic reactions or cancer induced by parasites from seafood. Infections can be caused by viruses, bacteria, protozoa, nematodes and trematodes. To date, cestodes, yeasts and fungi appear to be of minor relevance as aquatic food safety concerns, although the emergence of new aquaculture production systems, for example, for sea urchins *(Parcentrotus lividus)* and sea cucumbers (*Holothuria tuberosa*) may lead to emergence of new pathogens and biotoxins, including pathogenic yeasts such as *Candida albicans*, and mycotoxins (Kajikazawa et al., 2007; Viegas et al., 2019). Apart from yeasts and fungi, each group of pathogens will be discussed briefly, considering specific features as well as commonalities. One commonality is the increasing feasibility and affordability of detection of nucleic acids in seafood. There are many reports on the detection of pathogen RNA (ribonucleic acid) or DNA (deoxyribonucleic acid) in bivalves, snails, crustaceans, finfish or cephalopods without evidence of viability or infectivity of the pathogen. Such findings are often not presented as hazards or potential risks but as actual public health risks, even when there is little or no evidence of food-borne disease. Thus, caution is needed when interpreting pathogen prevalence data in terms of risk. Conversely, epidemiological studies looking at risk factors for disease may include data on exposure to seafood, but without evidence of pathogen presence. Outbreak studies are most likely to combine risk factor data with analysis of pathogen presence, but they are often limited in temporal and geographic scale and number of participants or samples. Thus, our knowledge of biological risk is a patchwork of studies of different kinds, with very strong evidence of major public health risks for some pathogens, and mere suggestions of evidence for others.

Viruses

Viruses are significant pathogens of shellfish and humans, and shellfish are significant sources of human viral disease. Viruses that cause disease in humans, however, are different from those that cause disease in shellfish. Crustaceans are highly susceptible to viral disease but not linked to food-borne disease. By contrast, bivalve molluscs are a prominent source of food-borne disease due to their location in surface waters, many of which are contaminated with human effluent, and their role as bioaccumulator of environmental contaminants, but the viruses causing food-borne disease are not known to affect the health of the bivalves (Figure 10.2).

Viruses and other pathogens present in water are filtered through the gills of shellfish during feeding and concentrated in the digestive glands or tissues, a process described as primary mechanical contamination. There is no amplification of viruses pathogenic to humans in shellfish (Stolle and Sperner, 1997). When shellfish are consumed raw or after preparation without heat treatment, as is often the case, they effectively close a human faecal–oral transmission cycle with an added pathogen-concentration step. Inadequate heating may be intentional, for example with oysters, or unintentional, as demonstrated for a norovirus outbreak caused by soup with carpet shell clams (*Tapes rhomboides*) (Lunestad et al., 2016). Norovirus is by far the most common seafood-borne virus. It is shed in high concentrations in human excreta and resistant to many environmental conditions. It was involved in 84% of some 360 shellfish-borne viral outbreaks reported globally from 1980 to 2012 (Bellou et al., 2013). Norovirus gastroenteritis is characterized by vomiting and watery diarrhoea and usually starts 1 to 2 days after exposure. It is short-lived but highly contagious and it can be severe enough to cause dehydration in children and the elderly. Outbreaks of norovirus disease from shellfish have been reported in East Asia, Europe, America, Oceania, Australia and Africa (Bellou et al., 2013). Although norovirus is the most common seafood-borne disease, most outbreaks of norovirus are the result of human-to-human transmission, for example, in care homes, restaurants or on cruise ships. The second most common shellfish-borne viral disease is hepatitis A (12.8% of outbreaks), which is also highly contagious but causes liver infection, accompanied by fever, nausea, abdominal discomfort, dark urine, and yellow skin and eyes (jaundice) (Bellou et al., 2013). Both viruses originate from faecal contamination, but their level in shellfish correlates poorly with other faecal indicators (coliform bacteria), as demonstrated for example in Mediterranean mussels (*Mytilus galloprovincialis*) (Ilic et al., 2017). This may explain why disease outbreaks associated with seafood continue to occur despite regulations based on bacterial indicators. Shellfish may also contain aichivirus, adenovirus, astrovirus, rotavirus and saprovirus, all of which can cause infectious gastroenteritis (Fusco et al., 2019). For some viruses, for example, aichivirus, the number of clinical reports is low relative to human exposure based on seroprevalence (90% at age 40) and human disease is not

Figure 10.2 (a) Potential transmission cycles of human, animal and agricultural pathogens, chemicals, toxins and pollutants via waste from domestic and industrial sources into water and seafood, toilet indicates human effluent; (b and c) plastic pollution of the Mekong River.

necessarily linked to shellfish consumption (Lodder et al., 2013). Compared with other food-borne diseases, those caused by viruses are generally less severe and rarely fatal, although their burden may be underestimated due to inadequate diagnostics or reporting (Stolle and Sperner, 1997).

Bacteria

As for viral pathogens, bacterial pathogens that cause disease in aquatic animals are largely different from seafood-derived pathogens that cause disease in humans. Reports of human infections with important fish pathogens, such as *Edwardsiella tarda* in catfish (*Pangasius* spp.) or *Streptococcus iniae* in flounder (*Paralichthys olivaceus*) or tilapia (*Oreochromis* spp.), are very rare and, for *S. iniae*, based on contact exposure during seafood preparation rather than food-borne exposure (Koh et al., 2009). The exception to this rule is *S. agalactiae* or Group B *Streptococcus* (GBS), a major pathogen of tilapia that also affects other fishes, frogs and crocodiles (Delannoy et al., 2013) (Figure 10.3). GBS has recently been recognized as significant cause of human invasive disease in Southeast Asia, with consumption of seafood without adequate heat treatment as most likely route of exposure (Barkham et al., 2019).

Aeromonas spp. can cause disease in aquatic species, including fishes (*A. salmonicida*, *A. hydrophila*), frogs (*A. hydrophila* subsp. *ranae*) and reptiles and *Aeromonas* has been described as an emerging

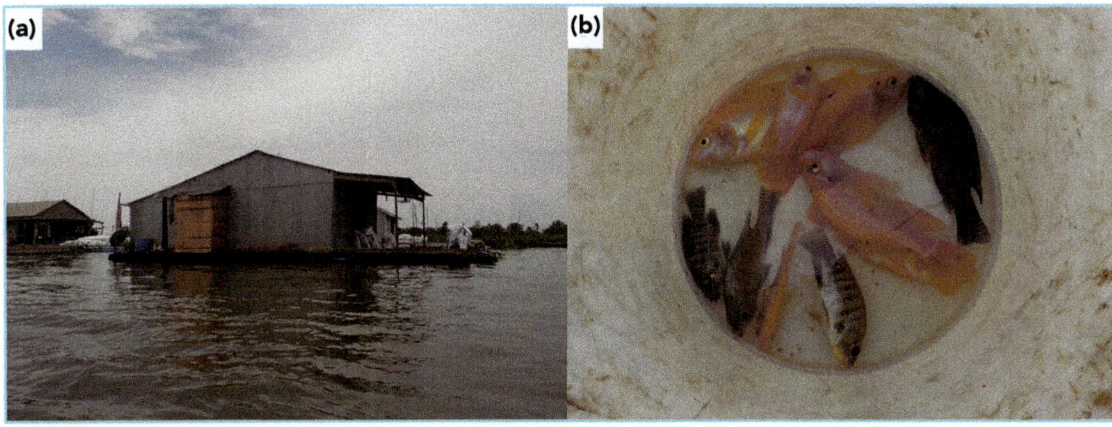

Figure 10.3 (a) Floating tilapia farm on the Mekong River in Vietnam, the outhouse on the side of the building being directly over the water and (b) tilapia with *Streptococcus agalactiae*.

food-borne pathogen that causes diarrhoea in children and travellers (Igbinosa et al., 2012). Unlike for GBS, there is no convincing evidence for *Aeromonas* as food-borne pathogen from seafood. Bacterial pathogens from seafood that do not cause disease in their aquatic host are categorized as naturally occurring (e.g. *Vibrio* spp.) or faecal contaminants (e.g. *E. coli*, *Salmonella*). Consistent with their distinct origins, counts of faecal coliforms, *E. coli* and *Salmonella* are correlated with each other, but not with those of *Vibrio* spp. (DePaola et al., 2010). *Vibrio* spp., notably *V. parahaemolyticus* (thermo-stable direct haemolysin producing strains) and *V. vulnificus*, are the major causes of bacterial seafood-borne disease in North America and they are of growing importance in China. The incidence and northerly range of *Vibrio* cases has expanded in the 21st century as a result of rising seawater temperatures, with outbreaks reported as far north as Canada and Alaska when water temperatures exceed 14°C (Konrad et al., 2017; McLaughlin et al., 2005). *Vibrio* food poisoning is primarily due to consumption of raw oysters and manifests as gastroenteritis, usually within 24 hours of infection, lasting only for a few days. Despite or due to the implementation of guidelines for shellfish quality based on coliform bacteria, there is limited evidence for salmonellosis or *E. coli* infection from seafood. This may also reflect underreporting in countries most affected by enteric food-borne pathogens, especially considering that important human pathogens such as *Shigella* and *Salmonella* are commonly found in oysters, for example, in Thailand, including the specific human pathogen *Salmonella parathyphi B* (Jeamsripong et al., 2018), in shrimp in Vietnam (Noor Uddin et al., 2015) and in clams, spiral shells, shrimp, fish and bullfrogs at retail in China (Zhang et al., 2015). With adequate depuration (shellfish) or processing (finfish), faecal pathogens that are present pre-harvest do not need to affect the post-harvest product (Dang and Dalsgaard, 2012). Other pathogens are primarily introduced post-harvest, e.g. *Bacillus cereus* (Kim et al., 2017), *Staphylococcus aureus* (Saito et al., 2011) and *Listeria monocytogenes* (Hansen et al., 2006). *Listeria* is of particular concern in ready-to-eat seafood because it can amplify at refrigeration temperature. Prevalence of food-borne bacteria may differ between retail systems, with high *Listeria* levels in imported fillets compared with locally purchased fillets, high *S. aureus* level in processed fish compared with fresh fish, and high *Vibrio* levels at fish markets compared with hypermarkets or online markets (Kim et al., 2017; Pao et al., 2008; Saito et al., 2011). Globally, there is growing concern about antimicrobial resistance, especially antibacterial resistance. Widespread use of antimicrobials in aquaculture may contribute to selection for resistant bacteria in aquatic environments and seafood. Some published studies show a shift rather than an increase in antibacterial resistance genes in fish farm sediments compared with control areas (Muziasari et al., 2016), whereas others raise alarm over high prevalence of antibacterial resistance genes in fish farms (Shah et al., 2012) or aquaculture products (Ye et al., 2013). It is not always clear whether presence of those genes is due to use of antibacterials in aquaculture or the result of run-off of human or agricultural waste into aquatic production systems, and the contribution of aquaculture or seafood to antimicrobial resistance as public health problem is yet to be quantified.

Protozoa

Protozoan parasites can cause disease and mortality in molluscs as well as people yet, again, the pathogen species that are detrimental to shellfish are not harmful to humans. Conversely, the pathogen species that pose a human health hazard are not harmful to shellfish. The main protozoan parasites of interest for aquatic food safety are *Toxoplasma gondii*, *Cryptosporidium* spp. and *Giardia duodenalis* (also known as *G. lamblia*). As for viruses and bacteria, excreta from other host species play a crucial role in the transmission cycle of the pathogen. Domestic cats and wild felids are the only definitive host species for *Toxoplasma*, while ruminants and dogs are important hosts for some species or assemblages of *Cryptosporidium* and *Giardia*, respectively. Humans shed both *Cryptosporidium* oocysts and *Giardia* cysts. Oocysts or cysts shed in faeces are transported into fresh and marine water via sewage, including flushed cat litter and surface run-off, and concentrated in filter feeders (Figure 10.4). *Toxoplasma* oocysts are known for their environmental robustness, contributing to long-term persistence in watersheds, ecosystems and seafood. Anthropogenic influences lead to increased pathogen contamination, with added spikes due to climate variability and flooding. *Toxoplasma* has been detected in clams, mussels, oysters and fishes in freshwater and marine habitats and, under experimental conditions, it can bioaccumulate in shrimp (Bigot-Clivot et al., 2016).

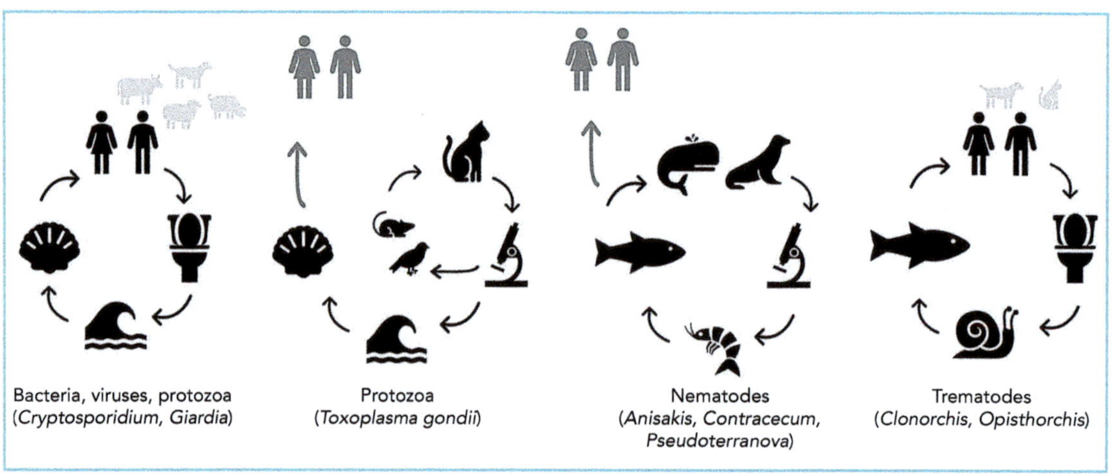

Figure 10.4 Simplified life cycles of pathogens (bacteria, viruses, protozoa, nematodes and trematodes) that can be acquired through consumption of seafood. Opportunities for control, for example, through prevention of contamination of aquatic environments with human effluent, depend on the role of humans and other host species as definitive (main figures) or paratenic (top row) host, and the existence of additional host species (shown in grey). Toilet indicates human effluent, microscope indicates various microscopic life stages.

Some studies, for example, in New Zealand, show seasonality in the prevalence of *Toxoplasma* in shellfish, whereas others, in Maine, USA, or Italy, do not (Coupe et al., 2018; Gomez-Couso et al., 2003; Marquis et al., 2019). The reported prevalence of *Toxoplasma* differs considerably between locations, with low prevalence in studies from e.g. southeast China (Zhang et al., 2015) and Sao Paulo State, Brazil (Esmerini et al., 2010) compared with the Atlantic and Pacific coast of the USA (Marquis et al., 2019; Staggs et al., 2015). Detection of *Toxoplasma* is often based on amplification of pathogen DNA by means of polymerase chain reaction (PCR) but only the sporozoite stage is infectious. The presence of viable sporozoites in bivalves has been demonstrated using reverse-transcriptase PCR, an elegant alternative to demonstration of viability using bioassays in mice (Coupe et al., 2018). Case-control studies have identified consumption of raw shellfish (e.g. oysters, clams or mussels) as a risk factor for toxoplasmosis in humans in geographically distant locations, including the USA (Jones et al., 2009) and Taiwan (Chiang et al., 2014). It is thought that *Cryptosporidium* retains its infectivity for prolonged periods in shellfish, whereas *Giardia* does not (Gomez-Couso et al., 2003; Robertson, 2007). Depuration processes designed to remove faecal coliforms from mussels, oysters, clams and cockles do not fully remove *Cryptosporidium* oocyst contamination (Gomez-Couso et al., 2003). Even so, there is no clear evidence that the hazard posed by presence of viable *Cryptosporidium* oocysts in shellfish translates into a significant health risk. Likewise, *Cryptosporidium parvum* has recently been detected in brown trout, a finding that is presented by the authors as a public health risk, but without any evidence of fish-borne infection in humans (Couso-Perez et al., 2019). To prioritize investment of resources in improvement of aquatic food safety, it is important that hazards are differentiated from risks, and that their impact on public health is quantified.

Nematodes

Anisakid nematodes in fish and shellfish are a major cause of human infections and allergies, responsible for an estimated 20,000 cases per year, whereas the rat lungworm *Angiostrongylus cantonensis* is a comparatively rare cause of aquatic food-borne disease. 'Anisakidosis' and 'anisakiasis' are used to describe zoonotic disease triggered by anisakid nematodes. Zoonotic fish-borne parasites within the Anisakidae belong to the genera *Anisakis, Contraceum* and *Pseudoterranova* (Buchmann and Mehrdana, 2016). Some authors also include the genus *Hysterothylacium*, although it formally belongs to the family Raphidascarididae (Aibinu et al., 2019; Shamsi et al., 2013). The life cycle of *Anisakis* ('herring worm' or 'whale worm') involves cetaceans as definitive host whereas *Contraceum* ('liver worm') and *Pseudoterranova* ('cod worm' or 'seal worm') use pinnipeds as definitive hosts (Figure 10.4). Invertebrates act as first intermediate host, with marine fish and cephalopods as secondary intermediate host. People are accidental hosts. Anisakiasis in people is due to ingestion of third-stage larvae and manifests as gastric, enteric, ectopic or allergic disease. Gastric and enteric forms are characterized by abdominal pain, with the peak of symptoms about 6 h and 48 h post-ingestion, respectively, and may be confused with or lead to bowel obstruction or perforation. Ectopic cases range from mild 'tingling throat' presentations to apparently life-threatening manifestations when clinical signs and symptoms are confused with other diseases. Allergic reactions are immunoglobulin E mediated and range from urticaria to anaphylaxis (Nieuwenhuizen and Lopata, 2013). The first infection with *Anisakis* was detected in the Netherlands in the 1960s. It was the first country to implement specific legislation to control 'herring worm disease' which includes freezing of raw herring, and the infection is now extremely rare (EFSA-BIOHAZ, 2010). Globally, 90% of anisakid cases are reported from Japan, with other supposed 'hotspots' including North America, South America, Northern Europe, Mediterranean Europe, South Korea and the Pacific islands, to name a few. Its global emergence is attributed to changes in consumer food habits, with increasing popularity of sashimi, sushi and other raw seafood (e.g. salmon tartare), followed by improved diagnostics and increased awareness. The distribution of *Anisakis* spp. in humans appears to reflect their distribution in fishes: *A. simplex* predominates in fishes on the Pacific side of Japan, and in people in Japan, which may also reflect that it is more likely to penetrate fish musculature and to survive in gastric juices than *A. pegreffii*. In the Sea of Japan and East China Sea, however, *A. pegreffii* predominates in fish, as it does in humans in South Korea. Human infections may be contracted from a wide range of aquatic species in different countries, including finfishes, cuttlefish and squid. As anisakids are of marine origin, they are largely absent from land-based aquaculture, unless such systems use marine lagoon water (Goffredo et al., 2019; Salati et al., 2013). In fish, the majority of L3 larvae are normally present in the body cavity and visceral submucosa so careful gutting can prevent human infection, as also described for bacterial pathogens. To mitigate the risk from larvae, present in musculature, freezing (below –20°C for at least 24 h) or heating (above 65°C for at least 10 minutes) are recommended.

In contrast to *Anisakis*, *A. cantonensis* is a freshwater nematode. It is also known as rat lungworm, based on its definitive host. Its life cycle relies on snails as intermediate hosts, including land snails and aquatic snails. Other terrestrial and aquatic species, including freshwater shrimp, land crabs, frogs, and lizards, are considered paratenic hosts. By 2008, *A. cantonensis*, which causes eosinophilic meningitis in humans, had been reported in over 2800 people from 30 countries, with the majority originating in Thailand, China and the Pacific (47%, 27% and *c.*12% of cases, respectively) (Wang et al., 2008). Its geographic range, the number of cases, and the number of reported intermediate hosts has increased since. Its food-borne origin reflects cultural culinary preferences, with cases in Thailand and Taiwan attributed to intentional consumption of raw snails, and cases in European travellers to Southeast Asia mostly attributed to freshwater prawns (Federspiel et al., 2020; Wang et al., 2008). Adult cases in Australia, the Americas and elsewhere are largely of non-aquatic origin and have been attributed to intentional ingestion of snails as part of bets or 'dares'. Severe paediatric cases are often due to unintentional ingestion by children licking (land)snails (Cowie, 2013; Epelboin et al., 2016).

Trematodes

Trematodes (flatworms) of aquatic origin include liver, lung and intestinal flukes. There are over 50 species of the lung fluke *Paragonimus* (Yoshida et al., 2019) and at least 74 species of intestinal flukes (Chai and Jung, 2020) but most fish-borne trematode infections are caused by a few liver fluke species. Some authors put the global number of people affected by fish-borne zoonotic trematodes at 18 million (Wiriya et al., 2013) whereas others attribute 35 million infections to 'the big 3', that is, the liver flukes *Clonorchis sinensis*, *Ophistorchis viverrini*, and *Ophistorchis felineus* (Lim, 2011), or even to *C. sinensis* alone (Lun et al., 2005), with another 8 million caused by *O. viverrini* in countries along the Mekong river, i.e. Thailand, Cambodia, Lao PDR and Vietnam (Forrer et al., 2012; Khamboonruang et al., 1997). *Clonorchis sinensis* is common in China, Korea, Japan and the eastern part of Russia, whereas *O. felineus* is also found further west, for example, in Western Siberia, Kazakhstan and Eastern Europe (Macpherson, 2005; Sultanov et al., 2014). Macpherson (2005) attributes a staggering 70–100 million human infections to parasites from fresh-water fish and another 20 million to lung fluke, which is contracted by eating undercooked crabs, crayfish, or shrimp, and has a distribution that includes Asia as well as parts of Africa and South America. Despite a case load that dwarfs all other aquatic food-borne pathogens combined, relatively little attention is paid to aquatic trematode infections in western countries. This may be because most cases occur in Asia, the scientific literature is largely published in Asian languages and clinical manifestations are not acute, unlike viral, bacterial or nematode infections. In some areas of Asia, e.g. in endemic parts of Thailand, prevalence of *O. viverrini* infection is as high as 85% and the country has a dedicated WHO Collaborating Centre for Research and Control of Opisthorchiasis (Saenna et al., 2017). Liver flukes may survive in human intrahepatic bile ducts for decades, causing chronic inflammation of the ducts and hepatobiliary disease, complicated by stone formation and cholangiocarcinoma (cancer) (Lim, 2011). An estimated 20% of Southeast Asian immigrants in Europe and North America have fish-borne trematode infection which they most likely contracted prior to migration (Macpherson, 2005). The role of aquaculture in prevention or dissemination of trematodes is variable, with some studies describing lower prevalence in farmed fish compared with wild-caught fish (Li et al., 2013; Wiriya et al., 2013) and other studies identifying the development of aquaculture as a driver of trematode dissemination, possibly through distribution of infected fish from hatcheries and nurseries (Sithithaworn et al., 2012). Although fish-borne trematodes are primarily a freshwater issue, marine species can also be infected and expansion of marine aquaculture may create a new food safety hazard (Thanh et al., 2009). In the 1990s, attempts were made to control the prevalence of trematodes in aquaculture fish through implementation of hazard analysis critical control points (HACCP) approaches (Khamboonruang et al., 1997). Due to the importance of human behaviour in transmission of the disease, subsequent control approaches have taken a more integrated One Health or ecohealth approach (Huang et al., 2017; Sripa et al., 2017). Significant reductions in infection levels in snails (primary intermediate host), fishes (secondary intermediate host) and humans (definitive host) were achieved for *C. sinensis* in South China and *O. viverrine* in the Lawa Lake region in Thailand. The integrated approach for *C. sinensis* included multiple factors: detection and treatment in people (chemotherapy with praziquantel), replacement of fish-pond toilets with upgraded toilets that avoid entry of live parasite eggs into the water, and health education in the community and schools covering the transmission route of the parasite as well as the role of raw fish consumption and the use of the same utensils for raw and cooked food (Huang et al., 2017). The Lawa model also included monitoring and anthelmintic treatment and community and school-based educational activities, with a strong focus on stakeholder engagement and long-term sustainability of activities (Sripa et al., 2017). Complete eradication of fish-borne trematodes is unlikely, in part because there is also a sylvatic (non-human) transmission cycle through piscivorous mammals including dogs, domestic and wild cats, foxes, pigs, marten and mink and in part because of long-term and intensive investment that is needed for successful control (Huang et al., 2017; Li et al., 2013; Sripa et al., 2017).

Biotoxins

Intoxications from seafood are relatively uncommon but likely underreported (Sobel and Painter, 2005). Of the seafood-borne outbreaks 80% are related to toxins produced by microalgae during harmful algal bloom events, by bacteria in spoiled food, or associated with consumption of raw molluscan shellfish (Huss et al., 2000).

Biotoxins produced by harmful algae bloom

Harmful algae bloom (HAB) occurs when dinoflagellate and diatom microalgae species producing biotoxins proliferate in a water body (Steidinger and Penta, 1999). Although the majority of marine microalgae are harmless, at least 80 species are toxic (Baden et al., 1995; Balmer-Hanchey et al., 2003). The number of HABs, their frequency and intensity are increasing in coastal areas worldwide (Glibert et al., 2005) and new species of toxic microalgae are emerging (Balmer-Hanchey et al., 2003; Grattan et al., 2016). This trend is considered largely the result of anthropogenic activity leading to water pollution, eutrophication, oxygen loss, climate change and ocean warming (Gobler, 2020). This trend is likely to continue, which is of growing concern for public health as the biotoxins can cause human illness by skin contact, aerosol inhalation or food ingestion (Balmer-Hanchey et al., 2003; Grattan et al., 2016). The toxins produced by the microalgae can be transferred throughout the food web and accumulate in aquatic organisms, mainly shellfish and finfish, but also crustaceans and cephalopods (Morabito et al., 2018; Lopes et al. 2013). The risk of having contaminated aquatic products is higher for the organisms caught by fisheries or farmed in open-water aquaculture than for those from land-based aquaculture. However, fish from land-based aquaculture can be contaminated if the water supply or feed is contaminated with toxins (DiNubile and Hokama, 1995). The role of the water supply to land-based aquaculture as driver of food safety risk has also been described for anisakids.

The five most commonly reported HAB-related illnesses are ciguatera food poisoning (CFP) caused by ciguatera toxins, paralytic shellfish poisoning (PSP) caused by saxitoxins, neurotoxic shellfish poisoning (NSP) caused by brevetoxins, amnesic shellfish poisoning (ASP) caused by domoic acid, and diarrhetic shellfish poisoning (DSP) due to okadaic acid and dinophysistoxins (Grattan et al., 2016). CFP is one of the most commonly reported aquatic food-borne diseases related to ingestion of toxins worldwide (Friedman et al., 2008). The annual number of people who suffer from CFP is estimated to range between 10,000 and 500,000 (Friedman et al., 2017). Ciguatera is a toxin mainly produced by dinoflagellate species from the genus *Gambierdiscus* occurring in tropical and subtropical areas, which are accumulated mostly in fish from coastal areas or shallow water (Balmer-Hanchey et al., 2003; Friedman et al., 2017). The reef fish scratch the surface of coral or algae on which the dinoflagellates are attached, and the toxins are then transferred throughout the food web to humans (Friedman et al., 2017; Mustafa et al., 2018; Yasumoto et al., 1977). More than 400 species of fish have been identified as carriers of ciguatera toxins (Huss et al., 2000). The symptoms of CFP include gastrointestinal, neurologic and cardiovascular effects (Sobel and Painter, 2005).

Shellfish containing saxitoxins, causing PSP, are primarily found in cold-water regions on the coasts of the northern USA, southern Chile, Japan and in the North Sea (Sobel and Painter, 2005). Risky levels of the toxins are encountered quite frequently but disease reports are rare, which is thought to reflect the effectiveness of control measures rather than underreporting, considering the affected regions (Sobel and Painter, 2005). Symptoms of PSP include numbness, tingling, headaches, breathing difficulties and gastrointestinal disorders (Etheridge, 2010; Sobel and Painter, 2005). If a high dose of saxitoxins is ingested death can occur (Garcia et al., 2004). NSP has been reported in western Florida, the Caribbean and New Zealand. The intoxication can result in gastrointestinal and neurological disorders in people (Sobel and Painter, 2005). ASP rarely occurs but has been reported in North America (Grattan et al., 2016). The symptoms of ASP are gastrointestinal and can include neurologic effects (Sobel and Painter, 2005). DSP has been reported in Japan, Europe, Canada, New Zealand, Africa and South America and the mains effects are gastrointestinal (Grattan et al., 2016; Sobel and Painter, 2005). Disease diagnosis relies on clinical presentation of the intoxicated person and the dietary history (Sobel and Painter, 2005). Toxicity varies according to the dose of the ingested toxin, the elimination rate of the toxin and the health of the consumer (Steidinger and Penta, 1999). Other than testing, there is no way to know that the aquatic product is contaminated as the toxins are tasteless and odourless (Balmer-Hanchey et al., 2003). Furthermore, most of the marine toxins responsible for these diseases have toxic effects at low concentrations, ranging from nanomoles to picomoles per kilogram of body weight (Baden et al., 1995), which can make their detection in aquatic food products challenging (Balmer-Hanchey et al., 2003). Intoxication cannot be prevented by improved storage, cooking, or cooling, as these toxins are temperature-resistant (Balmer-Hanchey et al., 2003; Sobel and Painter, 2005).

Biotoxins produced by cyanobacteria

Toxins can also be produced by certain phototrophic bacteria, the cyanobacteria, which have a broad distribution (Ibelings and Chorus, 2007; Zanchett and Oliveira-Filho, 2013). Food poisoning related to ingestion of cyanotoxins occurs mainly in Europe and the USA (Wood, 2016). The highest exposures to cyanobacterial toxins by humans are almost invariably associated with consumption of animals that are eaten whole such as bivalves, snails or planktivorous fish (Ibelings and Chorus, 2007) and rarely result in death (Wood, 2016). Some harmful cyanobacteria in freshwater environments are *Microcystis*, and *Planktothrix* (Ibelings and Chorus, 2007). The effects of toxins produced by cyanobacteria include liver and kidney damage, gastrointestinal disorders, cytotoxicity, neurotoxicity and skin toxicity (Zanchett and Oliveira-Filho, 2013).

In estuarine and marine environments, the cyanobacterium *Lyngbya majuscule* seems to be the cause of chelonitoxism, a rare type of food poisoning that is due to the consumption of turtle meat containing toxins. It is mainly reported in the Indo-Pacific region and intertropical area (Fussy et al., 2007). It is suggested that the causative agent is accumulated along the food chain, following the ingestion of the cyanobacterium by turtles (Fussy et al., 2007; Magnino et al., 2009). Chelonitoxins cause severe intoxication with a high mortality rate (Fussy et al., 2007). The symptoms may include gastrointestinal disorders, respiratory distress, and multi-organ failure, which can lead to coma and death (Magnino et al., 2009). The toxins can accumulate in breast milk and be transferred to breast-fed children.

Biotoxins produced by bacteria

Scombroid poisoning is also known as histamine fish poisoning and is due to the consumption of decaying or spoiled fish. It can result from mishandling of the product, e.g. storage under inadequate conditions, and can be prevented. Scrombroid poisoning is thought to be underreported. In spoiled fish it is primarily due to Gram-negative enteric bacteria that produce histamine from free histidine (Hungerford, 2010; Mustafa et al., 2018). As histamine is a common allergic agent, scombroid poisoning can be confused with other allergic reactions. The intoxicated victims respond well to antihistamines. The symptoms of scombroid poisoning include peppery or metallic taste, oral numbness, headache, dizziness, palpitations, low blood pressure, difficulty swallowing, thirst, rash, flushing, facial swelling, nausea, vomiting, abdominal cramps and diarrhoea. Sometimes, the nervous system is affected, and the consumer may feel anxious (Hungerford, 2010).

Another bacterial toxin, the tetrodotoxin, causes pufferfish (Tetraodontidae) poisoning, mainly reported in Japan, Taiwan, Hong Kong, and Cambodia, but also in Bangladesh and the USA (Haque et al., 2008). The fish contract bacteria producing the tetrotoxin via the food web and the bacteria subsequently stay in the fish as endosymbionts (Bane et al., 2014). Other aquatic species, including frogs, gastropods, starfish, sea slug and octopus may also contain tetrodotoxin (Bane et al., 2014). It can result in gastroenteritis and severe neurological disorders similar to PSP (Baden et al., 1995). The fact that consumer behaviour plays a crucial role in aquatic food-borne disease is illustrated by the fact that some people intentionally consume pufferfish because of the risk associated with it and the tingling sensation that can be caused by tiny amounts of toxin that are left in well-prepared pufferfish (Field, 1998). Food preparation practices are also an important contributor to seafood-derived botulism, or life-threatening paralytic illness caused by botulinum neurotoxin (BoNT). In the USA, several cases have been observed after consumption of a traditional Egyptian salt-cured fermented dish called fesikh, which is prepared from non-eviscerated fish containing BoNT type E and *Clostridium botulinum* type E. In response, the Centers for Disease Control (CDC) have suggested that this issue could be addressed by risk communication via public awareness campaigns, similar to the approach taken by the Egyptian Ministry of Health to discourage fesikh consumption (Ganapathiraju et al., 2019).

Chemical hazards

Another concern associated with the consumption of aquatic food is the presence of chemicals sourced from the environment, farming practices or food processing and the potential risk of chronic exposure and to a lesser extent, acute exposure to humans (Jennings et al., 2016; O'Keeffe and Kennedy, 1998). The presence of metals is increasing due to anthropogenic activities such as smelting, mining, burning of fossil fuels and metal input in agrochemicals (Bosch et al., 2016; Reichelt-Brushett et al., 2017). The use of agrochemicals on crops and livestock, such as pesticides, fertilizers and veterinary drugs, has resulted in pollution of compartments of the environment with worrying levels of residues, notably in soil, air, and fresh and marine ecosystems (Carvalho, 2017; Zhang et al., 2018). The use of these chemicals for the treatment of diseases or enhancement of production levels contributed to increased food production for a rapidly growing human population (Carvalho, 2017). However, this came at a cost: the pollution of terrestrial and aquatic ecosystems, especially in developing countries. This may cause environmental issues, and can directly affect aquatic food resources, fisheries and aquaculture, ultimately posing a threat to public health (Cabral et al., 2019; Carvalho, 2017; Reichelt-Brushett et al., 2017; Zhang et al., 2018). These threats have been considered by policymakers and led to altered regulation of agrochemical use in several countries. Examples include controlling the volume and type of chemicals applied, assessing toxicity levels, testing for residues in food products and controlling environmental pollution. All of these factors are likely to have improved food safety (O'Keeffe and Kennedy, 1998; Zhang et al., 2018).

Metals and persistent organic pollutants

One of the main pathways for human exposure to metals, polychlorinated biphenyls (PCBs) and organochlorine pesticides is via the consumption of contaminated fish and this can be an issue for human health if they are consumed at toxic concentrations above regulatory limits (Bosch et al., 2016; Marcotrigiano and Storelli, 2003; Prüss-Ustün et al., 2011). Contaminants can be transferred from the external environment to aquatic organisms when exposed to waterborne contaminants, contaminated sediment or via the food web, where they can bioaccumulate (Barron, 2003; Bushkin-Bedient and Carpenter, 2010). Harmful contaminants in aquatic systems that are prone to bioaccumulation are hydrophobic, persistent organic pollutants, such as organochlorine pesticides, PCBs and dioxins (Barron, 2003; Bushkin-Bedient and Carpenter, 2010; Streit, 1998). The skin, gill and digestive tract of aquatic organisms are potential sites of absorption of waterborne chemicals (Barron, 2003). Pollutants tend to accumulate in the muscles or elements of the digestive tract of fish (Marcotrigiano and Storelli, 2003).

Metals

Marine fish may contain high levels of metals, predominantly mercury, arsenic, cadmium, and lead (Bosch et al., 2016; FAO/WHO, 2012). A metal of great concern for human health related to consumption of fish and seafood from marine and freshwater environments is mercury, and more specifically its main stable organic form, the methylmercury (Bosch et al., 2016; Bushkin-Bedient and Carpenter, 2010).The contribution of methylmercury to total mercury increases throughout the aquatic food web from 11% in the water column, and 18% in phytoplankton, to 57% in zooplankton and up to 95% in small fish (Watras et al., 1998). Methylmercury binds to muscle of the animal which is the part consumed by humans, rather than fatty tissue and it is mostly absorbed through the human gastrointestinal tract (Bushkin-Bedient and Carpenter, 2010; Reichelt-Brushett et al., 2017). High levels of mercury were found in high-trophic-level fish predators in comparison with cephalopods, shellfish and crustaceans (Marcotrigiano and Storelli, 2003). The first case of methylmercury poisoning was reported in 1956 in Japan associated with the consumption of fish and shellfish by a fishing population located close to a chemical plant (Harada, 1995). Methylmercury poisoning in people affects the nervous system and is associated with a range of neurological symptoms including ataxia, dysarthria, visual and auditory disorders, tremor, and sensory disturbances and can impact on mental development or lead to death (Bushkin-Bedient and Carpenter, 2010; Harada, 1995). Exposure to mercury is a human health risk particularly to vulnerable sectors of the community including young children and pregnant or breastfeeding women as it passes from the mother to the foetus and breast milk (Bushkin-Bedient and Carpenter, 2010; Hicks, 2016). Thus, it is recommended that pregnant women limit their consumption of aquatic food and avoid specific fish species known to contain high levels of mercury such as shark, swordfish, king mackerel and tilefish (Bushkin-Bedient and Carpenter, 2010; Hicks, 2016). The maximum limits for mercury in fish set up by the FAO range from 0.5 mg kg^{-1} to 1 mg kg^{-1} of fresh weight (FW) (FAO, 2003).

Arsenic is another heavy metal of public health concern, even though in fish, 90% of any arsenic found is in the non-toxic organic form. Arsenic can affect human health even at quite low concentrations (Kapaj et al., 2006), resulting in a wide range of symptoms including gastrointestinal disorders, cardiovascular disease, neuropathy, muscle weakness, skin reactions and cancer (Bosch et al., 2016; Kapaj et al., 2006). Cadmium is also very toxic to human health and the main organ affected from cadmium exposure is the kidney, whereby continued and heavy exposure can lead to renal failure (Johri et al., 2010). Cadmium has carcinogenic effects and can weaken the skeleton (Bosch et al., 2016). The maximum limit for cadmium is 0.05 mg kg^{-1} FW in fishery products in general, with higher limits considered acceptable for a limited number of finfish species, bivalves and cephalopods (FAO, 2003). Finally, lead poisoning may cause neurological and renal disorders, haematologic issues and cancer (Bosch et al., 2016) in people. The maximum limits for lead are 0.2 mg kg^{-1} FW in fishery products in general, again with higher limits in some fin fish species, crustaceans, bivalves and cephalopods (FAO, 2003). Metals may become more available for bioaccumulation due to alteration of water chemistry, ocean acidification and increase of temperature resulting from anthropogenic activities, as shown under controlled conditions in edible marine bivalves (Barron, 2003; Cabral et al., 2019; Lemasson et al., 2019; Shi et al., 2016). Overall, further research is needed to understand how metals work individually and as a cocktail in aquatic organisms to better identify their safe levels (Bosch et al., 2016).

Persistent organic pollutants

Most fish are contaminated to some extent with persistent organic pollutants (POP) including dioxins, PCBs and chlorinated pesticides such as the organochlorine dichlorodiphenyltrichloroethane (DDT) (Bushkin-Bedient and Carpenter, 2010). The use of some of these POPs has been eliminated or restricted, for example, by the 2004 Stockholm Convention of the United Nations. Some studies showed that among finfish, cephalopods, crustaceans and shellfish higher concentrations of PCBs and DDTs were detected in liver and digestive glands in comparison to muscles due to the affinity of these chemicals to lipids (Katagi, 2010; Marcotrigiano and Storelli, 2003). POPS are endocrine disruptors and several studies associate diabetes and metabolic syndrome to human exposure to POPs, but the underlying causal relationship is not always clear (Bushkin-Bedient and Carpenter, 2010; Fierens et al., 2003; Lee et al., 2007; Rylander et al., 2005; Turyk et al., 2009; Uemura et al., 2009). Other functions or organs that might be affected by POPs include immune and reproductive systems, cardiovascular functions, blood, liver, kidney, muscles and skeleton, and they may be fetotoxic and carcinogenic (Huang et al., 2006). Consumption of fish contaminated with PCB by mothers before or during pregnancy can alter children's mental development and behaviour (Bushkin-Bedient and Carpenter, 2010).

Pharmaceutic products and veterinary drug residues

Veterinary pharmaceuticals include antimicrobials or pesticides that are used to treat a disease occurring in farmed aquatic animals. Many of the veterinary pharmaceuticals used in food animal production (e.g. aquaculture and livestock industries) are also available for use in humans. Antimicrobial use (and overuse) in food animal production is of global concern, although it is not always clear whether antimicrobials in aquatic production systems are of human, terrestrial animal, agricultural (crop production) or aquatic origin (Figure 10.2). Aquatic animal production systems present a unique environment for the administration of veterinary pharmaceuticals given that the environmental conditions can affect the bioavailability for the animal being treated and the fate of the compound in the environment. For example, half-life times are shorter in freshwater compared to seawater and in the presence of sediment as compared to without sediment (Kwon, 2016). The potential drawbacks associated with the use of veterinary pharmaceuticals in aquaculture are: (1) residual concentration left in the aquatic food; (2) the spread to non-target species and release to the environment; and (3) the development of resistance.

In aquatic animals, pharmaceuticals are commonly given through medicated feed, via an immersion bath (Figure 10.5) or by injection. Veterinary pharmaceuticals used according to the label instructions should not result in residues at slaughter that present a food safety issue for the human consumer (Okocha et al., 2018). Residues may include the non-altered parent compound as well as metabolites and/or conjugates that may have direct or indirect toxic effects on consumers (Mainero Rocca et al., 2017). To protect consumers, there are international food standard guidelines (e.g. Codex Alimentarius, European Union, US Food and Drug Administration) that set maximum residue limits (MRL) for veterinary pharmaceuticals in edible animal products. The MRL for three commonly used antimicrobials is 50 $\mu g\ kg^{-1}$, 200 $\mu g\ kg^{-1}$ and 1000 $\mu g\ kg^{-1}$ for amoxicillin, oxytetracycline and florfenicol, respectively (Codex Alimentarius, 2018; European Agency for the Evaluation of Medicinal Products, 2002). Regardless of the route of administration, veterinary pharmaceuticals (in particular, antibacterials) are known to accumulate as residues in tissues before they are completely metabolized and excreted from the body. The presence of residues in the edible tissue of aquatic animals can be attributed to not following the label directions or incorrect dose (e.g. off label use) or may include not following the recommended withdrawal periods prior to harvest, as well as failure to clean equipment used to mix or administer pharmaceuticals, mixing errors, and unaccounted for variations in water temperature (Okocha et al., 2018). Presence of antimicrobial residue or resistant bacteria in water and aquatic species may also result from human sewage or farm run-off (Baniga et al., 2020).

Figure 10.5 Grouper being returned to their sea cage following treatment for ectoparasites at a grow-out farm in Indonesia.

The biggest priority for the use of veterinary pharmaceuticals in food producing animals is the responsible use of antimicrobials to ensure safe food for human consumption and to minimize the development of antimicrobial resistance (AMR). Horizontal gene transfer (HGT) through mobile genetic elements (e.g. plasmids) and clonal selection due to antimicrobial selective pressure are the main drivers behind the development and spread of AMR (Preena et al., 2020). HGT can occur between unrelated bacterial species meaning that the resistance genes can be present in non-pathogenic, environmental bacteria and still pose a threat to animal and human health (Thornber et al., 2020). The unregulated use of antimicrobials in the aquaculture industry poses human health and food safety concerns that remain largely unaddressed in most developing nations (Okocha et al., 2018; Thornber et al. 2020). The issue is complicated by transmission of resistant bacteria or AMR determinants from humans, crops and animals to the aquatic environment via wastewater and run-off (Baniga et al., 2020). This is a prime example of AMR and aquatic food safety as One Health issues, with inextricable interconnectedness of human, animals, crops and their environment (Baniga et al., 2020). The vast majority of the world's aquaculture production is based in countries where there is a lack of regulation and enforcement of the responsible use of antibiotics (Chuah et al., 2016). The prophylactic use of antibiotics in aquaculture is still commonplace in shrimp (Thornber et al., 2019), as is the metaphylactic treatment of entire populations (e.g. the whole pond or sea cage) (Okocha et al., 2018; Watts et al., 2017). Further, the global trade in ornamental fish intensifies the risk of AMR associated with aquaculture with the widespread unregulated use of antimicrobials in this industry (Musto et al., 2006; Rose et al., 2013; Verner-Jeffreys et al., 2009). This use of antimicrobials goes largely unchecked, as ornamental fish do not fall under human food safety requirements and could be a hidden driver of AMR.

In addition to antimicrobials, the frequent use of other veterinary pharmaceuticals, such as parasiticides, has led to the development of resistance and spill-over to the environment. Pesticides are often used in farming salmon to control for parasitic copepods known as sea lice belonging to the genera *Caligus* and *Lepeophtheirus*. The extensive use of a variety of chemicals to combat sea lice infestations has resulted in leakage to the environment and ineffective treatments with the development of resistance (Aaen et al., 2015; Cheng et al., 2020; Lam et al., 2020; Quiñones et al., 2019; Tucca et al., 2017). Pesticide treatments are administered as an immersion bath (e.g. organophosphates dichlovos and azamethiphos or pyrethroids) or the preferred approach is combined in feed (e.g. SLICE® or emamectin benzoate) (Lam et al., 2020; Tucca et al., 2017). The development of resistant sea lice to a variety of pesticides has been documented for several years in many countries (Aaen et al., 2015) and often leads to industries cycling through treatment options to achieve clinical effectiveness. Pesticides (and other veterinary pharmaceuticals) can be excreted from the fish or enter the environment from uneaten feed. Pesticides have been found in sufficient concentrations to present a risk for benthic invertebrates in the sediments near salmon cages (Cheng et al., 2020; Quiñones et al., 2019; Tucca et al., 2017). Emamectin benzoate is considered highly persistent in marine sediments with a half-life of 175 days under aerobic conditions (Tucca et al., 2017). The persistent accumulation of emamectin benzoate in sediments has a twofold effect as it will interfere with ion transfer and nerve impulse transmission in many marine invertebrates (Tucca et al., 2017) and promote resistance to the drug with sub-therapeutic exposure (Cheng et al., 2020). As a green alternative to pesticides, the use of biological sea lice control using cleaner fish such as ballan wrasse (*Labrus bergylta*) and lumpfish (*Cyclopterus lumpus*) is promoted. As production and use of lumpfish for sea lice control intensifies, these fish are starting to develop their own health issues and associated disease control challenges (Powell et al., 2018).

Physical risks

Physical hazards include foreign objects found in food, including metal, plastic, glass, wood and stones (Huss et al., 2003), but also naturally occurring materials such as sand and, if you are lucky, pearls. Foreign items could be introduced in the aquatic products along the production chain, from the harvest to the manufacturing process. These foreign objects may cause injuries to the consumers such as choking, laceration and perforation in the mouth, throat, and organs along the digestive track, or even damage teeth and gums (Huss et al., 2003). Physical hazards are commonly reported by consumers immediately after product consumption as they are easy to identify (Huss et al., 2003).

Emerging risks

An emerging concern associated with seafood is the consumption of products contaminated with micro- and nanoplastics (Børresen, 2015; Hollman et al., 2013; Lusher et al., 2017; Smith et al., 2018). Microplastics (MP) and nanoplastics (NP) are plastics with a size smaller than 5 mm, and smaller than 1 μm, respectively (Smith et al., 2018). These plastics are directly produced by industry or result from the fragmentation of bigger plastics that humans have been increasingly using and discarding such as clothes, bags, bottles and cars (FAO, 2017). They reach aquatic ecosystems through run-off and waste (Figure 10.2) and they have been observed at different levels of the food web from plankton to higher trophic levels including invertebrates, fish, sea birds and marine mammals (Hollman et al., 2013). Plastic particles are found in the digestive tracts of aquatic organisms, but guts are commonly removed from finfish and cephalopods prior to consumption, reducing the direct exposure of consumers to plastic particles. This is not the case when consuming smaller finfish, crustaceans and bivalves, which are eaten whole (Lusher et al., 2017; Smith et al., 2018). A European study highlights that mussels from the wild were all contaminated with MP (Van Cauwenberghe et al., 2015). Uptake of NP was also reported in mussels (Ward and Kach, 2009; Wegner et al., 2012) and their translocation from algae to plankton to fish has been observed (Cedervall et al., 2012). The MP can absorb contaminants at their surface and further research on transfer through the food web and bioavailability is required (Andrady, 2011; Hollman et al., 2013). Plastic additives cause an additional concern because they can be endocrine disrupters (FAO, 2017). Finally, MP can bind microorganisms, including pathogens, at their surface (Hollman et al., 2013; Jongmans et al., 2018; Paul-Pont et al., 2018). There is a need to establish the tolerable daily intake for MP and to fully understand how MP affect human health (EFSA, 2016; FAO, 2017). In addition to plastic particles, other nanoparticles are of concern in the Anthropocene, as an increasing diversity of nanoparticles accumulates in marine environments and aquatic food. Accumulation of some nanoparticles, e.g. nanotubes of titanium dioxide (nTiO_2), may be exacerbated by ocean acidification, as already demonstrated in the three edible bivalve species *Meretrix meretrix*, *Cyclina sinensis*, and *Tegillarca granosa*, and in gastropods, amphipods, bivalves and fish (Shi et al., 2019). The impact of this accumulation in seafood on aquatic food safety and human health is unknown and further studies would be needed to determine whether this hazard translates into an actual risk.

Solutions

Since the 1990s, numerous food safety crises have made headlines, and this can reduce consumers' trust in the products they purchase. In order to minimize the occurrence of food-borne diseases, manage food safety and to protect human health there is a need to harmonize and standardize food safety strategies following an approach based on risk prevention along the entire food supply chain. This should include not just end point or product testing but also the adoption of good fisheries and aquaculture practices as well as increased consumer awareness. There are, however, many socio-economic and cultural barriers to uptake of potential solutions.

Regulation of food safety: the need for harmonization in response to the globalization of aquatic food trade

Food standards, guidelines and code of practice

Food travels around the globe and aquatic products are no exception. Fish and fish products, including shellfish and finfish, are among the most highly traded products and around 35% of global fish production was traded internationally in 2016 (FAO, 2018). As international food trade increased, so did the need to harmonize food safety criteria and standards, as standards from different countries created trade barriers (see Figure 10.6) (FAO/WHO, 2016). To address this, the Codex Alimentarius, also known as the Food Code, was developed by FAO and WHO in the early 1960s. It aims to provide 'safe and good food for everyone everywhere' and to ensure fairness in international food trade. The Food Code includes international food standards, guidelines and codes of practice. The Codex Alimentarius includes the 'Code of practice for fish and fishery products' whose target are the actors of the supply chain involved in the handling, production, storage, distribution, export/import and sale of fish and fishery products (FAO/WHO, 2012). This document covers finfish, shellfish, cephalopods, and crustaceans but not amphibians and aquatic reptiles. Other documents of interest include the 'Guidelines on the Application of General Principles of Food Hygiene to the Control of Pathogenic Vibrio Species in Seafood', which considers *V. parahaemolyticus*, *V. cholerae* or *V. vulnificus*, and the 'Application of General Principles of Food Hygiene to the Control of Viruses in Food' which considers hepatitis A virus and norovirus in bivalve molluscs (FAO/WHO, 2020).

Hazard analysis of critical control points

In 1959, the National Aeronautics and Space Administration (NASA) in the USA requested the development of a system to ensure the food safety of goods for astronauts' consumption, which gave rise to the concept of hazard analysis of critical control points (HACCP). The idea was to develop a method to ensure food safety based on prevention, implementing controls along the food supply chain rather than only at the end point (Miget, 2004). In the US aquatic sector, HACCP application in the 1990s reduced food-borne disease cases from seafood by between 20 and 60% (Reilly and Käferstein, 1997). HACCP has been adopted in most areas and countries to which aquaculture producers would be shipping, such as the European Union, Canada, Australia, New Zealand and Japan, and by the United Nations through its Codex Alimentarius program (see Figure 10.6) (Miget, 2004). HACPP is also used in low and middle income countries, despite barriers to implementation, particularly for small businesses, e.g. due to cultural or language problems, lack of education and training, and a shortage of experienced people to work in the aquaculture industry and associated regulatory authorities (Jirathana, 1998; Ropkins and Beck, 2000).

HACCP covers the 'fisheries/farm to table food safety continuum' (Reilly and Käferstein, 1997). It is based on a systematic approach which identifies the hazards (potential risks) and quantifies actual risks related to the production, distribution and use of the product (Reilly and Käferstein, 1997). The seven steps of HACCP consist of: (1) hazard analysis at each stage of the production chain (i.e. hazard identification, risk assessment and identification of measures to control the risk, including biological, chemical and physical hazards); (2) determination of critical control points (CCP) essential to eliminate or reduce the risk; (3) critical limits to ensure each CCP is effective; (4) monitoring of the CCP; (5) implementation of a corrective action when a CCP is not appropriate or effective enough; (6) validation of the HACCP over time; and (7) documentation of procedures and records (Garrett et al., 2000; Reilly and Käferstein, 1997). In addition to this HACCP approach, traceability is essential for food safety, including backward and forward tracing of products when a hazard has been identified (Garcia Martinez and Brofman Epelbaum, 2011; Stringer and Hall, 2007). Traceability can increase consumer trust, but it can be difficult to implement due to the complexity of market channels, particularly for wild-caught products (Stringer and Hall, 2007).

Food safety certification

To acknowledge good aquaculture practice and enhance consumer confidence in product quality, food safety certification can be used. The Global Food Safety Initiative (GFSI) was created in the early 2000s in a context where suppliers were audited by multiple retailers, which created duplication of effort. In response, the GFSI recognized different certification food safety schemes to avoid the suppliers being audited repeatedly. Recognized schemes include the BRC Global Standard for food safety, ISO (including the food safety related ISO 22000, Food Safety System Certification FSSC 22000), International Featured Standards (IFS) food, Safe Quality Food (SQF) Code, Best Aquaculture Practices, Global Read Meat Standards, and Canada GAP, to name a few (Sansawat and Muliyil, 2011). Certification and accreditation schemes aim to standardize and simplify production chains and to make the aquaculture industry more resilient, but certification can be challenging for fish farmers because of the proliferation of schemes, cost of certification, and lack of incentives. Rather than facilitating production and trade, certification schemes can result in additional cost to farmers and become barriers to trade, particularly for small-scale aquaculture. Attempts to overcome such barriers, for example, as led by the FAO's Technical Cooperation Programme, aim to establish and implement group certification for shrimp and tilapia farms in Thailand to spread the burden and cost and maximize benefit from the scheme through better access to profitable markets (Yamamoto, 2013). Using choice experiments, aquaculture producers' willingness to change and to adopt sustainable, eco-friendly or food safety practices can be assessed. As an example, a study in China showed that producers would require a premium of 2.5% per finfish to implement food safety practices for the national market based on the China Good Agricultural Practice (GAP), with vastly different attitudes towards international certification between groups of farmers (Ortega et al., 2014).

Because of the multitude of codes and standards, with differences between and within countries and over time, and rapid developments in detection methods, details for currently approved monitoring methods and associated thresholds are not included in this chapter.

Critical control points and associated measures

Pre-harvest

Whether aquatic organisms are farmed or harvested in the wild, they might be contaminated by pathogens, toxins and chemicals. With a few exceptions such as toxins from algal blooms, *Vibrio* spp. and fish-borne nematodes, most aquatic food-borne pathogens, toxins and contaminants are present in water and seafood as a result of human activity.

Some recycling of organic waste is intentional (e.g. use of human, pig, poultry or food waste as a method to recycle nutrients), whereas other human waste ends up in water due to lack of sanitation, reluctance to use sanitation or lack of sewage treatment infrastructure (see Figure 10.3). Industrial effluents may also contribute to environmental contamination with antimicrobials or toxins (Doron and Broom, 2019). Improvements to sewage treatment works can contribute to a significant decrease in contamination of shellfish with faecal bacteria (Campos et al., 2013). Likewise, methods and technology to re-use nutrients without recycling hazardous chemicals, toxins and pathogens could make a major contribution to aquatic food safety. When humans are the only or main definitive host species for a pathogen, control is also possible through elimination of human shedding. For example, vaccination of people against hepatitis A in Italy results in elimination of hepatitis A virus from shellfish, (Prato et al., 2013), and use of anthelmintic drugs for treatment of fish-borne trematode infections is an important component of holistic pathogen control strategies in China and Thailand (Huang et al., 2017; Sripa et al., 2017). Control of *T. gondii* in shellfish has been attempted through restrictions on flushing of cat litter to prevent environmental contamination (State of California).

To reduce the levels of aquatic food hazards associated with chemicals, changes may be needed in regulations and behaviours relating to the use of pesticides and fertilizers in agriculture or the use of medicines in people, terrestrial animals, and aquaculture. Options to reduce the use of chemicals on crops exist and include promotion of sustainable agriculture-aquaculture, with options ranging from organic farming to precision agriculture and development of food technology (Carvalho, 2017). There are considerably fewer registered veterinary products for use in aquaculture than for terrestrial animal production. This is due to biological factors, such as the fact that there are hundreds of species of aquatic animals each requiring efficacy and clinical testing, as well as economic factors. Because of the low value of an individual animal and the existence of species-specific industries, aquaculture is not considered a profitable market relative to the expense of getting veterinary products registered in each country, and limit registration of pharmaceuticals for aquaculture. To complicate matters, most aquaculture production takes place in countries where regulation around the sales and use of drugs may be difficult to enforce. Vaccination can reduce the reliance on antimicrobials for some diseases and has been used in Norway to significantly limit the use of antibacterials in the salmon industry (Anon, 2015; Klugman and Black, 2018). However, there are many biological, logistic, economic, social and regulatory barriers to development and uptake of vaccines in finfish, particularly in relatively low-value species such as tilapia, and in developing countries where access to infrastructure and resources can be limited. Oral (feed-based) administration would be the preferred route because it is cheap and does not require handling of fish. Engineered plants could potentially be used to produce oral vaccine ingredients, but they may not be acceptable to consumers (Clarke et al., 2013). Ironically, even vaccines themselves may meet with consumer reluctance, because they are perceived as foreign substance administered to fish (Engelstad, 2005). Vaccination is not possible for shellfish, including shrimp, which lack an adaptive immune system.

For meaningful implementation, regulations to reduce environmental contamination need to be accompanied by monitoring and surveillance of the hazards in areas used for production or harvest of aquatic products (Grattan et al., 2016; Huss et al., 2000). This can be done by screening the different compartments of aquatic ecosystems, such as the sediment, the water and sentinel species. In regions where coral reef fish, known to be contaminated with ciguatera toxins, are sold and consumed, it is recommended to monitor the sea surface temperature as this is one of the major triggers for the algal blooms (Yang et al., 2016). Temperature monitoring has also been recommended for prevention of *Vibrio* outbreaks from shellfish (Konrad et al., 2017). According to the results of the analyses from environmental samples, public authorities will be able to identify hazards, and to assess the potential risks at a specific time and to regulate the harvest and consumption of aquatic products. The harvest of fish and shellfish and subsequent consumption can be approved, restricted to a certain amount or frequencies at which the product may be eaten safely, or totally banned when alert levels of toxins are exceeded (Ibelings and Chorus, 2007).

Monitoring could also help in assessing the success of regulations. For example, monitoring of molluscs, fish, birds and mammals has shown that the concentrations of some organic compounds, such as DDT, PCB, hydrogen cyanide and polybrominated diphenyl ethers decreased due to regulations, while others, such as hexabromocyclododecane and per- and polyfluoroalkyl substances, have increased (Cabral et al., 2019). Currently, pesticide residues in North America and the European Union are thoroughly monitored. In less wealthy nations, monitoring may be absent or limited to products for export, with lower standards or limited or no testing for products for national or household consumption.

CCPs for shellfish farming are site selection, certification, maintenance, relaying contaminated product, harvest, bulk transport, storage and packaging (Garrett et al., 2008). The advantage of inland aquaculture over fisheries, shellfish harvest and off-land aquaculture, is the control over water and feed supplies, and the management of the growing steps, which constitute the three CCPs for fish farms (Garrett et al., 2000; Reilly and Käferstein, 1997). As new aquaculture systems and new feed sources are being developed, new hazards may arise. Examples include mycotoxins from plant-based ingredients used for marine salmon and gilthead seabream farming (Nacher-Mestre et al., 2015), pathogenic bacteria, resistance genes and integrase genes from fish meal (Han et al., 2019), and introduction of *Anisakis* in marine cage cultured cobia through chopped raw fish in their diet (Shih et al., 2010).

Post-harvest

Pre-consumer

Depending on the aquatic species and pathogen species, a variety of post-harvest, pre-consumer methods may be used to enhance aquatic food safety. For shellfish, depuration is used to reduce contamination with viral and bacterial pathogens, including norovirus and *Vibrio* species. Depuration times may differ between bivalve species and have been reported to be longer for cockles and clams than for mussels (Martinez et al., 2009). High salinity relay appears useful in the removal of *V. vulnificus* from oysters (Audemard et al., 2018).

Once the aquatic product is out of the water, it starts to deteriorate very quickly (Abbas et al., 2009). To limit the development of bacterial growth and toxin production, there is a need to keep the aquatic products at low temperatures, below 4°C if eaten quickly or stored at –18°C or below for later consumption (Abbas et al., 2009). This may not be feasible in rural or tropical areas, and seafood is often kept in water up to the point of sale or consumption in such environments. Freezing to –20°C for 72 h is a suitable method to kill *T. gondii* and *Anisakis* (Bouree et al., 1995).

When aquatic products are processed industrially, a way to minimize risks is to avoid using organs and tissues of aquatic organism where pathogens, toxins and contaminants may accumulate, or to avoid using organisms from specific places or harvested at a time of identified pollution or bloom events (Dang and Dalsgaard, 2012; Ibelings and Chorus, 2007; Lopes et al., 2013; Stringer and Hall, 2007). Proper gutting may also reduce the risk of exposure of consumers to bacterial pathogens (Dang and Dalsgaard, 2012), protozoa (Couso-Perez et al., 2019) and nematodes (Shih et al., 2010). Removal of fish skin may reduce the risk of *V. parahaemolyticus* exposure (Kim et al., 2017). Processing may also introduce new bacterial hazards, and contamination of consumption fish with a variety of fish- and human-derived bacterial species and strains at retail markets has been described. This includes bacterial species derived from human skin, such as *Staphylococcus aureus*, and the human gastrointestinal tract, such as *Streptococcus agalactiae*, suggesting a need for better personal hygiene during processing and retail (Chau et al., 2017; Saito et al., 2011). To avoid ciguatera poisoning, some markets avoid the sales of species known to be commonly contaminated with the toxin, or they refuse to sell fish that reach a certain weight or size, as bioaccumulation is higher in bigger fish (Yang et al., 2016). Information in the fish markets, with the location and the place of harvest of the fish, can help the consumer to consider if it is safe to buy and eat the fish (Yang et al., 2016).

A way to control physical hazards during food processing is to test the product using metal detection or to inspect equipment for signs of damage.

Consumer level

The final step in food safety is the consumer's responsibility, from purchase to preparation and handling of leftovers (Hicks, 2016). Proper handling, hygiene and preparation include maintenance at low temperature to avoid the development of pathogens and toxins produced by bacteria in the fresh product or leftovers, and prevention of cross-contamination with fish-borne or human pathogens or chemicals from cooking utensils, skin or faeces (Figure 10.6) (Stringer and Hall, 2007). Care should also be taken when aquatic product is purchased frozen, as it should not be defrosted and re-frozen (Hicks, 2016). Heat treatment destroys most pathogens, making heat treated products safer than raw or semi-cooked products. Cooking may not provide protection against thermostable toxins or chemicals and cooking is not always culturally acceptable. Alternative preparation methods using traditional recipes provide some reduction in microbiological hazards, as demonstrated for Thai 'koi-hoi' flavouring and Lab-Pla and it may be worth exploring if they provide a better balance between acceptability and hazard reduction (Chuboon et al., 2005; Eamsobhana et al., 2010). When discussing food safety, it is important to differentiate between food preparation, which is how 'cooking' may be understood in some language, and heat treatment, which is what cooking refers to in English.

Public awareness

Education to create public awareness complemented surveillance and regulation in decreasing the number of cases of shellfish poisoning in many countries (Figure 10.6) (Ibelings and Chorus, 2007). Thus, it is crucial to communicate the potential risks associated with the consumption of aquatic products to the public, especially the risks of eating raw molluscan shellfish and certain freshwater finfish species (Huss et al., 2000). However, the media tends to report health risks associated with seafood consumption without covering its health benefits (Raatz et al., 2013). There is a need to provide consumers with standardized and clear information on the quality and safety of aquatic food to improve their understanding of benefits and risks, and to increase confidence in food choices (Dickhoff et al., 2007).

A way to communicate about the risks associated with the consumption of shellfish or fish harvested recreationally is the use of 'toxin hotlines', a system that is in place in the USA, Europe and Australia (Ibelings and Chorus, 2007). Information campaigns can be conducted by authorities at local or regional scale (Ibelings and Chorus, 2007), in fish markets (Yang et al., 2016) or through the use of pre-packaged product labelling to convey information about responsible fisheries or farming practices. Different labels for aquatic products exist, such as the Best Aquaculture Practices labels, which captures food safety information. Other labels specific to seafood include those from the Aquaculture Stewardship Council (ASC), which promotes the use of responsible feeding in the farms and the use of veterinary products in case of diseases only, and Friend of the Sea, which forbids the use of antifouling products (surface coatings or treatments to control or prevent the attachment of unwanted organisms) and growth hormones. However, in order for this information to be effective, there is a need to educate the consumers on their meaning.

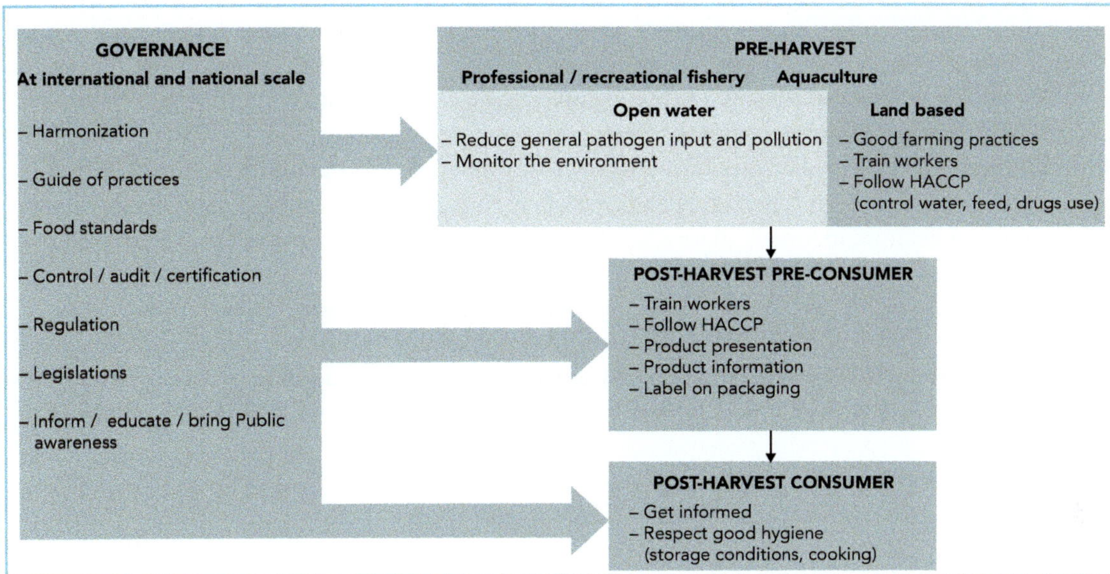

Figure 10.6 Measures to ensure food safety from the water to consumer's plate, the roles of governance bodies, the supply chain providers and public.

Cultural and socio-economic drivers of aquatic food safety

It might seem easy to dismiss many aquatic food-borne diseases as 'just' a problem of consumption of raw or undercooked seafood, but historical evidence shows that consumption of raw fish has been part of human culture for at least 2000 years in China (Huang et al., 2017), 1300 years in Japan (Japan News, 2022) and 700 years in Western Siberia (Slepchenko et al., 2015). Uncooked seafood is an integral part of food culture in many countries and sharing of food, including raw fish dishes, is a traditional practice that supports social cohesion (Saenna et al., 2017). The fact that health consequences are separate from the immediate satisfaction and social importance of seafood consumption, particularly for fish-borne trematodes in Asia, is somewhat reminiscent of the role played elsewhere by alcohol, which contributes significantly to social life in some parts of the world despite the associated risk of adverse long-term health impacts. In addition, the availability of effective treatment for fish-borne trematodes may result in complacency in terms of prevention, especially when prevention would mean abandoning deeply ingrained traditions around food. Consumption of raw or undercooked seafood in Asia is often described as a problem of the rural or uneducated poor, and publications on the subject appear in a journal like Infectious Diseases of Poverty (Forrer et al., 2012; Saenna et al., 2017). Somewhat ironically, consumption of raw oysters, sushi or sashimi in high-income countries, often in restaurants, is deemed fashionable and sophisticated. In China and Vietnam, eating in restaurants has been identified as risk factor for aquatic food-borne disease, emphasizing that food habits and associated food safety risks are not limited to poor rural households (Phan et al., 2011; Qian et al., 2013). The movement of people, (invasive) animals, seafood, parasites and culinary practices around the world, and the growing popularity of raw, undercooked, smoked, pickled or dried seafood is likely to result in an expansion of the geographic range of aquatic food-borne pathogens and an increase in number of people affected. In addition, seafood production from aquaculture is growing much faster than terrestrial agriculture, and seafood will play an increasingly important role in meeting the protein needs of the growing human population, which could potentially contribute to more cases of seafood-borne disease (Burket et al., 2018). More people also means more sewage, some 80% of which is released into the environment without adequate treatment, subsequently returning our pathogens and our chemicals to us via our seafood, particularly in rapidly expanding urban and peri-urban areas (see Figures 10.2 and 10.4) (Brooks and Conkle, 2019; Burket et al., 2018). As awareness of actual and perceived microbiological, parasitic and chemical food safety threats from aquaculture increases, there is a risk that the food supply and hence food security in Asia could be impacted (Khor et al., 2019). Aquaculture, however, has become too important as a source of nutrition, employment, livelihoods and foreign exchange to be abandoned. Moreover, aquaculture may provide opportunities to recycle nutrients, both from sewage and from food waste (Cheng and Lo, 2016; Koh et al., 2009), and the pressure exerted by human population growth on the planet's scarce resources makes that too good an opportunity to pass up.

Conclusion

In the Anthropocene, aquatic food safety is facing new challenges as pressure on waterways is increasing. These changes include increased water temperature, ocean acidification, and intensification of runoffs, which could support pathogen transmission and growth, favour harmful algae bloom, or lead to increasing input or availability of chemicals in aquatic ecosystems. In addition, we have created new problems such as micro- and nanoplastic pollution. To prevent these risks, monitoring and surveillance of the environment where animals are harvested or grown is essential. In addition, regulation and change in the use of sewage, waste products and agrochemicals by promoting sustainable agriculture-aquaculture should help in reducing the input of pollutants in the aquatic ecosystems and their presence in the long term. In addition, it is crucial to bring awareness to consumers of the health benefits and risks of seafood, and to promote good practices for food preparation and storage.

References

Aaen, S.M., Helgesen, K.O., Bakke, M.J., Kaur, K. & Horsberg, T.E. (2015) Drug resistance in sea lice: A threat to salmonid aquaculture. *Trends in Parasitology*, 31, 72–81. https://doi.org/10.1016/j.pt.2014.12.006.

Abbas, K.A., Saleh, A.M., Mohamed, A. & Lasekan, O. (2009) The relationship between water activity and fish spoilage during cold storage: A review. *Journal of Food, Agriculture and Environment*, 7, 86–90.

Abbott, D.W., Aasen, I.M., Beauchemin, K.A., Grondahl, F., Gruninger, R., Hayes, M., Huws, S., Kenny, D.A., Krizsan, S.J., Kirwan, S.F., Lind, V., Meyer, U., Ramin, M., Theodoridou, K., von Soosten, D., Walsh, P.J., Waters, S. & Xing, X. (2020) Seaweed and seaweed bioactives for mitigation of enteric methane: Challenges and opportunities. *Animals (Basel)*, 10, 2432. https://doi.org/10.3390/ani10122432.

Aibinu, I.E., Smooker, P.M. & Lopata, A.L. (2019) *Anisakis* nematodes in fish and shellfish from infection to allergies. *International Journal for Parasitology. Parasites and Wildlife*, 9, 384–393. https://doi.org/10.1016/j.ijppaw.2019.04.007.

Andrady, A.L. (2011) Microplastics in the marine environment. *Marine Pollution Bulletin*, 62, 1596–1605. https://doi.org/10.1016/j.marpolbul.2011.05.030.

Anonymous (2015) *Vaccinating Salmon: How Norway Avoids Antibiotics in Fish Farming*, Vol. 2020. World Health Organization: Geneva.

Audemard, C., Kator, H.I. & Reece, K.S. (2018) High salinity relay as a post-harvest processing method for reducing *Vibrio vulnificus* levels in oysters (*Crassostrea virginica*). *International Journal of Food Microbiology*, 279, 70–79. https://doi.org/10.1016/j.ijfoodmicro.2018.04.043.

Baden, D.A., Fleming, L.E. & Bean, J.A. (1995) Marine toxins. *Handbook of Clinical Neurology*, 21, 141–175.

Balmer-Hanchey, E.L., Jaykus, L.-A., Jaykus, L. & McClellan-Green, P. (2003) Marine biotoxins of algal origin and seafood safety. *Journal of Aquatic Food Product Technology*, 12, 29–53. https://doi.org/10.1300/J030v12n01_03.

Bane, V., Lehane, M., Dikshit, M., O'Riordan, A. & Furey, A. (2014) Tetrodotoxin: Chemistry, toxicity, source, distribution and detection. *Toxins*, 6, 693–755. https://doi.org/10.3390/toxins6020693.

Baniga, Z., Hounmanou, Y.M.G., Kudirkiene, E., Kusiluka, L.J.M., Mdegela, R.H. & Dalsgaard, A. (2020) Genome-based analysis of extended-spectrum β-lactamase-producing *Escherichia coli* in the aquatic environment and Nile perch (*Lates niloticus*) of Lake Victoria, Tanzania. *Frontiers in Microbiology*, 11, 108. https://doi.org/10.3389/fmicb.2020.00108.

Barkham, T., Zadoks, R.N., Azmai, M.N.A., Baker, S., Bich, V.T.N., Chalker, V., Chau, M.L., Dance, D., Deepak, R.N., van Doorn, H.R., Gutierrez, R.A., Holmes, M.A., Huong, L.N.P., Koh, T.H., Martins, E., Mehershahi, K., Newton, P., Ng, L.C., Phuoc, N.N., Sangwichian, O., Sawatwong, P., Surin, U., Tan, T.Y., Tang, W.Y., Thuy, N.V., Turner, P., Vongsouvath, M., Zhang, D., Whistler, T. & Chen, S.L. (2019) One hypervirulent clone, sequence type 283, accounts for a large proportion of invasive *Streptococcus agalactiae* isolated from humans and diseased tilapia in Southeast Asia. *PLOS Neglected Tropical Diseases*, 13, e0007421. https://doi.org/10.1371/journal.pntd.0007421.

Barron, M.G. (2003) Bioaccumulation and bioconcentration in aquatic organisms. In *Handbook of Ecotoxicology* (edited by D.J. Hoffman, B.A. Rattner, G.A. Burton and J. Cairns). Lewis Publisher.

Bellou, M., Kokkinos, P. & Vantarakis, A. (2013) Shellfish-borne viral outbreaks: A systematic review. *Food and Environmental Virology*, 5, 13–23. https://doi.org/10.1007/s12560-012-9097-6.

Bigot-Clivot, A., Palos Ladeiro, M., Lepoutre, A., Bastien, F., Bonnard, I., Dubey, J.P., Villena, I., Aubert, D., Geffard, O., François, A. & Geffard, A. (2016) Bioaccumulation of *Toxoplasma* and *Cryptosporidium* by the freshwater crustacean *Gammarus fossarum*: Involvement in biomonitoring surveys and trophic transfer. *Ecotoxicology and Environmental Safety*, 133, 188–194. https://doi.org/10.1016/j.ecoenv.2016.07.006.

Bosch, A.C., O'Neill, B., Sigge, G.O., Kerwath, S.E. & Hoffman, L.C. (2016) Heavy metals in marine fish meat and consumer health: A review. *Journal of the Science of Food and Agriculture*, 96, 32–48. https://doi.org/10.1002/jsfa.7360.

Bouree, P., Paugam, A. & Petithory, J.C. (1995) Anisakidosis: Report of 25 cases and review of the literature. *Comparative Immunology, Microbiology and Infectious Diseases*, 18, 75–84. https://doi.org/10.1016/0147-9571(95)98848-c.

Brooks, B.W. & Conkle, J.L. (2019) Commentary: Perspectives on aquaculture, urbanization and water quality. *Comparative Biochemistry and Physiology. Toxicology and Pharmacology*, 217, 1–4. https://doi.org/10.1016/j.cbpc.2018.11.014.

Buchmann, K. & Mehrdana, F. (2016) Effects of anisakid nematodes *Anisakis simplex* (s.l.), *Pseudoterranova decipiens* (s.l.) and *Contracaecum osculatum* (s.l.) on fish and consumer health. *Food and Waterborne Parasitology*, 4, 13–22. https://doi.org/10.1016/j.fawpar.2016.07.003.

Burket, S.R., Sapozhnikova, Y., Zheng, J.S., Chung, S.S. & Brooks, B.W. (2018) At the intersection of urbanization, water, and food security: Determination of select contaminants of emerging concern in mussels and oysters from Hong Kong. *Journal of Agricultural and Food Chemistry*, 66, 5009–5017. https://doi.org/10.1021/acs.jafc.7b05730.

Bushkin-Bedient, S. & Carpenter, D.O. (2010) Benefits versus risks associated with consumption of fish and other seafood. *Reviews on Environmental Health*, 25, 161–191. https://doi.org/10.1515/reveh.2010.25.3.161.

Børresen, T. (2015) Microplastics—A new threat to aquatic food safety? *Journal of Aquatic Food Product Technology*, 24, 415–416. https://doi.org/10.1080/10498850.2015.1048618.

Cabral, H., Fonseca, V., Sousa, T. & Costa Leal, M. (2019) Synergistic effects of climate change and marine pollution: An overlooked interaction in coastal and estuarine areas. *International Journal of Environmental Research and Public Health*, 16. https://doi.org/10.3390/ijerph16152737.

Campos, C.J., Acornley, R., Morgan, O.C. & Kershaw, S. (2013) Trends in the levels of *Escherichia coli* in commercially harvested bivalve shellfish from England and Wales, 1999–2008. *Marine Pollution Bulletin*, 67, 223–227. https://doi.org/10.1016/j.marpolbul.2012.11.030.

Carvalho, F.P. (2017) Pesticides, environment, and food safety. *Food and Energy Security*, 6, 48–60. https://doi.org/10.1002/fes3.108.

Cedervall, T., Hansson, L.-A., Lard, M., Frohm, B. & Linse, S. (2012) Food chain transport of nanoparticles affects behaviour and fat metabolism in Fis. *PLOS ONE*, 2.

Chai, J.Y. & Jung, B.K. (2020) Food-borne intestinal flukes: A brief review of epidemiology and geographical distribution. *Acta Tropica*, 201, 105210. https://doi.org/10.1016/j.actatropica.2019.105210.

Chau, M.L., Chen, S.L., Yap, M., Hartantyo, S.H.P., Chiew, P.K.T., Fernandez, C.J., Wong, W.K., Fong, R.K., Tan, W.L., Tan, B.Z.Y., Ng, Y., Aung, K.T., Mehershahi, K.S., Goh, C., Kang, J.S.L., Barkham, T., Leong, A.O.K., Gutiérrez, R.A. & Ng, L.C. (2017) Group B *streptococcus* infections caused by improper sourcing and handling of fish for raw consumption, Singapore, 2015–2016. *Emerging Infectious Diseases*, 23, 2002–2010. https://doi.org/10.3201/eid2312.170596.

Cheng, B., Van Smeden, J., Deneer, J., Belgers, D., Foekema, E., Roessink, I., Matser, A. & Van den Brink, P.J. (2020) The chronic toxicity of emamectin benzoate to three marine benthic species using microcosms. *Ecotoxicology and Environmental Safety*, 194, 110452. https://doi.org/10.1016/j.ecoenv.2020.110452.

Cheng, J.Y. & Lo, I.M. (2016) Investigation of the available technologies and their feasibility for the conversion of food waste into fish feed in Hong Kong. *Environmental Science and Pollution Research International*, 23, 7169–7177. https://doi.org/10.1007/s11356-015-4668-3.

Chiang, T.Y., Kuo, M.C., Chen, C.H., Yang, J.Y., Kao, C.F., Ji, D.D. & Fang, C.T. (2014) Risk factors for acute *Toxoplasma gondii* diseases in Taiwan: A population-based case-control study. *PLOS ONE*, 9, e90880. https://doi.org/10.1371/journal.pone.0090880.

Chuah, L.O., Effarizah, M.E., Goni, A.M. & Rusul, G. (2016) Antibiotic application and emergence of multiple antibiotic resistance (MAR) in global catfish aquaculture. *Current Environmental Health Reports*, 3, 118–127. https://doi.org/10.1007/s40572-016-0091-2.

Chuboon, S., Wongsawad, C., Ruamsuk, A. & Nithikathkul, C. (2005) Survival of *Haplorchis taichui* metacercariae in Lab-Pla, Thai traditional food preparation. *Southeast Asian Journal of Tropical Medicine and Public Health*, 36 (Supplement 4), 110–111.

Clarke, J.L., Waheed, M.T., Lössl, A.G., Martinussen, I. & Daniell, H. (2013) How can plant genetic engineering contribute to cost-effective fish vaccine development for promoting sustainable aquaculture? *Plant Molecular Biology*, 83, 33–40. https://doi.org/10.1007/s11103-013-0081-9.

Codex Alimentarius (2018) *Maximum Residue Limits (MRLs) and Risk Management Recommendations (RMRs) for Residues of Veterinary Drugs in Foods CX/MRL 2–2018*. Food and Agriculture Organization: Rome.

Coupe, A., Howe, L., Burrows, E., Sine, A., Pita, A., Velathanthiri, N., Vallée, E., Hayman, D., Shapiro, K. & Roe, W.D. (2018) First report of *Toxoplasma gondii* sporulated oocysts and *Giardia duodenalis* in commercial green-lipped mussels (*Perna canaliculus*) in New Zealand. *Parasitology Research*, 117, 1453–1463. https://doi.org/10.1007/s00436-018-5832-8.

Couso-Pérez, S., Ares-Mazás, E. & Gómez-Couso, H. (2019) First report of *Cryptosporidium molnari*-like genotype and *Cryptosporidium parvum* zoonotic subtypes (IIaA15G2R1 and IIaA18G3R1) in brown trout (*Salmo trutta*). *Journal of Parasitology*, 105, 170–179. https://doi.org/10.1645/18-83.

Cowie, R.H. (2013) Pathways for transmission of angiostrongyliasis and the risk of disease associated with them. *Hawai'i Journal of Medicine and Public Health: A Journal of Asia Pacific Medicine and Public Health*, 72 (Supplement 2), 70–74.

Dang, S.T. & Dalsgaard, A. (2012) *Escherichia coli* contamination of fish raised in integrated pig-fish aquaculture systems in Vietnam. *Journal of Food Protection*, 75, 1317–1319. https://doi.org/10.4315/0362-028X.JFP-11-501.

Delannoy, C.M.J., Crumlish, M., Fontaine, M.C., Pollock, J., Foster, G., Dagleish, M.P., Turnbull, J.F. & Zadoks, R.N. (2013) Human *Streptococcus agalactiae* strains in aquatic mammals and fish. *BMC Microbiology*, 13, 41. https://doi.org/10.1186/1471-2180-13-41.

DePaola, A., Jones, J.L., Woods, J., Burkhardt, W., III, Calci, K.R., Krantz, J.A., Bowers, J.C., Kasturi, K., Byars, R.H., Jacobs, E., Williams-Hill, D. & Nabe, K. (2010) Bacterial and viral pathogens in live oysters: 2007 United States market survey. *Applied and Environmental Microbiology*, 76, 2754–2768. https://doi.org/10.1128/AEM.02590-09.

Dickhoff, W., Collier, T. & Varanasi, U. (2007) The seafood 'dilemna' – A way forward. *Fisheries*, 5, 805–809.

DiNubile, M.J. & Hokama, Y. (1995) The ciguatera poisoning syndrome from farm-raised salmon. *Annals of Internal Medicine*, 122, 113–114. https://doi.org/10.7326/0003-4819-122-2-199501150-00006.

Doron, A. & Broom, A. (2019) The spectre of superbugs: Waste, structural violence and antimicrobial resistance in India. *Worldwide Waste*, 2. https://doi.org/10.5334/wwwj.20.

Eamsobhana, P., Yoolek, A. & Yong, H.S. (2010) Effect of Thai 'koi-hoi' food flavoring on the viability and infectivity of the third-stage larvae of *Angiostrongylus cantonensis* (Nematoda: Angiostrongylidae). *Acta Tropica*, 113, 245–247. https://doi.org/10.1016/j.actatropica.2009.11.004.

EFSA (2007) Scientific opinion of the panel on biological hazards on a request from the European Commission on public health risks involved in the human consumption of reptile meat. *EFSA Journal*, 578, 1–55.

EFSA Panel on Biological Hazards (BIOHAZ) (2010) Scientific opinion on risk assessment of parasites in fishery products. *EFSA Journal*, 8, 1543. https://doi.org/10.2903/j.efsa.2010.1543.

EFSA (2016) Presence of microplastics and nanoplastics in food, with particular focus on seafood. *EFSA* Journal, 14(6). https://doi.org/10.2903/j.efsa.2016.4501.

Engelstad, M. (2005) Vaccination and consumer perception of seafood quality. *Developments in Biologicals (Basel)*, 121, 245–254.

Epelboin, L., Blondé, R., Chamouine, A., Chrisment, A., Diancourt, L., Villemant, N., Atale, A., Cadix, C., Caro, V., Malvy, D. & Collet, L. (2016) *Angiostrongylus cantonensis* infection on Mayotte Island, Indian Ocean, 2007–2012. *PLOS Neglected Tropical Diseases*, 10, e0004635. https://doi.org/10.1371/journal.pntd.0004635.

Esmerini, P.O., Gennari, S.M. & Pena, H.F. (2010) Analysis of marine bivalve shellfish from the fish market in Santos city, São Paulo state, Brazil, for *Toxoplasma gondii*. *Veterinary Parasitology*, 170, 8–13. https://doi.org/10.1016/j.vetpar.2010.01.036.

Etheridge, S.M. (2010) Paralytic shellfish poisoning: Seafood safety and human health perspectives. *Toxicon*, 56, 108–122. https://doi.org/10.1016/j.toxicon.2009.12.013.

European Agency for the Evaluation of Medicinal Products (2002) *Florfenicol Extension to Fish*. European Union.

FAO (2003) Heavy metal regulations – FAOLEX. Legal Notice no. 66/2003.

FAO (2017) FAO aquaculture newsletter. *Newsletter*, 57.

FAO (2018) The state of world fisheries and aquaculture, 2018. *Meeting the Sustainable Development Goals*. FAO: Rome.

FAO/WHO (2012) *Codex Alimentarius – Code of Practice for Fish and Fishery Products*.

FAO/WHO (2016) *Understanding Codex*. Food and Agriculture Organization of the United Nations and World Health Organization: Rome.

FAO/WHO (2020) *Codex Alimentarius – Guidelines*. Food and Agriculture Organization/World Health Organization: Rome.

Federspiel, F., Skovmand, S. & Skarphedinsson, S. (2020) Eosinophilic meningitis due to *Angiostrongylus cantonensis* in Europe. *International Journal of Infectious Diseases*, 93, 28–39. https://doi.org/10.1016/j.ijid.2020.01.012.

Field, J. (1998) Puffer fish poisoning. *Journal of Accident and Emergency Medicine*, 15, 334–336. https://doi.org/10.1136/emj.15.5.334.

Fierens, S., Mairesse, H., Heilier, J.F., De Burbure, C., Focant, J.F., Eppe, G., De Pauw, E. & Bernard, A. (2003) Dioxin/polychlorinated biphenyl body burden, diabetes and endometriosis: Findings in a population-based study in Belgium. *Biomarkers*, 8, 529–534. https://doi.org/10.1080/1354750032000158420.

Forrer, A., Sayasone, S., Vounatsou, P., Vonghachack, Y., Bouakhasith, D., Vogt, S., Glaser, R., Utzinger, J., Akkhavong, K. & Odermatt, P. (2012) Spatial distribution of, and risk factors for *Opisthorchis viverrini* infection in southern Lao PDR. *PLOS Neglected Tropical Diseases*, 6, e1481. https://doi.org/10.1371/journal.pntd.0001481.

Friedman, M.A., Fleming, L.E., Fernandez, M., Bienfang, P., Schrank, K., Dickey, R., Bottein, M.Y., Backer, L., Ayyar, R., Weisman, R., Watkins, S., Granade, R. & Reich, A. (2008) Ciguatera fish poisoning: Treatment, prevention and management. *Marine Drugs*, 6, 456–479. https://doi.org/10.3390/md20080022.

Friedman, M.A., Fernandez, M., Backer, L.C., Dickey, R.W., Bernstein, J., Schrank, K., Kibler, S., Stephan, W., Gribble, M.O., Bienfang, P., Bowen, R.E., Degrasse, S., Flores Quintana, H.A., Loeffler, C.R., Weisman, R., Blythe, D., Berdalet, E., Ayyar, R., Clarkson-Townsend, D., Swajian, K., Benner, R., Brewer, T. & Fleming, L.E. (2017) An updated review of ciguatera fish poisoning: Clinical, epidemiological, environmental, and public health management. *Marine Drugs*, 15, 72. https://doi.org/10.3390/md15030072.

Fusco, G., Anastasio, A., Kingsley, D.H., Amoroso, M.G., Pepe, T., Fratamico, P.M., Cioffi, B., Rossi, R., La Rosa, G. & Boccia, F. (2019) Detection of hepatitis A virus and other enteric viruses in shellfish collected in the gulf of Naples, Italy. *International Journal of Environmental Research and Public Health*, 16. https://doi.org/10.3390/ijerph16142588.

Fussy, A., Pommier, P., Lumbroso, C. & de Haro, L. (2007) Chelonitoxism: New case reports in French Polynesia and review of the literature. *Toxicon*, 49, 827–832. https://doi.org/10.1016/j.toxicon.2006.12.002.

Ganapathiraju, P.V., Gharpure, R., Thomas, D., Millet, N., Gurrieri, D., Chatham-Stephens, K., Dykes, J., Luquez, C., Dinavahi, P., Ganapathiraju, S., Roger, S., Abbasi, D., Higgins, N., Loftus, F. & Trivedi, M. (2019) Notes from the Field: Botulism Type E after consumption of salt-cured fish – New Jersey, 2018. *MMWR. Morbidity and Mortality Weekly Report*, 68, 1008–1009. https://doi.org/10.15585/mmwr.mm6844a3.

García, C., del Carmen Bravo, M., Lagos, M. & Lagos, N. (2004) Paralytic shellfish poisoning: Post-mortem analysis of tissue and body fluid samples from human victims in the Patagonia fjords. *Toxicon*, 43, 149–158. https://doi.org/10.1016/j.toxicon.2003.11.018.

Garcia Martinez, M. & Brofman Epelbaum, F.M. (2011) The role of traceability in restoring consumer trust in food chains. In: *Food Chain Integrity*, Vol. 17 (edited by J. Hoorfar, K. Jordan, F. Butler & R. Prugger). Woodhead Publishing: Cambridge, UK, pp. 294–302.

Garrett, E.S., Jahncke, M.L. & Martin, R.E. (2000) Applications of HACCP principles to address food safety and other issues in aquaculture. *Journal of Aquatic Food Product Technology*, 9, 5–20. https://doi.org/10.1300/J030v09n01_02.

Garrett, T.R., Bhakoo, M. & Zhang, Z. (2008) Bacterial adhesion and biofilms on surfaces, *Progress in Natural Science*, 18(9). https://doi.org/10.1016/j.pnsc.2008.04.001.

Glibert, P.M., Anderson, D.M., Gentien, P., Granéli, E. & Sellner, K.G. (2005) The global, complex phenomena of harmful algal blooms. *Oceanography*, 18, 136–147. https://doi.org/10.5670/oceanog.2005.49.

Gobler, C.J. (2020) Climate change and harmful algal blooms: insights and perspective. *Harmful Algae*, 91, 101731. https://doi.org/10.1016/j.hal.2019.101731.

Goffredo, E., Azzarito, L., Di Taranto, P., Mancini, M.E., Normanno, G., Didonna, A., Faleo, S., Occhiochiuso, G., D'Attoli, L., Pedarra, C., Pinto, P., Cammilleri, G., Graci, S., Sciortino, S. & Costa, A. (2019) Prevalence of anisakid parasites in fish collected from Apulia region (Italy) and quantification of nematode larvae in flesh. *International Journal of Food Microbiology*, 292, 159–170. https://doi.org/10.1016/j.ijfoodmicro.2018.12.025.

Gómez-Couso, H., Freire-Santos, F., Martínez-Urtaza, J., García-Martín, O. & Ares-Mazás, M.E. (2003) Contamination of bivalve molluscs by *Cryptosporidium* oocysts: The need for new quality control standards. *International Journal of Food Microbiology*, 87, 97–105. https://doi.org/10.1016/s0168-1605(03)00057-6.

Grattan, L.M., Holobaugh, S. & Morris, J.G., Jr (2016) Harmful algal blooms and public health. *Harmful Algae*, 57, 2–8. https://doi.org/10.1016/j.hal.2016.05.003.

Han, Y., Wang, J., Zhao, Z., Chen, J., Lu, H. & Liu, G. (2019) Detecting antibiotic resistance genes and human potential pathogenic bacteria in fishmeal by culture-independent method. *Environmental Science and Pollution Research International*, 26, 8665–8674. https://doi.org/10.1007/s11356-019-04303-1.

Hansen, C.H., Vogel, B.F. & Gram, L. (2006) Prevalence and survival of *Listeria monocytogenes* in Danish aquatic and fish-processing environments. *Journal of Food Protection*, 69, 2113–2122. https://doi.org/10.4315/0362-028x-69.9.2113.

Haque, M. A., Islam, Q. T. & Ekram, A. S. (2009) Puffer fish poisoning. *TAJ: Journal of Teachers Association*, 21(2), 199–202. https://doi.org/10.3329/taj.v21i2.3806.

Harada, M. (1995) Minamata disease: Methylmercury poisoning in Japan caused by environmental pollution. *Critical Reviews in Toxicology*, 25, 1–24. https://doi.org/10.3109/10408449509089885.

Hicks, D.T. (2016) Seafood safety and quality: The consumer's role. *Foods*, 5, 71. https://doi.org/10.3390/foods5040071.

Hollman, P.C.H., Bouwmeester, H. & Peters, R.J.B. (2013) Microplastics in the aquatic food chain – Sources, measurements, occurrence and potential health risks. RIKILT Report 2013.003. Wageningen University and Research Center.

Hsieh, P.Y.-H., Leong, F.-M. & Rudloe, J. (2000) Jellyfish as food. *Hydrobiologia*, 451, 11–17. https://doi.org/10.1023%2FA%3A1011875720415.

Huang, X., Hites, R.A., Foran, J.A., Hamilton, C., Knuth, B.A., Schwager, S.J. & Carpenter, D.O. (2006) Consumption advisories for salmon based on risk of cancer and noncancer health effects. *Environmental Research*, 101, 263–274. https://doi.org/10.1016/j.envres.2005.08.011.

Huang, Y., Huang, D., Geng, Y., Fang, S., Yang, F., Wu, C., Zhang, H., Wang, M., Zhang, R., Wang, X., Wu, S., Cao, J. & Zhang, R. (2017) An integrated control strategy takes *Clonorchis sinensis* under control in an endemic area in South China. *Vector Borne and Zoonotic Diseases*, 17, 791–798. https://doi.org/10.1089/vbz.2017.2133.

Hungerford, J.M. (2010) Scombroid poisoning: A review. *Toxicon*, 56, 231–243. https://doi.org/10.1016/j.toxicon.2010.02.006.

Huss, H.H., Ababouch, L. & Gram, L. (2003) Physical hazards. In: *Assessment and Management of Seafood Safety and Quality*, Vol. 444 (edited by F.F.T. Paper). Food and Agriculture Organization of the United Nations: Rome.

Huss, H.H., Reilly, A. & Karim Ben Embarek, P. (2000) Prevention and control of hazards in seafood. *Food Control*, 11, 149–156. https://doi.org/10.1016/S0956-7135(99)00087-0.

Ibelings, B.W. & Chorus, I. (2007) Accumulation of cyanobacterial toxins in freshwater 'seafood' and its consequences for public health: A review. *Environmental Pollution*, 150, 177–192. https://doi.org/10.1016/j.envpol.2007.04.012.

Igbinosa, I.H., Igumbor, E.U., Aghdasi, F., Tom, M. & Okoh, A.I. (2012) Emerging *Aeromonas* species infections and their significance in public health. *TheScientificWorldJournal*, 2012, 625023. https://doi.org/10.1100/2012/625023.

Ilic, N., Velebit, B., Teodorovic, V., Djordjevic, V., Karabasil, N., Vasilev, D., Djuric, S., Adzic, B. & Dimitrijevic, M. (2017) Influence of environmental conditions on Norovirus presence in mussels harvested in Montenegro. *Food and Environmental Virology*, 9, 406–414. https://doi.org/10.1007/s12560-017-9298-0.

Japan News (2022) The history of sushi. https://jw-webmagazine.com/history-of-sushi/.

Jeamsripong, S., Chuanchuen, R. & Atwill, E.R. (2018) Assessment of bacterial accumulation and environmental factors in sentinel oysters and estuarine water quality from the Phang Nga estuary area in Thailand. *International Journal of Environmental Research and Public Health*, 15. https://doi.org/10.3390/ijerph15091970.

Jennings, S., Stentiford, G.D., Leocadio, A.M., Jeffery, K.R., Metcalfe, J.D., Katsiadaki, I., Auchterlonie, N.A., Mangi, S.C., Pinnegar, J.K., Ellis, T., Peeler, E.J., Luisetti, T., Baker-Austin, C., Brown, M., Catchpole, T.L., Clyne, F.J., Dye, S.R., Edmonds, N.J., Hyder, K., Lee, J., Lees, D.N., Morgan, O.C., O'Brien, C.M., Oidtmann, B., Posen, P.E., Santos, A.R., Taylor, N.G.H., Turner, A.D., Townhill, B.L. & Verner-Jeffreys, D.W. (2016) Aquatic food security: Insights into challenges and solutions from an analysis of interactions between fisheries, aquaculture, food safety, human health, fish and human welfare, economy and environment. *Fish and Fisheries*, 17, 893–938. https://doi.org/10.1111/faf.12152.

Jirathana, P. (1998) Constraints experienced by developing countries in the development and application of HACCP. *Food Control*, 9, 97–100. https://doi.org/10.1016/S0956-7135(98)00011-5.

Johri, N., Jacquillet, G. & Unwin, R. (2010) Heavy metal poisoning: The effects of cadmium on the kidney. *BioMetals*, 23, 783–792. https://doi.org/10.1007/s10534-010-9328-y.

Jones, J.L., Dargelas, V., Roberts, J., Press, C., Remington, J.S. & Montoya, J.G. (2009) Risk factors for *Toxoplasma gondii* infection in the United States. *Clinical Infectious Diseases*, 49, 878–884. https://doi.org/10.1086/605433.

Jongmans, E., Wichels, A. & Gunnar, C. (2018) Unravelling the pathogenic potential of microplastics in the Weser estuary (Germany). MICRO 2018, Fate and Impact of Microplastics: Knowledge, Actions and Solutions, 19–23 November, Lanzarote.

Kajikazawa, T., Sugita, T., Takashima, M. & Nishikawa, A. (2007) Detection of pathogenic yeasts from processed fresh edible sea urchins sold in a fish market. *Nihon Ishinkin Gakkai Zasshi*, 48, 169–172. https://doi.org/10.3314/jjmm.48.169.

Kapaj, S., Peterson, H., Liber, K. & Bhattacharya, P. (2006) Human health effects from chronic arsenic poisoning – A review. *Journal of Environmental Science and Health. Part A, Toxic/Hazardous Substances and Environmental Engineering*, 41, 2399–2428. https://doi.org/10.1080/10934520600873571.

Katagi, T. (2010) Bioconcentration, bioaccumulation, and metabolism of pesticides in aquatic organisms. *Reviews of Environment Contamination and Toxicology*, 204, 1–132.

Khamboonruang, C., Keawvichit, R., Wongworapat, K., Suwanrangsi, S., Hongpromyart, M., Sukhawat, K., Tonguthai, K. & Lima dos Santos, C.A. (1997) Application of hazard analysis critical control point (HACCP) as a possible control measure for *Opisthorchis viverrini* infection in cultured carp (*Puntius gonionotus*). *Southeast Asian Journal of Tropical Medicine and Public Health*, 28 (Supplement 1), 65–72.

Khor, D.X., Fernandez, C.J., Chee, D.M., Teo, X.H., Han, Z.Y., Jiang, J.H., Neo, H.S., Chang, S.F. & Yap, H.H. (2019) Food safety impacts of finfish and crustacean aquaculture on food security in Asia. *Revue Scientifique et Technique*, 38, 629–639. https://doi.org/10.20506/rst.38.2.3009.

Kim, H.W., Hong, Y.J., Jo, J.I., Ha, S.D., Kim, S.H., Lee, H.J. & Rhee, M.S. (2017) Raw ready-to-eat seafood safety: Microbiological quality of the various seafood species available in fishery, hyper and online markets. *Letters in Applied Microbiology*, 64, 27–34. https://doi.org/10.1111/lam.12688.

Klugman, K.P. & Black, S. (2018) Impact of existing vaccines in reducing antibiotic resistance: Primary and secondary effects. *Proceedings of the National Academy of Sciences of the United States of America*, 115, 12896–12901. https://doi.org/10.1073/pnas.1721095115.

Koh, T.H., Sng, L.H., Yuen, S.M., Thomas, C.K., Tan, P.L., Tan, S.H. & Wong, N.S. (2009) Streptococcal cellulitis following preparation of fresh raw seafood. *Zoonoses and Public Health*, 56, 206–208. https://doi.org/10.1111/j.1863-2378.2008.01213.x.

Konrad, S., Paduraru, P., Romero-Barrios, P., Henderson, S.B. & Galanis, E. (2017) Remote sensing measurements of sea surface temperature as an indicator of *Vibrio parahaemolyticus* in oyster meat and human illnesses. *Environmental Health: A Global Access Science Source*, 16, 92. https://doi.org/10.1186/s12940-017-0301-x.

Kwon, J.W. (2016) Environmental impact assessment of veterinary drug on fish aquaculture for food safety. *Drug Testing and Analysis*, 8, 556–564. https://doi.org/10.1002/dta.2007.

Lam, C.T., Rosanowski, S.M., Walker, M. & St-Hilaire, S. (2020) Sea lice exposure to non-lethal levels of emamectin benzoate after treatments: A potential risk factor for drug resistance. *Scientific Reports*, 10, 932. https://doi.org/10.1038/s41598-020-57594-7.

Landrigan, P.J., Stegeman, J.J., Fleming, L.E., Allemand, D., Anderson, D.M., Backer, L.C., Brucker-Davis, F., Chevalier, N., Corra, L., Czerucka, D., Bottein, M.D., Demeneix, B., Depledge, M., Deheyn, D.D., Dorman, C.J., Fénichel, P., Fisher, S., Gaill, F., Galgani, F., Gaze, W.H., Giuliano, L., Grandjean, P., Hahn, M.E., Hamdoun, A., Hess, P., Judson, B., Laborde, A., McGlade, J., Mu, J., Mustapha, A., Neira, M., Noble, R.T., Pedrotti, M.L., Reddy, C., Rocklöv, J., Scharler, U.M., Shanmugam, H., Taghian, G., van de Water, J.A.J.M., Vezzulli, L., Weihe, P., Zeka, A., Raps, H. & Rampal, P. (2020) Human health and ocean pollution. *Annals of Global Health*, 86, 151. https://doi.org/10.5334/aogh.2831.

Lee, D.H., Lee, I.K., Porta, M., Steffes, M. & Jacobs, D.R., Jr (2007) Relationship between serum concentrations of persistent organic pollutants and the prevalence of metabolic syndrome among non-diabetic adults: Results from the National Health and Nutrition Examination survey 1999–2002. *Diabetologia*, 50, 1841–1851. https://doi.org/10.1007/s00125-007-0755-4.

Lemasson, A.J., Hall-Spencer, J.M., Kuri, V. & Knights, A.M. (2019) Changes in the biochemical and nutrient composition of seafood due to ocean acidification and warming. *Marine Environmental Research*, 143, 82–92. https://doi.org/10.1016/j.marenvres.2018.11.006.

Li, K., Clausen, J.H., Murrell, K.D., Liu, L. & Dalsgaard, A. (2013) Risks for fishborne zoonotic trematodes in tilapia production systems in Guangdong Province, China. *Veterinary Parasitology*, 198, 223–229. https://doi.org/10.1016/j.vetpar.2013.08.011.

Lim, J.H. (2011) Liver flukes: The malady neglected. *Korean Journal of Radiology*, 12, 269–279. https://doi.org/10.3348/kjr.2011.12.3.269.

Lloret, J., Rätz, H.-J., Lleonart, J. & Demestre, M. (2016) Challenging the links between seafood and human health in the context of global change. *Journal of the Marine Biological Association of the United Kingdom*, 96, 29–42. https://doi.org/10.1017/S0025315415001988.

Lodder, W.J., Rutjes, S.A., Takumi, K. & de Roda Husman, A.M. (2013) Aichi virus in sewage and surface water, the Netherlands. *Emerging Infectious Diseases*, 19, 1222–1230. https://doi.org/10.3201/eid1908.130312.

Lopes, V.M., Lopes, A.R., Costa, P. & Rosa, R. (2013) Cephalopods as vectors of harmful algal bloom toxins in marine food webs. *Marine Drugs*, 11, 3381–3409. https://doi.org/10.3390/md11093381.

Lun, Z.R., Gasser, R.B., Lai, D.H., Li, A.X., Zhu, X.Q., Yu, X.B. & Fang, Y.Y. (2005) Clonorchiasis: A key food-borne zoonosis in China. *Lancet. Infectious Diseases*, 5, 31–41. https://doi.org/10.1016/S1473-3099(04)01252-6.

Lunestad, B.T., Maage, A., Roiha, I.S., Myrmel, M., Svanevik, C.S. & Duinker, A. (2016) An outbreak of Norovirus infection from shellfish soup due to unforeseen insufficient heating during preparation. *Food and Environmental Virology*, 8, 231–234. https://doi.org/10.1007/s12560-016-9245-5.

Lusher, A., Hollman, P. & Mendoza-Hill, J. (2017) Microplastics in fisheries and aquaculture: Status of knowledge on their occurrence and implications for aquatic organisms and food safety. FAO Fisheries and Aquaculture Technical Paper.

Macpherson, C.N. (2005) Human behaviour and the epidemiology of parasitic zoonoses. *International Journal for Parasitology*, 35, 1319–1331. https://doi.org/10.1016/j.ijpara.2005.06.004.

Magnino, S., Colin, P., Dei-Cas, E., Madsen, M., McLauchlin, J., Nöckler, K., Maradona, M.P., Tsigarida, E., Vanopdenbosch, E. & Van Peteghem, C. (2009) Biological risks associated with consumption of reptile products. *International Journal of Food Microbiology*, 134, 163–175. https://doi.org/10.1016/j.ijfoodmicro.2009.07.001.

Mainero Rocca, L., Gentili, A., Pérez-Fernández, V. & Tomai, P. (2017) Veterinary drugs residues: A review of the latest analytical research on sample preparation and LC-MS based methods. *Food Additives and Contaminants. Part A, Chemistry, Analysis, Control, Exposure and Risk Assessment*, 34, 766–784. https://doi.org/10.1080/19440049.2017.1298846.

Marcotrigiano, G.O. & Storelli, M.M. (2003) Heavy metal, polychlorinated biphenyl and organochlorine pesticide residues in marine organisms: Risk evaluation for consumers. *Veterinary Research Communications*, 27 (Supplement 1), 183–195. https://doi.org/10.1023/b:verc.0000014137.02422.f4.

Marquis, N.D., Bishop, T.J., Record, N.R., Countway, P.D. & Fernández Robledo, J.A. (2019) Molecular epizootiology of *Toxoplasma gondii* and *Cryptosporidium parvum* in the eastern oyster (*Crassostrea virginica*) from Maine (USA). *Pathogens*, 8. https://doi.org/10.3390/pathogens8030125.

Martínez, O., Rodríguez-Calleja, J.M., Santos, J.A., Otero, A. & García-López, M.L. (2009) Food-borne and indicator bacteria in farmed molluscan shellfish before and after depuration. *Journal of Food Protection*, 72, 1443–1449. https://doi.org/10.4315/0362-028x-72.7.1443.

McLaughlin, J.B., DePaola, A., Bopp, C.A., Martinek, K.A., Napolilli, N.P., Allison, C.G., Murray, S.L., Thompson, E.C., Bird, M.M. & Middaugh, J.P. (2005) Outbreak of *Vibrio parahaemolyticus* gastroenteritis associated with Alaskan oysters. *New England Journal of Medicine*, 353, 1463–1470. https://doi.org/10.1056/NEJMoa051594.

Miget, R. (2004) *The HACCP Seafood Program and Aquaculture*. Southern Regional Aquaculture Centre. SRAC Publication No. 4900.

Morabito, S., Silvestro, S. & Faggio, C. (2018) How the marine biotoxins affect human health. *Natural Product Research*, 32, 621–631. https://doi.org/10.1080/14786419.2017.1329734.

Mustafa, M., Yoshida, T. & Waheed, Z. (2018) Seafood poisoning, symptom, treatment and prevention. *Borneo Journal of Marine Science and Aquaculture*, 2, 64–69. https://doi.org/10.51200/bjomsa.v2i0.1252.

Musto, J., Kirk, M., Lightfoot, D., Combs, B.G. & Mwanri, L. (2006) Multi-drug resistant *Salmonella* Java infections acquired from tropical fish aquariums, Australia, 2003–04. *Communicable Diseases Intelligence Quarterly Report*, 30, 222–227.

Muziasari, W.I., Pärnänen, K., Johnson, T.A., Lyra, C., Karkman, A., Stedtfeld, R.D., Tamminen, M., Tiedje, J.M. & Virta, M. (2016) Aquaculture changes the profile of antibiotic resistance and mobile genetic element associated genes in Baltic Sea sediments. *FEMS Microbiology Ecology*, 92, fiw052. https://doi.org/10.1093/femsec/fiw052.

Nácher-Mestre, J., Serrano, R., Beltrán, E., Pérez-Sánchez, J., Silva, J., Karalazos, V., Hernández, F. & Berntssen, M.H. (2015) Occurrence and potential transfer of mycotoxins in gilthead sea bream and Atlantic salmon by use of novel alternative feed ingredients. *Chemosphere*, 128, 314–320. https://doi.org/10.1016/j.chemosphere.2015.02.021.

Nieuwenhuizen, N.E. & Lopata, A.L. (2013) *Anisakis* – A food-borne parasite that triggers allergic host defences. *International Journal for Parasitology*, 43, 1047–1057. https://doi.org/10.1016/j.ijpara.2013.08.001.

Noor Uddin, G.M., Larsen, M.H., Barco, L., Minh Phu, T. & Dalsgaard, A. (2015) Clonal occurrence of *Salmonella* Weltevreden in cultured shrimp in the Mekong Delta, Vietnam. *PLOS ONE*, 10, e0134252. https://doi.org/10.1371/journal.pone.0134252.

O'Keeffe, M. & Kennedy, O. (1998) Residues – A food safety problem? *Journal of Food Safety*, 18, 297–319. https://doi.org/10.1111/j.1745-4565.1998.tb00222.x.

Okocha, R.C., Olatoye, I.O. & Adedeji, O.B. (2018) Food safety impacts of antimicrobial use and their residues in aquaculture. *Public Health Reviews*, 39, 21–21. https://doi.org/10.1186/s40985-018-0099-2.

Ortega, D. L., Wang, H. H. & Olynk Widmar, N. J. (2014) Aquaculture imports from Asia: An analysis of U.S. consumer demand for select food quality attributes. *Agricultural Economics*, 45(5), 625–634. https://doi.org/10.1111/agec.12111.

Pao, S., Ettinger, M.R., Khalid, M.F., Reid, A.O. & Nerrie, B.L. (2008) Microbial quality of raw aquacultured fish fillets procured from Internet and local retail markets. *Journal of Food Protection*, 71, 1544–1549. https://doi.org/10.4315/0362-028x-71.8.1544.

Paul-Pont, I., Labreuche, Y., Chalopin, M., Dubreuil, C., Maignien, L., Cassone, A.L., Petton, B., Lambert, C., Reveillaud, J., Piel, D. et al. (2018) Are microplastics vehicles of bacteria or virus pathogens for marine bivalves? Micro 2018: Fate and Impact of Microplastics, 19–23 November, Lanzarote.

Phan, V.T., Ersbøll, A.K., Do, D.T. & Dalsgaard, A. (2011) Raw-fish-eating behavior and fishborne zoonotic trematode infection in people of northern Vietnam. *Food-borne Pathogens and Disease*, 8, 255–260. https://doi.org/10.1089/fpd.2010.0670.

Powell, A., Treasurer, J.W., Pooley, C.L., Keay, A.J., Lloyd, R., Imsland, A.K. & Garcia de Leaniz, C. (2018) Use of lumpfish for sea-lice control in salmon farming: Challenges and opportunities. *Reviews in Aquaculture*, 10, 683–702. https://doi.org/10.1111/raq.12194.

Prato, R., Martinelli, D., Tafuri, S., Barbuti, G., Quarto, M., Germinario, C.A. & Chironna, M. (2013) Safety of shellfish and epidemiological pattern of enterically transmitted diseases in Italy. *International Journal of Food Microbiology*, 162, 125–128. https://doi.org/10.1016/j.ijfoodmicro.2012.12.025.

Preena, P.G., Swaminathan, T.R., Kumar, V.J.R. & Singh, I.S.B. (2020) Antimicrobial resistance in aquaculture: A crisis for concern. *Biologia*, 75, 1497–1517. https://doi.org/10.2478/s11756-020-00456-4.

Prüss-Ustün, A., Vickers, C., Haefliger, P. & Bertollini, R. (2011) Knowns and unknowns on burden of disease due to chemicals: A systematic review. *Environmental Health: A Global Access Science Source*, 10, 9. https://doi.org/10.1186/1476-069X-10-9.

Qian, M.B., Chen, Y.D., Fang, Y.Y., Tan, T., Zhu, T.J., Zhou, C.H., Wang, G.F., Xu, L.Q. & Zhou, X.N. (2013) Epidemiological profile of *Clonorchis sinensis* infection in one community, Guangdong, People's Republic of China. *Parasites and Vectors*, 6, 194. https://doi.org/10.1186/1756-3305-6-194.

Quiñones, R.A., Fuentes, M., Montes, R.M., Soto, D. & León-Muñoz, J. (2019) Environmental issues in Chilean salmon farming: A review. *Reviews in Aquaculture*, 11, 375–402. https://doi.org/10.1111/raq.12337.

Raatz, S.K., Silverstein, J.T., Jahns, L. & Picklo, M.J. (2013) Issues of fish consumption for cardiovascular disease risk reduction. *Nutrients*, 5, 1081–1097. https://doi.org/10.3390/nu5041081.

Reichelt-Brushett, A.J., Stone, J., Howe, P., Thomas, B., Clark, M., Male, Y., Nanlohy, A. & Butcher, P. (2017) Geochemistry and mercury contamination in receiving environments of artisanal mining wastes and identified concerns for food safety. *Environmental Research*, 152, 407–418. https://doi.org/10.1016/j.envres.2016.07.007.

Reilly, A. & Käferstein, F. (1997) Food safety hazards and the application of the principles of the hazard analysis and critical control point (HACCP) system for their control in aquaculture production. *Aquaculture Research*, 28, 735–752. https://doi.org/10.1046/j.1365-2109.1997.00939.x.

Robertson, L.J. (2007) The potential for marine bivalve shellfish to act as transmission vehicles for outbreaks of protozoan infections in humans: A review. *International Journal of Food Microbiology*, 120, 201–216. https://doi.org/10.1016/j.ijfoodmicro.2007.07.058.

Ropkins, K. & Beck, A.J. (2000) Evaluation of worldwide approaches to the use of HACCP to control food safety. *Trends in Food Science and Technology*, 11, 10–21. https://doi.org/10.1016/S0924-2244(00)00036-4.

Rose, S., Hill, R., Bermudez, L.E. & Miller-Morgan, T. (2013) Imported ornamental fish are colonized with antibiotic-resistant bacteria. *Journal of Fish Diseases*, 36, 533–542. https://doi.org/10.1111/jfd.12044.

Rylander, L., Rignell-Hydbom, A. & Hagmar, L. (2005) A cross-sectional study of the association between persistent organochlorine pollutants and diabetes. *Environmental Health: A Global Access Science Source*, 4, 28. https://doi.org/10.1186/1476-069X-4-28.

Saenna, P., Hurst, C., Echaubard, P., Wilcox, B.A. & Sripa, B. (2017) Fish sharing as a risk factor for *Opisthorchis viverrini* infection: Evidence from two villages in north-eastern Thailand. *Infectious Diseases of Poverty*, 6, 66–66. https://doi.org/10.1186/s40249-017-0281-7.

Saito, E., Yoshida, N., Kawano, J., Shimizu, A. & Igimi, S. (2011) Isolation of *Staphylococcus aureus* from raw fish in relation to culture methods. *Journal of Veterinary Medical Science*, 73, 287–292. https://doi.org/10.1292/jvms.10-0198.

Salati, F., Meloni, M., Cau, M. & Angelucci, G. (2013) Presence of *Contracaecum* spp. in teleosts cultured and fished in Sardinia. *Veterinary Parasitology*, 196, 382–387. https://doi.org/10.1016/j.vetpar.2013.03.014.

Sansawat, S. and Muliyil, V. (2011) Comparing Global Food Safety Initiative (GFSI) recognised standards. White paper. www.SGS.com.

Shah, S.Q., Colquhoun, D.J., Nikuli, H.L. & Sørum, H. (2012) Prevalence of antibiotic resistance genes in the bacterial flora of integrated fish farming environments of Pakistan and Tanzania. *Environmental Science and Technology*, 46, 8672–8679. https://doi.org/10.1021/es3018607.

Shamsi, S., Gasser, R. & Beveridge, I. (2013) Description and genetic characterization of *Hysterothylacium* (Nematoda: Raphidascarididae) larvae parasitic in Australian marine fishes. *Parasitology International*, 62, 320–328. https://doi.org/10.1016/j.parint.2012.10.001.

Shi, W., Zhao, X., Han, Y., Che, Z., Chai, X. & Liu, G. (2016) Ocean acidification increases cadmium accumulation in marine bivalves: A potential threat to seafood safety. *Scientific Reports*, 6, 20197. https://doi.org/10.1038/srep20197.

Shi, W., Han, Y., Guo, C., Su, W., Zhao, X., Zha, S., Wang, Y. & Liu, G. (2019) Ocean acidification increases the accumulation of titanium dioxide nanoparticles (nTiO2) in edible bivalve mollusks and poses a potential threat to seafood safety. *Scientific Reports*, 9, 3516. https://doi.org/10.1038/s41598-019-40047-1.

Shih, H.H., Ku, C.C. & Wang, C.S. (2010) *Anisakis simplex* (Nematoda: Anisakidae) third-stage larval infections of marine cage cultured cobia, *Rachycentron canadum* L., in Taiwan. *Veterinary Parasitology*, 171, 277–285. https://doi.org/10.1016/j.vetpar.2010.03.023.

Sithithaworn, P., Ziegler, A.D., Grundy-Warr, C., Andrews, R.H. & Petney, T.N. (2012) Changes to the life cycle of liver flukes: Dams, roads, and ponds. *Lancet. Infectious Diseases*, 12, 588. https://doi.org/10.1016/S1473-3099(12)70174-3.

Slepchenko, S.M., Gusev, A.V., Ivanov, S.N. & Svyatova, E.O. (2015) *Opisthorchiasis* in infant remains from the medieval Zeleniy Yar burial ground of XII-XIII centuries AD. *Memórias do Instituto Oswaldo Cruz*, 110, 974–980. https://doi.org/10.1590/0074-02760150156.

Smith, M., Love, D.C., Rochman, C.M. & Neff, R.A. (2018) Microplastics in seafood and the implications for human health. *Current Environmental Health Reports*, 5, 375–386. https://doi.org/10.1007/s40572-018-0206-z.

Sobel, J. & Painter, J. (2005) Illnesses caused by marine toxins. *Clinical Infectious Diseases*, 41, 1290–1296. https://doi.org/10.1086/496926.

Sripa, B., Tangkawattana, S. & Sangnikul, T. (2017) The Lawa model: A sustainable, integrated *opisthorchiasis* control program using the EcoHealth approach in the Lawa Lake region of Thailand. *Parasitology International*, 66, 346–354. https://doi.org/10.1016/j.parint.2016.11.013.

Staggs, S.E., Keely, S.P., Ware, M.W., Schable, N., See, M.J., Gregorio, D., Zou, X., Su, C., Dubey, J.P. & Villegas, E.N. (2015) The development and implementation of a method using blue mussels (*Mytilus spp.*) as biosentinels of *Cryptosporidium spp.* and *Toxoplasma gondii* contamination in marine aquatic environments. *Parasitology Research*, 114, 4655–4667. https://doi.org/10.1007/s00436-015-4711-9.

State of California (n.d.) *Fish and Game Code*, section 4501. https://codes.findlaw.com/ca/fish-and-game-code/fgc-sect-4501.html.

Steidinger, K.A. & Penta, H.L.M. (1999) Harmful microalgae and associated public health risks in the Gulf of Mexico. Florida Marine Research Institute.

Stolle, A. & Sperner, B. (1997) Viral infections transmitted by food of animal origin: The present situation in the European Union. *Archives of Virology. Supplementum*, 13, 219–228. https://doi.org/10.1007/978-3-7091-6534-8_21.

Streit, B. (1998) Bioaccumulation of contaminants in fish. In: *Fish Ecotoxicology*, 86th edn (edited by T. Braunbeck, D. E. Hinton & B. Streit). Springer: Dortrecht, pp. 353–387.

Stringer, M.F. & Hall, M.N. (2007) A generic model of the integrated food supply chain to aid the investigation of food safety breakdowns. *Food Control*, 18, 755–765. https://doi.org/10.1016/j.foodcont.2006.01.008.

Sultanov, A., Abdybekova, A., Abdibaeva, A., Shapiyeva, Z., Yeshmuratov, T. & Torgerson, P.R. (2014) Epidemiology of fishborne trematodiasis in Kazakhstan. *Acta Tropica*, 138, 60–66. https://doi.org/10.1016/j.actatropica.2014.04.030.

Tacon, A.G.J. & Metian, M. (2013) Fish matters: Importance of aquatic foods in human nutrition and global food supply. *Reviews in Fisheries Science*, 21, 22–38. https://doi.org/10.1080/10641262.2012.753405.

Thanh, B.N., Dalsgaard, A., Evensen, O. & Murrell, K.D. (2009) Survey for fishborne zoonotic metacercariae in farmed grouper in Vietnam. *Food-borne Pathogens and Disease*, 6, 1037–1039. https://doi.org/10.1089/fpd.2008.0230.

Thornber, K., Verner-Jeffreys, D., Hinchliffe, S., Rahman, M.M., Bass, D. & Tyler, C.R. (2020) Evaluating antimicrobial resistance in the global shrimp industry. *Reviews in Aquaculture*, 12, 966–986. https://doi.org/10.1111/raq.12367.

Tucca, F., Moya, H., Pozo, K., Borghini, F., Focardi, S. & Barra, R. (2017) Occurrence of antiparasitic pesticides in sediments near salmon farms in the northern Chilean Patagonia. *Marine Pollution Bulletin*, 115, 465–468. https://doi.org/10.1016/j.marpolbul.2016.11.041.

Turyk, M., Anderson, H., Knobeloch, L., Imm, P. & Persky, V. (2009) Organochlorine exposure and incidence of diabetes in a cohort of Great Lakes sport fish consumers. *Environmental Health Perspectives*, 117, 1076–1082. https://doi.org/10.1289/ehp.0800281.

Uemura, H., Arisawa, K., Hiyoshi, M., Kitayama, A., Takami, H., Sawachika, F., Dakeshita, S., Nii, K., Satoh, H., Sumiyoshi, Y., Morinaga, K., Kodama, K., Suzuki, T., Nagai, M. & Suzuki, T. (2009) Prevalence of metabolic syndrome associated with body burden levels of dioxin and related compounds among Japan's general population. *Environmental Health Perspectives*, 117, 568–573. https://doi.org/10.1289/ehp.0800012.

Van Cauwenberghe, L., Claessens, M., Vandegehuchte, M.B. & Janssen, C.R. (2015) Microplastics are taken up by mussels (*Mytilus edulis*) and lugworms (*Arenicola marina*) living in natural habitats. *Environmental Pollution*, 199, 10–17. https://doi.org/10.1016/j.envpol.2015.01.008.

Verner-Jeffreys, D.W., Welch, T.J., Schwarz, T., Pond, M.J., Woodward, M.J., Haig, S.J., Rimmer, G.S.E., Roberts, E., Morrison, V. & Baker-Austin, C. (2009) High prevalence of multidrug-tolerant bacteria and associated antimicrobial resistance genes isolated from ornamental fish and their carriage water. *PLOS ONE*, 4, e8388. https://doi.org/10.1371/journal.pone.0008388.

Viegas, C., Esteves, L., Faria, T., Pombo, A., Caetano, L.A., Quintal-Gomes, A., Twarużek, M., Kosicki, R., Grajewski, J. & Viegas, S. (2019) Fungal diversity and mycotoxin distribution in echinoderm aquaculture. *Mycotoxin Research*, 35, 253–260. https://doi.org/10.1007/s12550-019-00352-0.

Wang, Q.P., Lai, D.H., Zhu, X.Q., Chen, X.G. & Lun, Z.R. (2008) Human angiostrongyliasis. *Lancet. Infectious Diseases*, 8, 621–630. https://doi.org/10.1016/S1473-3099(08)70229-9.

Ward, J.E. & Kach, D.J. (2009) Marine aggregates facilitate ingestion of nanoparticles by suspension-feeding bivalves. *Marine Environmental Research*, 68, 137–142. https://doi.org/10.1016/j.marenvres.2009.05.002.

Watras, C.J., Back, R.C., Halvorsen, S., Hudson, R.J.M., Morrison, K.A. & Wente, S.P. (1998) Bioaccumulation of mercury in pelagic freshwater food webs. *Science of the Total Environment*, 219, 183–208. https://doi.org/10.1016/s0048-9697(98)00228-9.

Watts, J.E.M., Schreier, H.J., Lanska, L. & Hale, M.S. (2017) The rising tide of antimicrobial resistance in aquaculture: Sources, sinks and solutions. *Marine Drugs*, 15. https://doi.org/10.3390/md15060158.

Wegner, A., Besseling, E., Foekema, E.M., Kamermans, P. & Koelmans, A.A. (2012) Effects of nanopolystyrene on the feeding behavior of the blue mussel (*Mytilus edulis* L.). *Environmental Toxicology and Chemistry*, 31, 2490–2497. https://doi.org/10.1002/etc.1984.

Wells, M.L., Potin, P., Craigie, J.S., Raven, J.A., Merchant, S.S., Helliwell, K.E., Smith, A.G., Camire, M.E. & Brawley, S.H. (2017) Algae as nutritional and functional food sources: revisiting our understanding. *Journal of Applied Phycology*, 29(2), 949–982. https://doi.org/10.1007/s10811-016-0974-5.

Wiriya, B., Clausen, J.H., Inpankaew, T., Thaenkham, U., Jittapalapong, S., Satapornvanit, K. & Dalsgaard, A. (2013) Fish-borne trematodes in cultured Nile tilapia (*Oreochromis niloticus*) and wild-caught fish from Thailand. *Veterinary Parasitology*, 198, 230–234. https://doi.org/10.1016/j.vetpar.2013.08.008.

Wood, R. (2016) Acute animal and human poisonings from cyanotoxin exposure — A review of the literature. *Environment International*, 91, 276–282. https://doi.org/10.1016/j.envint.2016.02.026.

Yang, Z., Luo, Q., Liang, Y. & Mazumder, A. (2016) Processes and pathways of ciguatoxin in aquatic food webs and fish poisoning of seafood consumers. *Environmental Reviews*, 24(2), 144–150. https://doi.org/10.1139/er-2015-0054.

Yamamoto, K. (2013) *Small-Scale Aquaculture in Thailand: Farmer Groups and Aquaculture Certification.* Vol. 31. Food and Agriculture Organization: Rome.

Yasumoto, T., Nakajima, I., Bagnis, R. & Adachi, R. (1977) Finding of a dinoflagellate as a likely culprit of ciguatera. *Nippon Suisan Gakkaishi*, 43, 1021–1026. https://doi.org/10.2331/suisan.43.1021.

Ye, L., Lu, Z., Li, X., Shi, L., Huang, Y. & Wang, H.H. (2013) Antibiotic-resistant bacteria associated with retail aquaculture products from Guangzhou, China. *Journal of Food Protection*, 76, 295–301. https://doi.org/10.4315/0362-028X.JFP-12-288.

Yoshida, A., Doanh, P.G., Maruyama, H. (2019) Paragonimus and paragonimiasis in Asia: An update. *Acta Tropica*, 199. https://doi.org/10.1016/j.actatropica.2019.105074.

Zanchett, G. & Oliveira-Filho, E.C. (2013) Cyanobacteria and cyanotoxins: From impacts on aquatic ecosystems and human health to anticarcinogenic effects. *Toxins*, 5, 1896–1917. https://doi.org/10.3390/toxins5101896.

Zhang, J., Yang, X., Kuang, D., Shi, X., Xiao, W., Zhang, J., Gu, Z., Xu, X. & Meng, J. (2015) Prevalence of antimicrobial resistance of non-typhoidal *Salmonella* serovars in retail aquaculture products. *International Journal of Food Microbiology*, 210, 47–52. https://doi.org/10.1016/j.ijfoodmicro.2015.04.019.

Zhang, L., Yan, C., Guo, Q., Zhang, J. & Ruiz-Menjivar, J. (2018) The impact of agricultural chemical inputs on environment: Global evidence from informetrics analysis and visualization. *International Journal of Low-Carbon Technologies*, 13, 338–352. https://doi.org/10.1093/ijlct/cty039.

11

Quality issues in our global seafood

Jørgen Lerfall, Anita N. Jakobsen, Ioannis S. Boziaris, Ram C. Bhujel, Eirin Marie S. Bar and Amaya Albalat

Introduction

To secure sustainable use of marine resources and to comply with the United Nations (UN) Sustainable Development Goals (SDGs), we must prioritize increased utilization of the available biomass and actions to reduce spoilage and loss of seafood quality. The Food and Agriculture Organization (FAO) of the UN (FAO, 2011) set out five system boundaries that are crucial to maintaining quality and reducing nutrient loss in the food value chain. Adapting these boundaries to the seafood sector, we find that diseases (Lerfall et al., 2012) and utilization of feed resources in aquaculture (Gatlin III et al., 2007; Naylor et al., 2009) together with reducing losses related to by-catch and discards in the fishery sector are highly relevant (Hazen et al., 2018). Pre- and post-harvest handling of fish (Erikson et al., 2016) are other areas that need more attention. The industry needs to implement technology improvements along the whole value chain (Harmon, 2009; Refaey et al., 2017; Sampaio and Freire, 2016) to allow for appropriate temperature control, for example, use of wet-ice, refrigerated seawater (RSW) or superchilling (Magnussen et al., 2008; Piñeiro et al., 2004), packaging (Church and Parsons, 1995; Hansen et al., 2012; Sivertsvik et al., 2002), storage (Cortesi et al., 2009) and transportation after landing (Mai et al., 2012). However, the sector has been and is still dominated by traditional preservation technologies such as salting, drying, smoking and canning (Horner, 1997). Changes in consumer demand for more natural and minimally preserved seafood products drive the development of more lightly processed products (Abel et al., 2022; Carlucci et al., 2015; Speranza et al., 2009). Hurdle technology, which is a method of controlling food pathogens in food, needs to be implemented to meet quality, safety and sustainability issues, and several new interesting combinations of technologies have been tested/developed in recent years (Abel et al., 2019a; Lerfall et al., 2018b; Rosnes et al., 2011; Rosnes and Skipnes, 2018; Singh and Shalini, 2016).

To avoid quality loss during distribution, improvements in the cold-chain are necessary (Wu et al., 2014). The distribution of seafood is varied and includes markets, supermarkets, retailers and wet markets. During distribution, the cold-chain is typically ensured by using wet-ice for fresh seafood, whereas preserved products are distributed in refrigerated conditions. Supply of wet-ice can be challenging under tropical- and sub-tropical conditions, and actions to avoid solar heating must

Margaret Crumlish and Rachel Norman (eds) **Aquatic Food Security**
DOI: 10.52517/9781789183368.011,

be considered. However, the cold-chain can be improved by lowering the temperature 1–2°C below the initial freezing point of the muscle (Kaale and Eikevik, 2016; Kaale et al., 2011). This technology, known as superchilling, is performed using liquid nitrogen or carbon dioxide (dry ice) to improve shelf-life by inhibiting psychrophilic spoilage bacteria (Kaale et al., 2011; Wu et al., 2014).

Marked consumer changes have increased the demand for convenient food options (Anetta, 2018). This includes seafood products that are portion-sized, ready-to-eat, ready-to-cook, sustainable and convenient for consumers to handle. For producers offering such products, it is essential to focus on sustainable side-streams to increase the utilization of available resources. To ensure sustainable side-streams, technology that secures high retention of nutritional components is important.

Seafood quality is a multi-faceted phenomenon defined either objectively, according to its composition, or subjectively, related to consumers' preferences. In the research literature, this is often referred to as primary or secondary quality (Nortvedt et al., 2007). More specifically, primary quality refers to the intrinsic/biological properties (e.g. species, size, season and health status), professionalism during handling and production and product quality (e.g. hygienic, sensory, technological, nutritional and ethical), whereas the secondary quality focus more on marked quality (delivery according to specifications) and perceived quality. However, the quality of a specific product is often defined according to specific government regulations (import and exporting countries), marked demands and consumer trends.

This chapter will introduce the main aquatic species leading seafood production and quality issues in the seafood sector and consider how innovative technologies and control systems can avoid losses due to reduced quality. Moreover, the importance of product shelf-life in the utilization of marine resources is reviewed.

The seafood value chain

Fisheries

Global fish production peaked at about 178 mt in 2020, with global capture fisheries production representing 53% (90.3 mt). Fisheries in marine waters provided 87.2% of the global total, indicating the importance of the marine versus inland waters for this sector (FAO, 2020). Despite the challenging situation with many stocks fished at biologically unsustainable levels, some progress has been reported, especially in fisheries managed by countries with high governance capacity. For instance, the proportion of fishing stocks exploited within biologically sustainable levels increased from 53% in 2005 to 74% in 2016 in the USA, while in the Northeast Atlantic, the percentage of stocks where fishing mortality is below the maximum sustainable yield increased from 34% in 2003 to 60% in 2015 (FAO, 2018b). The major species taken by marine capture fisheries are Peruvian anchoveta (*Engraulis ringens*), Alaskan pollock (*Theragra chalcogramma*), Skipjack tuna (*Katsuwonus pelamis*), Atlantic herring (*Clupea harengus*), Yellowfin tuna (*Thunnus albacares*) and blue whiting (*Micromesistius poutassou*). One of the key traits of the production of capture fisheries is its temporal variability. For example, annual catches of Peruvian anchoveta ranged from 94,000 t to 12.3 mt during 1970 to 2004, with changes in the El Niño-Southern Oscillation being the main causative agent. This is important as our capacity to control the variability in supply is limited, making it difficult to predict future production (Brander, 2007). Within this context, it is necessary to mention the effect that climate change is having and is expected to have on future fisheries production patterns, especially shifting production levels and changes in the net marine primary production (Cheung et al., 2010). From a nutritious value perspective, fishery products can be perceived as superior to their farmed counterparts (Verbeke et al., 2007). Several studies have reported differences in fat content, which tend to be higher in farmed fish than in the wild, while the percentage of omega-3 fatty acids (FA) can be higher in wild fish (González et al., 2006; Henriques et al., 2014). However, from a nutritional perspective, fishery product composition is affected by seasonal and geographical factors (Celik, 2008; Hyldig et al., 2012). Variations in proximate composition are closely linked to feeding intake, which can vary due to external factors such as food availability and internal factors, such as spawning or migration.

Alaska pollock (*Theragra chalcogramma*)

Alaska pollock, also known as walleye pollock, is a semipelagic schooling fish widely distributed in the North Pacific Ocean with the largest concentrations in the eastern Bering Sea. The Alaska pollock fishery is managed by the National Marine Fisheries Service (NOAA Fisheries) and the North Pacific Fishery Management Council and is mainly caught by a pelagic trawl. Other gears used include bottom trawl, jig, longlines and pots. The fish is primarily sold as fillets or surimi. In general, products derived from cold-water fish are characterized by high activity of endogenous enzymes at a lower temperature, translating in a reduction of fish quality during refrigerated storage even before frozen storage (Simpson, 2000). For this reason, post-harvest storage and processing conditions of Alaska pollock are essential to minimize protein denaturation. For both products, the effects of refreezing have been studied in detail (Abe et al., 2019; Lee et al., 2016), while in the case of surimi, the effects of several shelf-extension technologies have been tested (Cando et al., 2016; Cao et al., 2018; Zhang et al., 2018).

Peruvian anchoveta (*Engraulis ringens*)

Peruvian anchoveta is probably the most heavily exploited fish species in history. The modern anchoveta fishery started in Peru around 1955. During the 1960s, the fleet and fishery increased until, in 1970, peaking with a harvest of 12.3 mt (Chavez et al., 2008). Two years later, the stock collapsed probably due to a combination of fishing pressure and a strong El Niño-Southern Oscillation. From that point onwards, the fishery has been characterized by significant fluctuations in production. From a product perspective and according to PRODUCE statistics, the industrial fleet landings for indirect human food consumption account >99% of the total catches, while small- and medium-scale fleet landings directed to direct human consumption (fresh, frozen, canning and curing) represent <1% of total catches. The low volumes of anchoveta destined to direct human consumption are attributed to several factors such as regulatory limitations (only artisanal not industrial purse seiners allowed to land for direct human consumption), consumer preferences, lack of a cold-chain for fish in Peru and the shelf price (Avadí et al., 2014). Therefore, the production and export of fishmeal and fish oil are the main drivers for the anchoveta industry. Although the inclusion of fishmeal and fish oil in aquaculture feeds has decreased due to the inclusion of vegetable protein sources (Ytrestøyl et al., 2015), the demand for both fishmeal and fish oil has remained constant due to the expansion of aquaculture.

Tuna (*Katsuwonus pelamis* and *Thunnus* pp.)

Tuna fisheries with a production of 4.6 mt and a landed value of US$12.2 billion, represent a significant source of food and trade commodity for both developed and developing countries. As tuna stock spans multiple national jurisdictions, they are regulated by regional fishery management organizations responsible for harvest control measures within their regional jurisdictions. There are mainly two primary harvest strategies for tuna. Fleets using purse seine, gillnet, and pole and line or trolling gear traditionally target skipjack, yellow tuna and some albacore (*Thunnus alalunga*) that are marketed as canned or pouched products while fleets that use longline and handline gear traditionally target yellow tuna, bigeye (*T. obesus*), bluefin (*T. orientalis*; *T. maccoyii*) and albacore, which supply the fresh and frozen steak and sashimi markets (McCluney et al., 2019). The tuna industry is an excellent example of how post-harvest value-added processes affect the ecological, economic and community performance when using fishery performance indicators (FPIs). Indeed, the international sashimi market of the tuna fisheries scores higher than international canned tuna and tuna fisheries supplying local markets in local states, reflecting that value chains that can preserve the quality and transport of fish to high-value markets outperform others (McCluney et al., 2019).

Aquaculture

In general, farmed fish and seafood are excellent raw materials for consumers and the processing industry due to reasonable availability and even quality. Today Atlantic salmon (*Salmo salar*), tilapia (*Oreochromis* spp.), catfish (*Pangasius* spp.), seabass (*Dicentrarchus labrax*), seabream (*Sparus aurata*), carp (all freshwater cyprinids) and warm-water shrimp (*Penaeus vannamei* and *P. monodon*) are important species worldwide. Among these species, Atlantic salmon is unique because of its size and anatomy, which make the production of boneless fillets easy. Boneless fillets are convenient for consumers and are an example of a value-added product (VAP) that simplifies logistics, distribution and the rest of the raw material side-stream processing.

Although farmed fish are known for their consistent quality, large biological variations occur within and between populations. For instance, several studies have shown genetic variations in feed intake, growth and feed utilization, meat quality, disease resistance and age at sexual maturation (Gjerde et al., 1994, 2011; Kolstad et al., 2004, 2005; Powell et al., 2008; Thodesen et al., 2001a, 2001b; Vieira et al., 2007; Wetten et al., 2007). Other important factors that influence the biological variation are; seasonality (Mørkøre and Rørvik, 2001; Shearer, 1994; Shearer et al., 1994) and geographic and local environments such as temperature, light and water quality (Austreng et al., 1987; Boeuf and Le Bail, 1999; Brett, 1979; Lerfall et al., 2017; Maria Poli, 2009; Smith et al., 1993; Ørnholt-Johansson et al., 2017). In comparison, the variable effects of feed in commercial farming are relatively small due to the similarity of feed management practices among the farmers (Einen and Roem, 1997).

Atlantic salmon (*Salmo salar*)

Today's generation of farmed Atlantic salmon is a descendant of wild Atlantic salmon collected in the early 1970s (Gjedrem et al., 1991) and reared over decades with a focus on rapid growth, limited lipid deposition, intense fillet colour and resistance against diseases. The life cycle of farmed Atlantic salmon can be divided into broodstock facility, hatchery and sea sites. The broodstock is carefully selected, and eggs and milk are mixed and incubated. After 230 degree days, eggs get eyes, and between 350–400 degree days, they change to yolk-sac fryes. After this stage, the fry 'float up,' and the feeding can begin. The hallmark of salmon (called parr in this stage) is clear black marks in the skin, but these disappear during smoltification. At approximately 20 g, the salmon are now called parrs and are induced to light manipulation to speed up smoltification (Solbakken et al., 1994). Smoltification is the process where the fish undergoes physiological, morphological and behavioural changes that prepare the fish to live in seawater (Björnsson et al., 2011; Hoar, 1988). Once fish have gone through the smoltification process, smolts can be transferred from the fresh to the seawater facilities. With the use of light, salmon parr smoltifies after one year (0+ smolts). Farmers may also use 1+ smolts, which take a year and a half to produce. After the smolts are transferred to sea facilities, a commercial size of 3–8 kg is reached after 12 to 24 months of growth. The inclusion level of plant oils in salmon feeds varies according to the price and its availability. Hence, there may be significant variations in the FA composition of salmon fillets as the FA composition in the fish is reflected by the FA acid composition of the feed (Lerfall et al., 2016; Torstensen et al., 2005). In addition, diseases may reduce the flesh quality of Atlantic salmon (Dawson-Coates et al., 2003; Lerfall et al., 2012; Moran et al., 1999; Whitaker and Kent, 1991). Starvation (Einen and Thomassen, 1998; Mørkøre et al., 2008) and handling due to well-boat transportation, slaughtering (Lerfall et al., 2015; Mørkøre et al., 2008; Skare et al., 2021) and thereafter processing also affect food quality (Chan et al., 2020a; Chan et al., 2020b; Lerfall and Rotabakk, 2015; Rotabakk et al., 2018). Highly developed slaughtering and processing plants increase the development of VAPs. Salmon VAPs include fresh and frozen fillets, portioned products (sashimi grade fresh loins, cutlets, etc.), together with preserved products such as salt-cured (Russian markets), cold- and hot-smoked salmon and Gravadlax. Technological improvements have made early processing (pre-rigor) possible, but approximately 80% of the whole production is still exported on ice as head-on-gutted (HOG) salmon and processed 3–6 days post-slaughtering.

In 2017, 17.5 billion meals of Atlantic salmon were produced worldwide (ISFA, 2018). Norway was the leading producer with an approximate production of 1.3 mt, followed by Chile (0.6 mt) and Scotland (UK) (0.2 mt) (FAO, 2018a). Other important producing countries are Canada, the USA, Ireland, New Zealand and Tasmania (Australia). The global salmon farming industry was in 2017 worth US$15.4 billion, inclusive of down-stream ripple effects that significantly strengthen rural communities in respective countries (ISFA, 2018).

Tilapia (*Oreochromis* spp.)

Among the many species of tilapias, Nile tilapia (*Oreochromis niloticus*) along with its hybrid (red tilapia, *Oreochromis aureus x Oreochromis mossambicus*) are one of the most widely cultured species playing a critical role in global food security. Global tilapia production doubled in each decade during the last three decades reaching over 6.3 mt in 2018 (FAO, 2019a). Originally from Africa, 75% of production now comes from Asia. The flesh is suitable to be processed as fillets for export as its meat is free of small Y-bones and has a mild flavour. In western countries, tilapia is considered a cheap fish, and its quality has been questioned owing to its low omega-3 fatty acid content. However, it contains the same level of protein and minerals as other species (Zajdband, 2012). Tilapias raised in green water ponds have higher omega-3 than those fed with commercial fish pellets (Karapanagiotidis et al., 2006). Some research has demonstrated that omega-3 can be supplemented through natural food or commercial pellets (dos Santos et al., 2014; Stoneham et al., 2018).

Tilapias are traditionally grown using earth ponds making the water green; therefore, a high-protein diet is unnecessary (Bhujel, 2014). However, most tilapias exported to Europe and the USA are grown in cages (5 m × 5 m or 6 m × 4 m in area and 2 m deep) in canals, rivers, lakes and reservoirs. Recently, large circular cages (e.g. 10–20 m diameter), generally used for salmon farming, are also being used as the demand for tilapia is increasing. Farmers stock 30–40 g size fish and feed 5–8% of biomass/day with floating pellets (25–30% crude protein (CP)) 2 to 3 times daily (Bhujel, 2014). Some farmers use vaccines against bacterial infections from *Streptococcus* spp., while others use antibiotics, for example, oxytetracycline, amoxillin or sulfa-drugs, formalin and copper sulphate, to control disease outbreaks and parasites (Rico et al., 2014). Farms, which are producing for export, are scrutinized for chemical use and other environmental safeguards. Tilapia are harvested between 0.5–1 kg size.

About 90% of tilapia are sold in the domestic market in most Asian countries (Ponte et al., 2014). After harvesting, they are transported by trucks directly to the fresh market or to the processing plants (Ponte et al., 2014). For domestic markets, tilapia is available in every form (live, whole fresh, fresh fillet, and even pre-cooked), although live tilapia is the most common. However, the low filleting yield (29–33%) of tilapia is challenging (Peterman and Phelps, 2012; Suresh and Bhujel, 2019). Only gills and guts should be discarded as they may contain heavy metals, parasites and bacteria if they are exposed to polluted environments (Osman and El-Khateeb, 2016).

The most critical parts of tilapia farming are technical and managerial (human resources), especially for hatcheries, nurseries and farms in rural areas, which impacts how tilapias are produced and influences subsequent product quality. Most countries, including those who were previously reluctant to accept tilapia as food fish, are becoming more positive and are promoting tilapia farming and developing guidelines, for example, India (NFBD, 2015).

Catfish (*Pangasius* spp.)

Pangasius is well-known worldwide, and is farmed in Asia, especially in Vietnam's Mekong Delta. Other countries also produce pangasius, but their production is lower and is mainly consumed locally. Vietnamese pangasius is available worldwide in over 160 countries as 'cheap' and white meat with the mild taste preferred in western countries. Its annual production increased 36-fold during 1998–2008, reaching a peak of 1.2 mt in 2008. Staggering production has seen annual production rising each year from 2016, 2017 to 2018, reaching values of 1.2, 1.25 and 1.3 mt, respectively, securing a value of *c.*US$2 billion in 2018 (FAO, 2019a). Production has increased due to high demand from China. In 2019, production was expected to reach 1.5 mt and a value of US$2.2 billion (FAO, 2019a). The EU was the largest importer, followed by the USA, but recent demand from Asian countries has seen them overtaking them in terms of market share (FAO, 2019a).

Pangasius is grown at high densities (40–100 fish/m^2), making ponds deep (4–6 m). As an air-breather, they can survive at high densities exchanging water daily (up to 30% daily). The average yield is approximately 400 t/ha, where the fish needs 6–7 months to reach the marketable size of around 1 kg. The fish prefer dark and murky water with high organic matter, which can affect flesh quality. Farm-made feed was common in the past, but nowadays, most farmers use commercial pellets containing 27% crude protein with low fishmeal (3%) and more vegetable protein, for example, soybean meal ranged 45–60% (Khoa et al., 2013). The use of the right feed ingredients and culture systems is necessary to minimize yellowish colour flesh, high fat content and the strong fishy flavour in pangasius.

Pangasius grow-out farmers usually buy fingerlings from traders in nearby areas who receive them from hatcheries in other provinces/districts. Hatcheries maintain their broodstock in ponds. Injection of hormones to induce breeding in captivity normally occurs in circulation tanks. Pangasius is a freshwater fish often grown near low-lying areas. There is a need to develop salinity tolerant strains that survive if salinity intrusion occurs (Khoa et al., 2013).

As already mentioned, Vietnam is the top producer of pangasius. Over 90% of production is exported as frozen fillets or other processed products. Several processing factories (e.g. 90 in Vietnam) follow the standard procedure for processing (Thi et al., 2014). Most fillets are frozen using blast freezers. Frozen fillets will have acceptable quality for up to 6 months. The processing is mainly manual, meaning that there is a need for a large workforce for filleting, skinning, cleaning and packing.

In some countries, agencies such as the Aquaculture Stewardship Council (ASC) certify pangasius farms based on feed ingredients used, wastewater treatment, employees and community welfare. Other agencies certify the products in terms of food safety and traceability, for example, GMP (good manufacturing practices), SSOP (sanitation standard operating procedure) and HACCP (hazard analysis and critical control point).

European seabass (*Dicentrarchus labrax*) and gilthead seabream (*Sparus aurata*)

European seabass and gilthead seabream are the main farmed species in the Mediterranean region with Greece, Turkey, Spain, Italy and Israel being the primary producers (Basurco et al., 2011; FAO, 2007). The estimated production of seabass and seabream in 2018 was 172,500 t (+6%) and 200,500 t (+7%), respectively. Most seabass production derives from sea cage farming, with production in seawater ponds and lagoons making up the rest. Around half the production of both species is exported within the EU and with the rest going worldwide. Italy absorbs approximately 50% of the seabass and around 40% of seabream production. However, a 10% rise in Italian production has been forecasted for 2018 (FAO, 2019b).

The hatcheries for seabass consist of two central units: the production unit and the service unit, which supports the correct function of the production unit. In order to ensure the constant supply of best quality fertilized eggs, every hatchery has its own broodstock unit where natural maturation, initiation of ovulation by photoperiod manipulation, triggering of the last phase of egg maturation by hormonal treatments, fertilization in spawning tanks, and incubation in an open-water circulation system, take place. Growth of seabass larvae into 2 g fry lasts about 4 months (at water temperature 18–20°C). The grow-out and fattening of fry supplied from hatcheries occur, in intensive production, mainly in a cage or in tank systems. Fry starts from a size of 1.5–2.5 g and in 18–24 months weighs 400–450 g (distribution of 2–15 g of feed every 10–15 minutes) (FAO, 2005a).

Similarly, for seabream, every hatchery has its broodstock unit where breeders are kept, and at the start of the spawning season, they are transferred to the spawning tanks. The most crucial factor that should be considered is controlling the sex ratio in the tanks because sex reversal occurs. A seabream larval rearing system can be small scale (<10 m^3, producing many juveniles) or large scale (~200 m^3, better larval quality but producing a smaller number of juveniles). Seabream larvae feed on only rotifers at first, and then (after 10–11 days) they also feed on artemia enriched with commercial lipid preparations until they achieve metamorphosis (32–35 days post-hatching). After this stage, juveniles are moved into tanks where weaning occurs, and feed is presented at 2 h intervals from 8:00 to 20:00. The grow-out of gilthead seabream can take place in extensive, semi-intensive and intensive systems. The primary fattening system used in the Mediterranean basin is in sea cages (lower densities than tanks but more profitable) (FAO, 2005b).

Organic production of seabream and seabass has gained attention in the last decade (Mente et al., 2011, 2012). The requirements for organic aquaculture are described in EU Regulation (EC, 2009). The most important aspect is the maximum stocking densities for organic production in coastal lagoons and open-water systems.

Harvesting of both European seabass and gilthead seabream can occur after the correct period of starvation has been observed, either by using dipnets or by vacuum pumps in very high densities

(70–100 kg/m^3), and killing occurs by thermal shock and asphyxiation in tubs with iced water. Extended crowding of fish before harvesting should be avoided, and the killing methods should ensure instant and irreversible loss of consciousness securing fish welfare and high quality of the product.

The main distributed product is the whole fish kept on ice and distributed to the market in no longer than 4–5 days. The shelf-life of whole fish stored exclusively in ice can be up to 16 days (Mokrani et al., 2018; Parlapani and Boziaris, 2016), which is often not fully needed as supermarkets usually clear stock within 10–12 days maximum after catch. Other products seeing increasing demand recently, especially for export, are the gutted and filleted pieces of seabass and seabream packed aerobically or in modified atmospheric packaging (MAP). Products diversification by mild processing, such as smoking or marination, is another market segment showing increasing interest, but the number of producers offering such products is currently low (Kurt Kaya and Baştürk, 2014; Vasiliadou et al., 2005; Vidal et al., 2017).

Carp (all freshwater cyprinids)

There are many species of carp cultured in China and India. Nine major species of farmed carp account for 33 mt, slightly over 40% of the total global aquaculture production (FAO, 2018b, 2019). Carp are important sources of protein and other nutrients, especially for widely distributed rural families, whereas pangasius, Atlantic salmon and shrimp contribute to the food security of mostly urban elites as they are sold mainly in the form of VAP such as fresh or frozen fillets.

Out of nine different cultured carp, seven were originally from and are mainly produced in China. Those are known as Chinese major carp. Among them, grass carp (*Ctenopharyngodon idella*) has the highest production volume, which accounts for more than 6 mt per year, followed by silver carp (*Hypophthalmichthys molitrix*) with over 5 mt. Other carp are common carp (*Cyprinus carpio*) and bighead carp (*Aristichthys nobilis*), which are produced each at the levels of 4 mt per year. Although by volume, grass carp and silver carp rank highest, common carp is the most commonly distributed fish among the cyprinids as it also can grow across the temperate regions of Europe and Asia (Woynarovich et al., 2011). Other carp with a significant production are the Indian major carp catla (*Catla catla*) and rohu (*Labeo rohita*), which are produced at the level of 2–3 mt per year. Today, they are primarily produced in India and its neighbouring countries such as Bangladesh, Pakistan, Myanmar, Nepal.

Carp have a long history of being farmed, traditionally in earth ponds, which are green in colour due to the high production of phytoplankton. Some carp, such as silver carp, eat phytoplankton, while others, such as big head and catla, prefer zooplanktons. Grass carp are often fed with grasses, but due to poor digestibility, their excreta contains a high level of nutrients that become good sources of nutrients for the pond water, resulting in high algae and phytoplankton production. Therefore, carp are mostly stocked together in a polyculture system to utilize different types of natural food organisms, including grasses, detritus, phytoplankton and zooplanktons.

Carp are very important in terms of food and nutrition security, especially in Asia and Europe, as they are found ubiquitously from family-level farms in rural communities to extensive corporate farming. They contribute to the nutrition and health of hundreds of millions of people. Carp have been heavily promoted worldwide by UN organizations and respective country governments with the objectives of supporting low-income families to achieve food and nutrition security and income generation. Most carp are bred using induced spawning methods; either providing a suitable natural environment only (e.g. common carps) or injecting artificial hormones such as extracts of the pituitary gland (PG) of other adult carps, luteinizing hormone-releasing hormone analogy (LHRHa), human chorionic gonadotropin (HCG) together with motilium or domperidone (Jhingran and Pullin, 1985; Khatun et al., 2017). However, due to growing concerns about the use of chemicals, the use of pituitary gland is considered the best practice management (BPM) (Woynarovich et al., 2011).

From 1960 to the 1980s, carp farming was heavily promoted through the provision of training and consultancy services to professionals in developing countries, and hatchery facilities were also built to breed and nurse at many government facilities supported by various development agencies, for example, the Japanese International Cooperation Agency (JICA). Eggs are hatched in circular tank systems, and hatchlings are fed with mostly egg-yolk prepared as a suspension from hard-boiled chicken eggs (Woynarovich et al., 2011). They were traditionally reared in the same tanks where females hatch. Nursing is done in small ponds or tanks, popularly called nursery ponds or in outdoor tanks where natural foods, especially zooplanktons, such as cladocerans and copepods, were found. During nursing, they were often fed with fine powders of cheap crop by-products, such as pure rice bran or a mixture with oil cake, soybean meal and fish meal (Jhingran and Pullin, 1985). More recently, fry are nursed in hapas and fed with commercial starter diets, such as powdered feed or small pellets, that ensure higher survival rates and better growth. Generally, hatchery operators do the breeding only and then supply to nursery farmers to grow for 1–3 months before being sold on to grow-out farmers. Carp fry or fingerlings are transported using plastic bags, oxygenated under pressure. During packing, it is common to use common salt, acriflavine or anaesthesia to reduce stress and avoid pathogen transfer. During grow-out, farmers may opt not to feed completely or to feed intermittently with commercial pellets. Commercial pellets are generally low in protein ranging from 15% crude protein to 30%. Most farmers grow carp in earth ponds. Ponds are adequately drained and dried. Lime is used to neutralize or balance the water pH level, reducing levels of human pathogens (e.g. *C. botulinum* and *E. coli*). After filling, plankton are grown, causing the green water colour. The use of animal manures or compost and/or chemical fertilizers commonly used in agriculture (i.e. urea, triple super-phosphate (TSP) or diammonium phosphate (DAP)), is common in carp farming. These are applied once before stocking and then weekly, depending upon how green the water is. Water that is too green can cause oxygen depletion, especially during the early morning due to increased oxygen consumption by plankton.

As they are grown at low densities and with low feeding rates disease is less of a problem than with some species and carp feed does not habitually contain antibiotics. Incidence of bacterial, protozoan, viral and fungal diseases will rise with poor water quality. In such a situation, some common antibiotics are applied to reduce losses. More recently, probiotics are adapted to enhance immunity and help digestion to improve growth and survival rates. The use of probiotics is especially effective during early stages (Jha et al., 2015).

Shrimp (*Penaeus monodon* and *P. vannamei*)

Shrimp is one of the most attractive high-quality, and high-value aquatic food items. Its annual global production from farming has reached almost 4 mt; out of which 3 mt were traded internationally (FAO, 2019). Nearly the same amount is also captured from the wild. Over three-fourths of the shrimp is produced in Asia. Extensive shrimp farming uses wild seed and natural food producing 100–200 kg/ha per year. In semi-intensive systems, *P. monodon* are stocked at 5–10 post larvae/m^2 in small ponds (1–8 ha) and production reaches 1000 kg/ha/year (i.e. 10 times more). With a stocking density of 20–40 post larvae/m^2, production can reach 4000–10,000 kg/ha/year with two crops in a year. After disease outbreaks, farmers started to grow Pacific white shrimp

(*P. vannamei*), with production increasing from 2003 onwards. In intensive systems stocking 100–200 post larvae/m^2 can lead to a harvest from 8000 up to 30,000 kg/ha/year. In 2018, the Pacific white shrimp accounted for 65% of total crustacean production (FAO, 2019). However, early mortality syndrome (EMS) or acute hepatopancreatic necrosis disease (AHPND) caused by a toxin produced *Vibrio parahaemolyticus* reduced shrimp production in Thailand by 50%. The effect was so significant that production has still not recovered fully, though it has been improving since 2015. Antibiotics are commonly used at low levels, increasing risks for antibiotic resistance (Hoang et al., 2011; Pham et al., 2018).

The frequent ups and downs in shrimp farming provide a stark lesson in how a booming industry could be devastated by viruses, in turn jeopardizing global food security. Shrimp farming has tremendous economic importance, but it also has been blamed for environmental problems. Good aquaculture practices (GAP) and other certification schemes are making headway. According to the (FAO, 2019) the EU, the USA, Vietnam, China, Japan, Korea and Canada are the top markets for shrimp in addition to domestic markets. Some producers are producing VAP, such as ready-to-cook and ready-to-eat shrimps destined for retail chains worldwide.

The role of technology in improving utilization and quality

Innovation in fisheries

Catching technology

A significant challenge within fisheries is unwanted catches (Kelleher, 2005). Unwanted catches are defined as the portion of the catch that cannot be marketed because they have little or no market value or because of legal requirements (Maeda et al., 2017). Those commercial and non-commercially valuable fish returned at sea during commercial fishing operations, dead or alive, are referred to as discards (Catchpole et al., 2005). As highlighted by Feekings et al. (2012), the reasons for discarding include the fact that some individuals might be under the legal minimum conservation reference size, individuals are from species that have low or no market value, they are damaged or the quota for this species has been reached and therefore cannot be legally landed. While the practice of discarding can be attributed to social, economic and legislative reasons, one of the fundamental causes is the use of unselective fishing techniques coupled with environmental factors affecting the distribution of species (Johnsen and Eliasen, 2011). The degree of discards as a problem is very fishery-dependent, and it has been particularly challenging in mixed-species demersal trawl fisheries, which are currently responsible for most of the discards, at least in the European context (Johnsen and Eliasen, 2011). Several countries established discard bans previously (Norway, Iceland, Chile and New Zealand). The EU landing obligation, also known as the discard ban, was fully implemented in early 2019 by the latest reform of the Common Fisheries Policy (EU 2013, Article 5). One of the aims of this ban is the development and deployment of more selective gear. So, from a catching technology perspective, the need for increased selectivity has been a main driver for innovation. From a quality perspective the type of gear used can influence the size and quality of the animals caught. This is particularly the case in crustacean species, such as Norway lobster (*Nephrops norvegicus*) with creel-caught *Nephrops* being in better condition (physically and physiologically) and serving a premium market compared to trawl-caught *Nephrops* (Albalat et al., 2009). Similarly, trawl-caught fish are frequently associated with more inferior quality, and in this respect, selective gear and gear modifications have been shown to improve catch quality (Brinkhof et al., 2018b) and reduce catch sorting times (Cosgrove et al., 2019). Besides gear type, practices such as 'buffer towing' have also been shown to harm Atlantic cod quality. In this practice, if the gear is full before the catch from the previous haul has been processed, the trawl is lifted off the seabed and towed at low speed until the vessel's production capacity can process the catch. Data indicates that this practice leads to poor exsanguination and an increased probability of fillet redness. Indeed, quality for both whole cod and cod fillets was negatively affected by 'buffer towing' (Brinkhof et al., 2018a). Practices such as 'buffer towing' highlight the lack of fish welfare awareness within the context of capture fisheries. Nevertheless, research interest in this area is increasing as reviewed recently by Veldhuizen et al. (2018). Within this context, the implementation of gear and fishing practices that reduce injuries and mortality in commercial capture fisheries will require an in-depth evaluation of welfare benefits, and ecological and economical consequences (Veldhuizen et al., 2018).

On-board processing

Capture can affect aquatic animals in terms of external damage and increased stress, both impacting seafood quality. Together with the increased awareness of fish welfare in capture fisheries, this fact makes both capture and on-board processing key factors to consider from both quality and an ethical animal welfare perspective. The Norwegian Council for Animal Ethics has stated that both the durations of harvesting and the length of time that fish experience high levels of stress, fear or pain should be shortened, aiming for gentle handing and minimal damage during capture. On-board processing is important as by the time animals have been hauled on board the vessel, fish are often stressed due to the capture process (Olsen et al., 2013). This is shown by low pH and elevated blood lactate levels. On-board processing after capture will depend on the species and markets being supplied. For example, Atlantic cod and haddock (*Melanogrammus aeglefinus*) need to be bled just after capture to minimize bleed-out and minimize fillet discolorations (Botta et al., 1986; Olsen et al., 2014). In horse mackerel, Ando et al. (1999) reported bleeding by gill cut to delay muscle softening by 6 h during storage at 5°C. Similarly, early bleeding moderated the rate of rigor mortis development and K-value in horse mackerel during ice storage for up to 7 days (Mochizuki et al., 1998). However, when the catch is large, it can be very challenging to keep the fish alive before they can be bled, and therefore blood coagulation is an issue. Several authors have evaluated whether the use of live holding tanks could be used as a strategy to keep the animals before being bled, thus minimizing fillet discoloration (Olsen et al., 2013). However, in this case, holding conditions should be optimized to ensure fish welfare (Digre et al., 2017).

On-board refrigeration systems

Although fishing vessels are intended to catch fish, the reality is that they must also provide adequate conditions for processing and adequate cold-chain storage conditions until the fish is landed. Of the total estimated number of fishing vessels in the world, 4.4 million, 80% are less than 12 m long, and industrialized fishing vessels of 24 m or more only represent 2% of all motorized fishing vessels (UNEP, 2016). Only medium to large commercial-scale vessels going out for mid-distance (*c.*2 week trips) will have refrigerated seawater tanks and ice-making machines onboard, while only large industrialized vessels going out for long distances will have freezing capability. Depending on the species being caught and processes deployed, these large industrialized vessels will have different blast freezers, plate freezers, refrigerated sea water tanks and ice-making machines (UNEP, 2017).

On-board freezing systems

Freezer trawlers are generally large commercial fishing vessels that operate far from shore with freezing equipment on board. Most used freezing systems in on-board processing vessels are freezing tunnels (air blast) and plate freezers, which can be installed simultaneously in the same vessel, and freezing tanks (immersion in salt solution at low temperature), which can be used for large whole fish and pelagic fish used for bait (Ruize, 2012). Ruize (2012) explained that freezing tunnels are defined as any refrigerated space that works at a temperature range of –25 to –32°C, and is used for fast freezing. In a plate freezer, seafood products, such as fillets, are frozen by contact between two plates with some pressure. Studies have shown that in Indian mackerel (*Rastrelliger kanagurta*) the reduction of freezing time recorded in plate freezers can significantly result in higher taste and overall acceptability scores after 3 months of storage (Lakshmisha et al., 2008).

Processing technology

Downstream processing of seafood includes an array of traditional processes such as, slaughtering, filleting, chilling, freezing, thawing, curing, drying, smoking and heating. We now discuss some innovative technologies with the potential to retain or improve seafood quality.

Superchilling

Superchilling is a food processing method in which the temperature of the product is lowered to an interval between frozen and chilled, typically 1 to 2°C below the initial freezing point of the product (Kaale et al., 2011). The reduced storage temperature, generally at –0 to –4°C, will have a more significant inhibitory effect on microbial activity and growth than conventional refrigeration and refrigerated ice storage (at usually 0 to 4°C). While frozen storage at –18 to –40°C will have a greater preservative effect than superchilling in terms of microbiological quality and safety, however, the formation of ice crystals throughout the product will often have detrimental effects upon other quality parameters. This is undoubtedly the case with muscle foods, where intracellular and extracellular ice crystals lead to the rupturing of cellular membranes and de-ordering of myofibrillar structures (Kaale and Eikevik, 2016; Kaale et al., 2011, 2014; Magnussen et al., 2008; Strasburg et al., 2007).

Since sub-freezing temperatures are employed both during storage and initial chilling of a superchilled product, some ice crystal formation will occur (Kaale et al., 2011, 2014). The size and prevalence of these ice crystals will depend not only on product type but also on the chilling process. Faster chilling processes will generally lead to the formation of smaller and more uniformly distributed ice crystals. Conversely, a slower decrease in temperature leads to large crystals, which are more likely to cause damage to the cell membranes and highly ordered protein networks present in muscle tissues. The partial freezing of water in the products also leads to an increased concentration of solutes, such as ions and sarcoplasmic proteins, in the remaining fluid fraction. The concentration of solutes will increase osmotic pressure on cell membranes and may affect protein–water interactions. As substrate and cofactor concentration are important regulators of enzyme activity, an increase in autoproteolytic degradation may occur; the low temperature (and thus reduction in enzyme activity) will work to balance this effect (Strasburg et al., 2007). The impact of these effects is well known from conventional freezing.

As discussed, ice crystal formation is affected mainly by the chilling process (Kaale et al., 2011; Strasburg et al., 2007). Simultaneously, products may be placed directly into a freezer set to its superchilled storage temperature to obtain increased microbiological shelf-life. Sivertsvik et al. (2003) showed that such quality changes would be acceptable in Atlantic salmon fillets using appropriate superchilling technology. Kaale et al. (2014) demonstrated that drip loss and WHC (water holding capacity) parameters could be improved by including the freezing step in an equivalent product: in this case, by impingement freezing.

Three freezing technologies are commonly employed in the food processing industry: mechanical freezers, cryogenic freezers and impingement freezers (Kaale et al., 2011). Mechanical freezers use a circulating refrigerant to chill the product by air-to-product heat exchange; they are thus limited by their low heat-transfer coefficient, resulting in high processing times and unfavourable ice crystal formation. They have the benefit of being cheap and straightforward to operate. Cryogenic freezers utilize a cryogen to chill the product. Within the confines of food processing, cryogen commonly refers to liquid nitrogen (–196°C) or solid carbon dioxide (–78°C). These low temperatures combined with a high rate of surface heat transfer from the boiling/sublimation of the cryogen allows cryogenic freezers to chill products quickly, allowing for favourable ice crystal formation. While these freezers are simple in use, the high amount of cryogen required makes operating costs high, limiting their usefulness to high-value products. Impingement freezers separate a continuous belt freezer into zones of progressively colder chambers; each chamber has jets with individually adjustable velocity. This design allows for quick temperature reductions in the product by creating a series of high heat-transfer coefficients in the consecutive chambers, resulting in lower processing times than continuous belt mechanical freezers and favourable ice crystal formation. Impingement freezers have comparable processing times to cryogenic freezers at a lower operating cost, making them ideal choices for use in superchilling (Kaale et al., 2011; Kaale et al., 2014).

A downside to superchilling technology is the requirement to maintain a temperature slightly below the initial freezing point during storage of the product (Kaale et al., 2011). The average consumer may have access to a traditional consumer-grade refrigerator and freezer unit but not a freezer dedicated to keeping a temperature relevant to superchilling. A stable temperature must be kept to maintain maximum quality in a superchilled product. Any fluctuation may result in melting and recrystallization of the frozen fraction of the liquid content, leading to ice crystal growth and quality reduction. The cost of such precision instruments makes superchilled storage an unlikely proposition for the private consumer, but the technology is still highly relevant for use in industry and during transport.

Ultra rapid freezing systems

Freezing is a long-established food preservation process that produces high-quality, nutritious foods with a long storage life. The formation of ice crystals during freezing and frozen storage causes physical changes to the structure of foods with a subsequent reduction in the quality of the thawed material. In meat and fish, the main result is increased drip on thawing. Drip loss on thawing can be reduced with fast freezing rates (e.g. pressure shift freezing) known to reduce the size of the present ice crystals (Chevalier et al., 2000). Established food freezing technologies encompass air blast, contact/plate, immersion and spray freezing systems. Although often considered a separate technology, cryogenics can be considered a subfield of blast, immersion or spray, depending on how the cryogen is used and the application method (James et al., 2015). In recent decades, there has been considerable scientific interest in developing novel systems that can improve the organoleptic quality or economics of the production of frozen foods. Novel freezing technologies have been reviewed by James et al. (2015) and Li and Sun (2002), pointing out several promising technologies to either improve quality or reduce costs (or both) of food products.

The use of cryogenics has a huge potential to increase freezing rates and retain the quality of seafood products. The technology is beneficial since it allows the product to directly contact the refrigerant (Khan and Mittal, 2017). Using a conveyor belt, the product will be exposed to the counter-current flow of nitrogen gas (at a temperature of about –50°C). As the product passes to the pre-cooling stage of the freezer, the gaseous nitrogen partially freezes it, and up to 50% of the product's heat is extracted. The food product then passes below the spray of liquid nitrogen, where the freezing process is completed. The end-stage in the freezer provides some time for the product temperature to reach equilibrium. The main advantage of this system is that freezing is very rapid, and the size of the freezer is comparatively small. Also, the compressor and condenser are not an integral part of this freezer. Liquid nitrogen must be retained in a vacuum insulated pressure vessel with continuous venting to keep the contents cool and the internal pressure down. The limitation of cryogenic freezing technology is the cost of nitrogen, energy losses due to poor insulation (transport and storage of nitrogen), and the need for regular nitrogen gas supply.

The use of impingement technology to increase the surface heat transfer in air and cryogenic freezing systems has been one of the few recent innovations to be truly commercially realized (Sarkar and Singh, 2004; Sarkar et al., 2004). Impingement is the process of directing a jet or jets of fluid at a solid surface to effect a change. The very high velocity (up to 50 m s^{-1}) of impingement jets are used to 'break up' the static surface boundary layer of gas that surrounds a food product (James et al., 2015). The resulting medium around the product is more turbulent, and the heat exchange through this zone becomes much more effective and can produce substantially faster freezing than conventional blast systems (James et al., 2015).

Impingement freezing is best suited to products with a high surface area to weight ratio (i.e. flat products, such as burgers or fish fillets). Testing has shown that products less than 20 mm thick freeze most effectively in an impingement heat-transfer environment (Newman, 2001). The process is also attractive for products that require very rapid surface freezing (i.e. crust freezing).

Radio-frequency thawing

As frozen seafood is widely used, thawing also needs attention. Thawing of fish is traditionally performed in air or in circulating water. Those technologies are, however, not optional for retaining quality due to limited heat-transfer potential. The heat-transfer potential can be increased by using, for example, microwaves, ultrasound or radio-frequency (RF) heating (Nagarajarao, 2016). RF thawing has the highest potential (Altemimi et al., 2019; Gambuteanu and Alexe, 2015; Nagarajarao, 2016) and is nowadays widely used in the fish industry. After using RF for food heating and blanching in 1947, attempts were made to use RF energy for thawing frozen products (Altemimi et al., 2019). The thawing time depends on multiple factors, including the uniformity of the blocks used and size and dielectric properties. In general, studies have demonstrated that thawing time using RF is much shorter than conventional methods (Altemimi et al., 2019).

High-pressure processing

High hydrostatic pressures can be utilized as a food processing step to inactivate or injure bacteria, viruses and proteolytic enzymes (Duranton et al., 2014; Strasburg et al., 2007). High-pressure processing (HPP) units typically work by applying pressure (by hydraulics, for example) to a confined water bath that contains the product that is to be processed. The pressures typically utilized in food processing are in the 100–600 MPa range, and they are commonly held for 5–30 minutes (Duranton et al., 2014; Murchie et al., 2005). Some lessons learnt from heat processing are applicable to HPP, such as utilizing a milder process (low temperature or low pressure, respectively) for a long time to achieve a high reduction in microorganisms while preserving organoleptic qualities. Pressure is uniformly and instantly distributed throughout the entirety of the product; this is an advantage of HPP compared to other bacteriostatic processes, such as heating, in which a thermal gradient will be established throughout the product with variable effects (Baldwin, 2012; Rosnes et al., 2011).

HPP requires the product to be packaged in a snugly fitting and robust pouch. It can be considered particularly suitable for products intended for vacuum storage. No repacking would be required, and thus there is no chance of recontamination. It would also be efficient to combine with sous vide processing if a heat treatment is desirable, as this technology also requires vacuum packaging (Baldwin, 2012).

Since pressure can be built up quickly and instantly released in an HPP unit, it is possible to use such equipment to perform an oscillating high-pressure treatment (Mor-Mur et al., 2014). This has been shown to induce germination of spore-forming bacteria into vegetative cells, which have reduced resistance to pressure and can be injured or inactivated by the following pressure cycles. This may be a valuable process to inactivate toxin-producing *C. botulinum* spores, which otherwise requires intense heat processing to inactivate.

Impact of shelf-life on utilization of marine food resources

Seafood shelf-life

Shelf-life is generally defined as the length of time a food product remains acceptable to consumers and safe (Steel, 2004). According to EC (2011) No. 1169/2011, article 2, a food's shelf-life or date of minimum durability is the date until which the food retains its specific properties when properly stored. Thus, both chemical, physical, nutritional and sensory properties should be retained, and the growth of microorganisms should not reduce quality or safety. Food-borne pathogens, causing infections or intoxications, can be present in the raw material or be introduced during processing. Several pathogenic bacteria, viruses, parasites and biotoxins are linked to seafood (Huss et al., 2014) and are reviewed by Løvdal (2015), Buchanan et al. (2017), Elbashir et al. (2018) and Hoel et al. (2019).

Quality loss of seafood occurs throughout the value chain, from harvesting, processing, distribution and during storage by the end consumer. Microorganisms are the main cause of spoilage in fresh and minimally processed seafood (FAO, 2018b; Gram and Dalgaard, 2002; Gram and Huss, 1996). Microbiological criteria give guidance on the acceptability of foodstuffs and their manufacturing processes.

Total viable counts (TVC) are commonly used to indicate shelf-life, and the maximum acceptable microbial concentration for seafood is generally considered 10^6–10^7 CFU/g (Stannard, 1997). However, the levels of TVC reported as indicative of the end of shelf-life vary considerably, between 10^5 and 10^9 CFU/g (Boziaris et al., 2011; Dalgaard et al., 1997b; Parlapani et al., 2014; Robson et al., 2007) and sensory rejection typically occurs at microbial levels between 10^6–10^9 CFU/g (Kuuliala et al., 2018). Microbiological testing alone cannot guarantee the safety of a foodstuff tested, but these criteria provide objectives and reference points to assist food businesses and competent authorities in their activities to manage and monitor the safety of foodstuffs, respectively.

Worldwide, post-harvest seafood losses are of significant concern and occur throughout the value chain. According to FAO (2018b), 27% of landed fish worldwide is lost or wasted between landing and consumption. In some small-scale fisheries in Africa, the losses are up to 50%, and deterioration of quality can account for more than 70% of the loss (Akande and Diei-Ouadi, 2010). To ensure sustainable seafood production, an important goal is to reduce these losses. There are significant differences in the utilization of seafood between developed and developing countries (FAO, 2018b). Whereas more than two-thirds of the production of fish used for human consumption is distributed in frozen, prepared or preserved forms in Europe and North America, most fish are commercialized in live or fresh forms in Asia and Africa (FAO, 2018b). An unbroken cold-chain worldwide is essential to reduce food losses and meet the challenges of food insecurity and a growing world population. Especially, are improvements necessary in developing countries to improve the cold-chain globally (Mercier et al., 2017).

The higher degree of post-harvest processing in developing countries can reduce seafood spoilage and thus allow distribution to a worldwide market. In the western world, the assortment of minimally processed and ready-to-eat seafood (e.g. sashimi, sushi, smoked fish) in retail stores has grown in the last 5–10 years, which is in accordance with the customers' preferences for natural, healthy and convenient food (Carlucci et al., 2015; Contini et al., 2015). The combination of convenience and high sensory quality is challenging for the producers and distributors of seafood. It is, however, important to never compromise with the microbiological quality and food safety of such products.

As fresh seafood is highly perishable, attention towards post-harvest handling, processing, preservation, packaging, storage and transportation is crucial for shelf-life extension and to maintain quality, safety and nutritional attributes, and to avoid waste losses. Fresh seafood (e.g. finfish, crustaceans and molluscs) are in general more perishable than meat products mainly due to their higher post-mortem pH (~6.5 to above 7) (Gram and Huss, 1996; Huss et al., 2014). Furthermore, high concentrations of non-protein nitrogen fractions, for example, trimethylamine oxide (TMAO) and amino acids present in seafood, support microbiological growth (Gram and Huss, 1996; Huss et al., 2014). Thus, intrinsic factors of the seafood support microbiological growth, and the microbial metabolism leads to decomposition of nutrients and production of specific metabolites (Jakobsen et al., 2020; Shumilina et al., 2017) causing unpleasant and unacceptable off-flavours (Gram and Dalgaard, 2002; Gram and Huss, 1996). These compounds, produced by specific spoilage organisms (SSO), include organic acids, alcohols, aldehydes, amines, ketones, sulphides, ammonia and hydrogen sulphide (Parlapani et al., 2017; Ryder et al., 2014). Microbial spoilage may also result in the production of slime, discolouration, the appearance of microbial colonies or changes in texture properties of the product. In newly harvested fish and shellfish, SSOs are usually present in low concentrations and constitute only a minor part of the total microbiota. During storage, the SSOs have a higher growth rate than the remaining microbiota and are the dominant part of the microbiota at the timepoint of sensory rejection. The SSOs often consist of a single or a few microbial species (Gram and Dalgaard, 2002), and the development of SSOs during storage is affected by type of seafood (marine or freshwater species, tropical or cold-water species) and process and storage conditions (Boziaris and Parlapani, 2017).

Although seafood spoilage is primarily caused by microbiological activity, enzymatical and chemical reactions (Hong et al., 2017; Kassemsarn et al., 1963) and physical damage also contribute. At the time of death, the initial loss of fish freshness is caused by autolytic reactions catalysed by endogenic enzymes. Post-mortem glycolysis results in the accumulation of lactic acid, which lowers the pH of the muscle. The end pH is related to species, for example, in cod the pH is reduced from 6.8 to about 6.1–6.5 (Huss, 1995). The level also depends on the amount of glycogen in the tissue at the time of death. Stressed fish have a lower level and will be affected by a smaller pH reduction (Daskalova, 2019; Lerfall et al., 2015). Furthermore, autolytic degradation of adenosine triphosphate (ATP) is one of the first autolytic changes to occur after fish deaths (Figure 11.1).

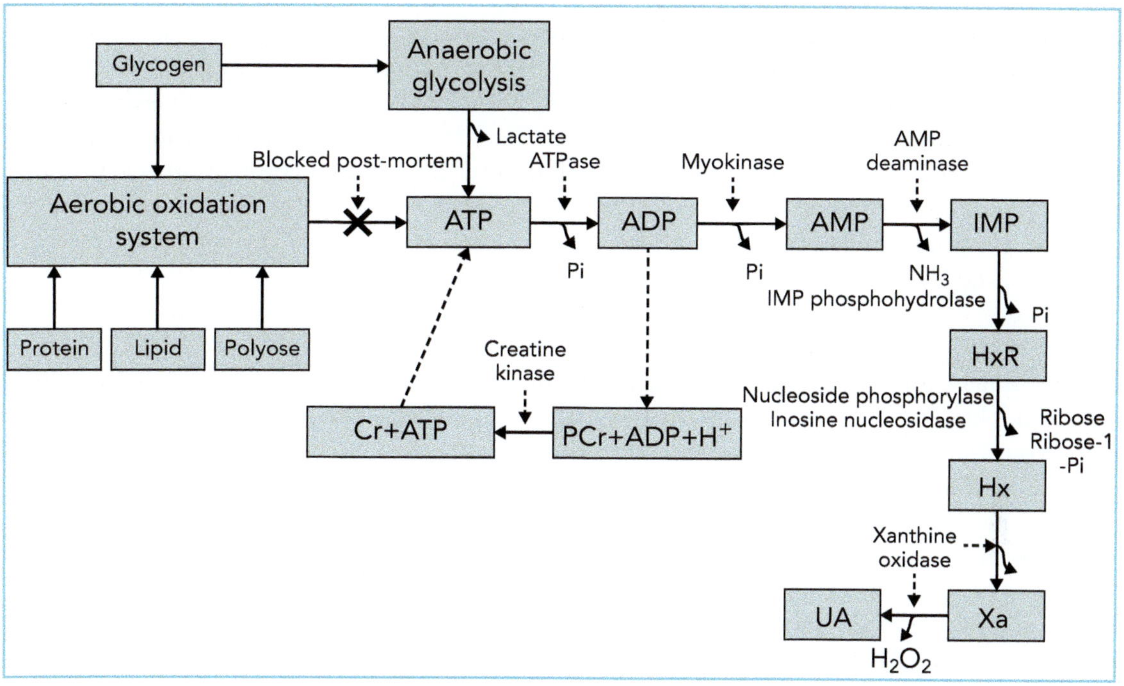

Figure 11.1 Post-mortem ATP metabolism in fish muscle (modified from Huss (1995); adapted from Hong et al. (2017). ATP = adenosine triphosphate, ADP = adenosine diphosphate, AMP = adenosine monophosphate, IMP = inosine monophosphate, HxR = inosine and Hx = hypoxanthine.

The total amount of ATP-related compounds is usually constant. The ratio (%) of the amounts of inosine (HxR) and hypoxanthine (Hx) to their total amount was defined as the K-value or 'freshness' index by Saito et al. (1959).

The K-value gives a relative freshness rating based primarily on the autolytic changes in the muscle during storage (Figure 11.1). The importance of ATP-related compounds for the freshness and flavour of fish and shellfish is extensively reviewed by Hong et al. (2017), highlighting species-specific differences in ATP degradation. Autolytic changes involving proteolytic enzymes cause softening and textural changes of the fish tissue. Examples of proteases contributing to the deterioration of fish tissue are cathepsins, calpains, collagenases (Huss, 1995).

Factors affecting shelf-life and spoilage

Shelf-life and the typical SSO present are affected by several factors, including capture methods, onboard handling, initial natural microbiota load and contamination throughout the value chain, seafood species, processing technology, processing environment and storage conditions (Boziaris and Parlapani, 2017).

Capture technology and onboard handling

Capture technology, hygienic handling and chilled storage during primary processing are essential for good quality and extended shelf-life (Anders et al., 2020). Rough handling or delayed gutting will result in a faster spoilage (Huss, 1995). Several complex parameters determine the choice of fishing gear, for example, historical aspects, fuel prices, politically set fishing quotas and the marked price of frozen versus fresh raw material (Rotabakk et al., 2011). However, as demonstrated by Esaiassen et al. (2004), Rotabakk et al. (2011) and Anders et al. (2020) the chosen gear can significantly affect quality. Esaiassen et al. (2004) found that the pH of the cod caught by gillnet was significantly lower, and the K-value was higher than the cod caught by longline. Another study also demonstrated an overall better quality of Atlantic cod was caught by longline than caught by trawl. The primary quality parameters affected were texture, protein denaturation, colour and sensory score (Rotabakk et al., 2011).

Microbial contamination and control throughout the value chain

In newly caught fish bacteria are mainly present in gills (10^3–10^7 CFU/g), guts (10–10^8 CFU/g) and on the skin (10^2–10^5 CFU/g) (Ryder et al., 2014). Gram-negative psychotropic bacteria (e.g. *Pseudomonas, Photobacterium, Shewanella, Psychrobacter, Moraxella, Acinetobacter, Flavobacterium, Vibrio, Aeromonas* and Enterobacteriaceae) predominate the microbiota on the skin, outer shell and in gills of newly caught or harvested finfish, crustaceans and shellfish. The most common Gram-positive genera are *Micrococcus*, lactic acid bacteria (LAB), *Bacillus* and *Clostridium* (Gram and Huss, 1996; Huss et al., 2014; Parlapani et al., 2015a; Svanevik and Lunestad, 2011).

The composition of the indigenous microbiota of fresh seafood is affected by geographic origin, water composition and temperature, type of species and catching methods. However, few conclusions have so far been drawn about how these factors affect the distribution of microbial genera present in the indigenous microbiota (Huss et al., 2014). Parlapani et al. (2018a), using the 16S Next Generation Sequencing (NGS) approach, reported a different microbiota composition in seabream originating from fish farms from two geographically distinct areas of Greece (Ionian and Aegean seas), indicating that geographical origin affects the initial and spoilage microbiota of chill-stored fish.

The flesh of healthy fish does not contain microorganisms, but it can be contaminated in the slaughter process and throughout the value chain. In a 10 year spot sampling program (2005–2014) of pelagic fish species harvested in Norway, 51% of all fishing vessels and fish processing factories did not comply with the limits proposed concerning hygiene and food safety (Svanevik et al., 2015). During processing, seafood can be contaminated from the environment, food contact surfaces and workers handling the food. Langsrud et al. (2015) found a diverse microbiota dominated by *Pseudomonas* spp. after regular sanitation in three salmon processing plants. They further demonstrated that a cocktail of bacterial isolates from conveyer belts was able to form stable biofilms. Biofilm formation in the seafood industry constitutes a severe economic, quality and health issue (Galié et al., 2018; Møretrø et al., 2016). Møretrø et al. (2016) also reported the processing environment as a source of contamination of salmon fillets with *Pseudomonas* spp. and *Shewanella* spp. Preventative actions, such as applying good hygiene and manufacturing practices (GHP, GMP) and the hazard analysis critical control point (HACCP) principles, are crucial for controlling both safety and spoilage of seafood throughout the value chain. Møretrø et al. (2016) found ensuring optimum hygene conditions could reduce the initial levels of bacteria on salmon fillets by about 90%. Novel technology to prevent biofilm formation has been reviewed by Galié et al. (2018).

TVC is often used as a shelf-life indicator due to convenience, although the quantification of SSO may be a better shelf-life indicator (Gram and Dalgaard, 2002). Fogarty et al. (2019) concluded that hydrogen sulphide-producing bacteria, LAB, *Pseudomonas* spp., *Brochothrix thermosphacta* and *Photobacterium* spp. were a better indicator of fish spoilage in Atlantic salmon.

Processing technology

Minimal processing of seafood aims to retain sensory and nutritional attributes of the raw material to the greatest extent possible. To meet consumer demands, minimal processing technology (e.g. minimal heat processing, non-thermal processing or modified atmospheric packaging) has been developed and applied in the food industry to produce fresh-like products with longer shelf-life and increased product safety (Abel et al., 2019a, 2019b, 2022; Ohlsson, 1994; Ohlson and Bengtsson, 2000; Rosnes et al., 2011; Rosnes and Skipnes, 2018; Sivertsvik et al., 2002). Minimal processing of food often requires multiple sub-lethal processes, or hurdles, to achieve a similar level of microbial control as are obtained in traditional processing methods using a single lethal process (Rosnes et al., 2011; Singh and Shalini, 2016; Abel et al., 2022). For example, proper rice acidification combined with low storage temperature (≤4°C) was required to prevent the growth of *Aeromonas* spp. in retail sushi (Hoel et al., 2018).

Innovative processing technology and new combinations of technology that focus on minimal heating or rapid heating (Rosnes et al., 2011; Stormo et al., 2018), the combination of modified atmosphere and minimal heating (Abel et al., 2019a; Lerfall et al., 2018b), chilling (Margeirsson et al., 2012) and superchilling technology (Kaale et al., 2011; Magnussen et al., 2008), high-pressure technology (Duranton et al., 2014), natural preservatives (Olatunde and Benjakul, 2018) and biopreservation (Ghanbari et al., 2013; Singh, 2018; Wiernasz et al., 2017) have therefore gained increased attention lately.

Minimal thermal processing of food aims to minimize thermal damage, for example, loss of nutrients or sensory quality (Abel, et al., 2022). Sous vide technology is often applied for retail seafood products and are based on a safety level of 6D reduction of non-proteolytic *Clostridium botulinum* type E (Rosnes et al., 2011). Surface pasteurization is an efficient method to enhance the shelf-life of many foodstuffs (Baldwin, 2012; Peng et al., 2017; Rosnes et al., 2011). However, Stormo et al. (2018) found limited effects on fresh cod loins. Microwave pasteurization is another exciting technology that offers rapid heating and enhanced product quality (Lerfall et al., 2018b; Thostenson and Chou, 1999).

HPP can be utilized as a food processing step to inactivate or injure bacteria, viruses, and proteolytic enzymes (Duranton et al., 2014). Pressures typically applied are 100–600 MPa for 5–30 minutes (Duranton et al., 2014; Murchie et al., 2005). The main quality challenge related to HPP technology is sensory quality. Much HPP-treated seafood is reported to have adverse textural and visual change (Cheftel and Culioli, 1997; Murchie et al., 2005; Tintchev et al., 2009, 2010).

To reduce the rate of spoilage, low temperature is essential. The seafood industry and the distribution chain employ refrigerated and/or ice storage at 0–4°C and superchilled storage at –1 and –4°C for fresh seafood (Kaale et al., 2011; Magnussen et al., 2008). Superchilling technology (see page 222) has shown several advantages for the microbiological quality of seafood, as the growth of psychotropic spoilage bacteria is significantly reduced compared to conventional chilled products (Hansen et al., 2009).

Natural preservative options include the use of plant extracts, chitosan, bioactive peptides, essential oil, among others, to enhance shelf-life and quality of food due to their antimicrobial and antioxidant properties (Olatunde and Benjakul, 2018). Natural preservatives can serve as an alternative to synthetic additives and can be used in either pure or crude form. Biopreservation aims to extend shelf-life and enhance food safety by applying naturally occurring microorganisms and/or their inherent antibacterial compounds of defined quality and at certain quantities (Ghanbari et al., 2013). LAB are excellent candidates as they are generally regarded as safe and are naturally present in many food commodities (Ghanbari et al., 2013). LAB produces several antimicrobial components, for example, organic acids (lactic, acetic and propionic acids), several bacteriocins with nisin as the best defined, hydrogen peroxide, carbon dioxide, diacetyl and FAs (Ghanbari et al., 2013). The antimicrobial effect is demonstrated against Gram-negative and Gram-positive bacteria (Ghanbari et al., 2013; Stupar et al., 2021). To use LAB as a tool in food biopreservation, several criteria and requirements must be fulfilled (Ghanbari et al., 2013): (1) be safe for the consumers; (2) have demonstrable antimicrobial effects against a broad spectrum of relevant food-borne pathogen and/or spoilage bacteria and fungi; (3) not contribute to the quality loss (e.g. spoilage or nutritional degradation); and (4) LAB must survive and grow in the processing and storage conditions used. Another approach in biopreservation is applying genetically modified bacteriophages. However, their safety issues must be adequately addressed before selecting a biopreservative agent (Singh, 2018).

Temperature control and storage conditions

Temperature control and keeping an unbroken cold-chain is the crucial factor for maintaining the microbiological quality of perishable seafood. Nevertheless, weaknesses in the cold-chain occur frequently throughout the value chain (Mercier et al., 2017). For example, examination of temperature-dependent microbiota in retail sushi confirmed that strict temperature control throughout the entire shelf-life significantly impacted the bacterial growth rates of different bacterial groups. That spoilage was delayed or inhibited at low temperatures (Hoel et al., 2017). Hoel et al. (2015) suggested poor temperature control as the main contributor to loss of quality in retail sushi.

Modified atmospheric packaging (MAP) combined with low storage temperature is a well-studied and established technology with several benefits, such as the possibility to design a specific atmosphere to inhibit the growth of SSOs (Sivertsvik et al., 2002). The assortment of MAP seafood in the retail stores has increased in line with consumer interest for convenient and portion packaged fresh food (Carlucci et al., 2015). Extended shelf-life and increased quality attract both the processing industry and trendy consumers (Carlucci et al., 2015). The effectiveness of MAP is generally determined by the amount of available carbon dioxide to dissolve into the food, as carbon dioxide is the main gas for inhibiting microbial activity (Devlieghere and Debevere, 2000; Sivertsvik et al., 2002). However, the optimal gas mixture to achieve extended shelf-life with optimal quality depends on the type of seafood.

Hydrogen sulphide-producing *Shewanella* spp. and *Pseudomonas* spp. are predominant SSOs in fresh refrigerated or iced aerobically stored seafood (Gram and Dalgaard, 2002; Gram and Huss, 1996). The use of 16S NGS analysis of fish tissue also revealed other potential spoilers such as *Psychrobacter,* in seabream, cuttlefish, hake and plaice (Parlapani et al., 2018a, 2018b; Zotta et al., 2019), while *Shewanella* spp. was found to constitute a low percentage of the total spoilage microbiota of fish using NGS methods (Parlapani et al., 2018a, 2018b; Zotta et al., 2019). In a modified atmosphere (MA), the carbon dioxide-resistant *P. phosphoreum* is often detected as SSO in cold-water marine species (e.g. cod) (Dalgaard et al., 1997a, 1997b; Gram and Dalgaard, 2002; Leroi, 2010). In a study of MA-packed raw saithe filets (Lerfall et al., 2018a), the specific gas mixture (67% or 33% carbon dioxide, balanced with either oxygen or nitrogen) affected the overall quality of the product. A study of the microbial ecology revealed *P. phosphoreum* and *Pseudomonas* spp. as predominant under all gas-conditions tested, while *Shewanella* spp. was not detected in saithe packaged in a gas mixture of low carbon dioxide/high oxygen levels. Thus, the main discriminant was the use of oxygen as part of the gas mixture. The use of oxygen (33% or 67%) in combination with carbon dioxide (67% or 33%) reduced the proliferation of psychotropic bacteria and inhibited hydrogen sulphide-producing bacteria. Using oxygen as part of the gas mixture was beneficial against the accumulation of hypoxanthine and biogenic amines, and a reduced drip loss during storage was achieved compared with saithe packaged in carbon dioxide/nitrogen or vacuum. Vacuum-packaged saithe stand out negatively compared with all investigated MAP-solutions with higher drip loss and reduced shelf-life. Other studies (Kuuliala et al., 2018; Sivertsvik, 2007) have reported that growth of hydrogen sulphide-producing bacteria in cod are reduced or completely inhibited as oxygen is included in the package gas. Lactic acid bacteria *and B. thermosphacta* often acts as spoilers in warm-water species and in lightly processed and chilled products (Gram and Huss, 1996).

Lactobacillus, *Carnobacterium* and *B. thermosphacta* have been found to dominate in chilled fish from the Mediterranean Sea stored under a modified atmosphere (Drosinos and Nychas, 1996; Koutsoumanis et al., 2000; Parlapani et al., 2015b). However, it has to be noted that in packages where the oxygen concentration is at high levels, *Pseudomonas* can be the dominant organism although the LAB are favoured (Kostaki et al., 2009; Parlapani et al., 2015a, 2015b; Pournis et al., 2005). Parlapani et al. (2015a) observed *Pseudomonas* to be dominant in chill seabream fillets stored both under air and a modified atmosphere (60/10/30% $CO_2/O_2/N_2$). Simultaneously, bacteria such as *Carnobacterium* (*C. maltaromaticum*, *C. divergens*), *Vagococcus*, *Lactobacillus* and *Shewanella* were also found to co-dominate with Pseudomonas in MA-stored fillets at 5°C. Macé et al. (2014) evaluated the spoilage potential of isolates belonging to five bacterial groups/species (*Shewanella baltica*, *Carnobacterium maltaromaticum*, *Aeromonas salmonicida*, *Vibrio* sp., 'other Gamma-Proteobacteria' [containing one strain of *Pseudoalteromonas* sp. and one strain of *Psychrobacter* sp.]) isolated from spoiled cooked and whole tropical shrimp stored under a modified atmosphere. They concluded that the major spoilage bacterial isolate groups were *C. maltaromaticum* and *S. baltica*.

In MAP, the practical upper carbon dioxide level is limited by the degree of filling (DF) which is the volume of product vs. volume of package. A percentage between 25 and 33 is recommended for seafood to ensure bacteriostatic carbon dioxide availability and preventing packaging collapse (Sivertsvik et al., 2002). To increase the level of dissolved carbon dioxide, enough carbon dioxide can be dissolved into the product for 2 h in pure carbon dioxide before retail packaging. This method is called soluble gas stabilization (SGS) (Esmaeilian et al., 2021; Sivertsvik, 2003). SGS has shown promising results on fresh Atlantic salmon fillets (Sivertsvik, 2003), on cooked, peeled shrimp (Sivertsvik and Birkeland, 2006), and Atlantic halibut fillets (Rotabakk et al., 2008). The SGS-technology has also successfully been combined with heat treatment to prolong the shelf-life of lightly processed fish patties (Abel et al., 2019a) and lightly processed salmon (Abel et al., 2019b).

Increased availability for consumers

Fisheries and aquaculture products are among the most important foods consumed by people worldwide. The fish nutrient content is responsible for the positive nutritional impact on the human diet. Fish are products of great nutritional value with inherent functionality, containing valuable proteins, FAs, vitamins and minerals with significant benefits for human health. However, the availability of fish for consumers depends not only on the fish quantity produced by the fishery sectors but also on the shelf-life. In general, the shelf-life of fresh fish is limited, comprising a significant shortcoming for their commerce and distribution globally. The extension and estimation of the shelf-life of the fresh fish can lead to the excellent management of fish batches, which batches can be sent directly in the local, national or international markets.

Meanwhile, the fresh fish is very perishable, and its freshness deteriorates fast due to bacterial action. Spoilage microorganisms consume nutrients of fish flesh and produce metabolites responsible for the development of off-flavours/off-odours, thus cause the sensory rejection of the product

(Gram and Huss, 1996). Thus, the implementation of various preservation strategies for the extension of shelf-life of fresh fish, mainly based on low temperature and appropriate packaging, is necessary for international commerce. Besides distributing fresh fish, other preservation methods, such as freezing, salting, etc., can be employed to aid fish trade in long distances. Indeed, a significant percentage of fish is distributed frozen (Jessen et al., 2014), while salted fish, especially Gadidae species, are traded worldwide (Arason et al., 2014).

The conditions, mainly temperature, maintained during storage and distribution of fresh fish markedly affect its quality retention and shelf-life. Most high-quality fresh fish usually are distributed to the markets while stored in ice. The shelf-life of ice-stored farmed fish varies from 10 to 24 days for species such as gilthead seabream, European seabass, turbot, Atlantic salmon, grass carp, shortfin pompano, sutchi catfish, pacu and tambaqui (Borges et al., 2014; Cakli et al., 2007; Erikson et al., 2019; Parlapani and Boziaris, 2016; Roth et al., 2014; Sveinsdottir et al., 2002; Viji et al., 2015). Whereas, the shelf-life of ice-stored wild fish is shorter, primarily due to the higher initial microbial load. Shelf-life of wild fish varies from 10–15 days for Mediterranean horse mackerel and blue jack mackerel (Panguila et al., 2018; Tzikas et al., 2007), while a shorter shelf-life (<4–7 days) has been recorded for smaller pelagic fish, such as sardines and anchovy (Ariano et al., 2014; Erkan and Özden, 2008). From those, fish species with short shelf-life usually distributed for local (< 5 days) or national (5–10 days) consumption, while fish with a longer shelf-life (10–20 days or more) move forward to the international market. Superchilling is another option to extend the shelf-life of fresh fish. By this method, the shelf-life is markedly increased, however from a practical point of view is rather difficult for various reasons to keep fish at such temperatures throughout the distribution chain, from processor to consumer table.

MAP combined with chill storage prolongs the shelf-life of fish products compared with those packaged under air, mainly due to the use of a high concentration of carbon dioxide in the package headspace, which has bacteriostatic effects against the most important spoilage bacteria act on the fish tissue. The extension of fish shelf-life depends on the fish species, initial microbial population level, fat content, gas mixture, gas/product ratio, and storage temperature (Sivertsvik et al., 2002). Thus, an extension of 20–100% or more has been recorded in various chill-stored fish products under MAP conditions compared to aerobic packaging (Parlapani et al., 2014, 2015b; Poli et al., 2006; Silbande et al., 2018; Sivertsvik et al., 2002). Alternatively, vacuum packaging is also used in the fish processing industry, especially for fatty species and before freezing. The use of vacuum packaging prolongs fish products' shelf-life; however, it has to be used in combination with additional preservation methods to ensure food safety, especially against anaerobic psychotropic pathogens such as *C. botulinum* type E (Anonymous, 2011).

In contrast to traditional packaging, an alternative, and environmentally friendly technology, using edible films or coatings can increase food quality and safety of such products, and it has gained much attention in the last decade (Kapetanakou and Manios, 2014; Kapetanakou and Skandamis, 2016). Carbohydrates and proteins such as chitosan and gelatine are some of the most common ingredients used to produce edible films or coatings. The use of such films and coating can improve the quality of fish products extending their shelf-life for an extended period and also contribute to their safety (Alsaggaf et al., 2017; Feng et al., 2016; López de Lacey et al., 2014; Nowzari et al., 2013).

The application of processing hurdles and technologies extend the shelf-life of fish, enabling exports worldwide, thus increasing availability for consumers.

Control systems to improve quality and to reduce waste

Consumers increasingly demand convenient, high-quality and safe products based on fresh seafood (Carlucci et al., 2015). Today's food industry mainly focuses on traceability, where specific quality and food safety parameters are proved at fixed points in processing facilities and or along the value chain from producer to consumer. To meet these challenges, additional measures, more than traceability alone, is needed. From a futuristic perspective, it is possible to extend today's level of traceability further through a digital transformation that makes it possible to implement a chain of custody, which refers to chronological documentation of physical or electronic shreds of evidence. Through a digital transformation of the chain of custody, the real impact starts when essential knowledge about a specific processing condition and or a specific raw material is connected to an Internet of Things (IoT) platform. If we consider that the IoT refers to a combination of large number of physical devices being connected, collating and sharing data via the internet, we can see how handling all kinds of data may be used to predict food quality, shelf-life and safety in a more 'real-time' concept. Thereby making dynamical shelf-life and prize systems possible. Such models have recently being developed by using artificial intelligence (AI) and data based on abiotic- and biotic-omics of the raw material and online non-destructive measures applied throughout the processing line and value chains. The use of sensors and on-line technology is significant to collect information and might make it possible to build systems to improve food safety, food quality, and thereby contribute towards improved food security (availability and safety). Metagenomics can be used taxonomically and help understand the properties and interactions within the 'in-house' microbial community (Cocolin et al., 2017; den Besten et al., 2017). The formation of diverse biological metabolites in fish muscle is affected by the specific production protocol (Ørnholt-Johansson et al., 2017) and the environment (Enders and Boisclair, 2016) is essential and can be quantified and monitored through advanced chemical methods. The introduction of sequencing data and metabolomics in developing more comprehensive prediction models and manufacturing data requires new knowledge and infrastructures in the sector.

In this direction, quite a few publications exist in the literature regarding multi-omics approaches to assess fish quality, in particular the connection of the microbiome with volatilome and their link with the sensory attributes and microbiological data (microbial populations) of fish during storage at refrigeration temperatures under air (Parlapani et al., 2018b, 2019b) or MAP (Kuuliala et al., 2018). Multi-omics approaches have also been used to assess fish quality fed with various fish meals (Parlapani et al., 2019a). These data might be used to rapidly assess fish freshness/quality through novel approaches in the future.

New (multi-omics) technologies to decode the metabolic and microbial complexity in food matrixes will generate a diverse collection of differential data that need to be addressed to realize the suggested IoT platform. Developing predictive models based on this complex platform will raise possibilities to introduce dynamical prize- and shelf-life systems that further can reduce food waste and, in the future, give better utilization of marine raw materials.

Sustainability

Today, around one-third of the global carbon footprint is related to food production and its indirect support systems, such as the production of fertilizers, production equipment and distribution.

In addition, the food sector consumes 70% of the world's freshwater resources and 20% of the energy produced (Bacenetti et al., 2018; Hoolohan et al., 2013). Considering a growing global population estimated to reach 9 billion by 2050, these problems will only worsen in the coming years. Therefore, reducing the carbon footprint from food production is crucial in limiting climate change (Vermeulen et al., 2012).

Food waste or loss is measured only for products directed to human consumption, excluding feed and parts of products that are not edible. Food loss refers to a decrease in edible food intended for human consumption throughout the food supply chain, from production to post-harvest operation, while food waste is connected to the actions and behaviour of retailers and consumers that causes the food intended for human consumption to spoil (FAO, 2011; Parfitt et al., 2010).

According to a study by the FAO (2015), 54% of the world's food losses and waste occurs upstream during production, post-harvest handling, and storage, while the remaining occurs downstream at the processing, distribution, and consumption stages. Roughly 30% of the edible parts of food produced for human consumption gets lost or wasted globally, which amounts to 1.3 bt per year. In the case of fisheries and aquaculture, 35% of global catch are lost annually, either because they are thrown back to sea or they become spoiled before they reach the consumer (FAO, 2018b). In addition to ethical and economic concerns, food waste also contributes to the depletion of natural resources. Most of retail food waste occurs because of seafood's high level of perishability (Chintagari et al., 2017; Gutierrez et al., 2017).

Several drivers influence loss at the retail level, such as a customer's experience of perishable products in the store, the aesthetic quality of the items, forcing retailers to display products of ideal appearance and discarding food with minor flaws or sign of ripening/maturing (Gutierrez et al., 2017). Keeping food products fresh longer maintains its aesthetic quality for an extended period and allows retailers to present it for sale for longer periods without downgrading. Focusing on shelf-life extension methods and technologies has been shown to reduce food loss and waste not only at retail but throughout the food supply chain (Bacenetti et al., 2018).

Reduction in food loss and waste is directly linked to a reduction in resource consumption, economic loss and environmental impact. Less will need to be produced to provide the same amount of product to the customer. Thus, the environmental effect of extended shelf-life is related to the magnitude of reduction in production as well as the effectiveness within the value chain. However, increased efforts need to be taken to understand the economic and environmental impact of shelf-life extension and reduced losses. However, in most cases, an increase in shelf-life has considerable environmental and economic benefits (Amani and Gadde, 2015; Bacenetti et al., 2018; Gutierrez et al., 2017; Silvenius et al., 2014).

Conclusion

In order to solve the quality issues related to our global seafood, a deeper and wider understanding of the mechanisms affecting marine resources, fisheries and aquaculture, and technological solutions for downstream processing is necessary. Moreover, it is crucial to understand the worldwide seafood market, including macro-economic and consumer forces and how they affect trading. To meet the UN's SDGs, it will be important to reduce losses from fisheries and, in general, improve the flow of high-quality nutrients in the seafood value chain. To meet this, actions on technological development are important, especially by improving catching technology and systems that improve handling and chilling of fish and thereby improve shelf-life. Maintaining the excellent quality of the raw material throughout processing and logistics is the key to improving seafood shelf-life and reducing losses due to reduced quality and spoilage.

References

Abe, S., Asada, T. & Kajiwara, K. (2019) Effects of freeze–thaw cycles on gel-forming ability and protein denaturation in Alaska pollock frozen surimi. *Journal of Food Quality*, 2019, 1–9. https://doi.org/10.1155/2019/3760368.

Abel, N., Rotabakk, B.T. & Lerfall, J. (2019a) Effect of heat treatment and packaging technology on the microbial load of lightly processed seafood. *LWT*, 101, 123–129. https://doi.org/10.1016/j.lwt.2018.11.025.

Abel, N., Rotabakk, B.T., Rustad, T., Ahlsen, V.B. & Lerfall, J. (2019b) Physiochemical and microbiological quality of lightly processed salmon (Salmo salar L.) stored under modified atmosphere. *Journal of Food Science*, 84, 3364–3372. https://doi.org/10.1111/1750-3841.14852.

Abel, N., Rotabakk, B.T., Lerfall, J. (2022) Mild processing of seafood – a review. *Comprehensive Reviews in Food Science and Food Safety*, 21 (1), 340-370. https://doi.org/10.1111/1541-4337.12876.

Akande, G. & Diei-Ouadi, Y. (2010) *Postharvest Losses in Smallscale Fisheries: Case Studies in Five Sub-Saharan African Countries*. FAO Fisheries and Aquaculture Technical Paper no. 550. Food and Agriculture Organization: Rome.

Albalat, A., Gornik, S.G., Atkinson, R.J.A., Coombs, G.H. & Neil, D.M. (2009) Effect of capture method on the physiology and nucleotide breakdown products in the Norway lobster (Nephrops norvegicus). *Marine Biology Research*, 5, 441–450. https://doi.org/10.1080/17451000802603637.

Alsaggaf, M.S., Moussa, S.H. & Tayel, A.A. (2017) Application of fungal chitosan incorporated with pomegranate peel extract as edible coating for microbiological, chemical and sensorial quality enhancement of Nile tilapia fillets. *International Journal of Biological Macromolecules*, 99, 499–505. https://doi.org/10.1016/j.ijbiomac.2017.03.017.

Altemimi, A., Aziz, S.N., Al-Hilphy, A.R.S., Lakhssassi, N., Watson, D.G. & Ibrahim, S.A. (2019) Critical review of radio-frequency (RF) heating applications in food processing. *Food Quality and Safety*, 3, 81–91. https://doi.org/10.1093/fqsafe/fyz002.

Amani, P. & Gadde, L.E. (2015) Shelf life extension and food waste reduction. In: *Proceedings of the in Food System Dynamics*, pp. 7–14.

Anders, N., Eide, I., Lerfall, J., Roth, B. & Breen, M. (2020) Physiological and flesh quality consequences of pre-mortem crowding stress in Atlantic mackerel (Scomber scombrus). *PLOS ONE*, 15, e0228454. https://doi.org/10.1371/journal.pone.0228454.

Ando, M., Nishiyabu, A., Tsukamasa, Y. & Makinodan, Y. (1999) Post-mortem softening of fish muscle during chilled storage as affected by bleeding. *Journal of Food Science*, 64, 423–428. https://doi.org/10.1111/j.1365-2621.1999.tb15056.x.

Anetta, B. (2018) Millennial consumers in the convenience food market. *Management*, 22, 251–264.

Anonymous (2011) *Fish and Fishery Products Hazards and Controls Guidance*. Food and Drug Administration, Center for Food Safety and Applied Nutrition.

Arason, S. Nguyen, M. & Thorarinsdottir, K. (2014) Preservation of fish by curing. *Seafood Processing: Technology, Quality and Safety*.

Ariano, A., Scarano, L., Mormile, A., Barile, M., Palma, G. & Murru, N. (2014) Shelf life of anchovy products (Engraulis encrasicolus): Evaluation of sensory, microbiological and chemical properties. *Italian Journal of Food Safety*, 3, 1678. https://doi.org/10.4081/ijfs.2014.1678.

Austreng, E., Storebakken, T. & Åsgård, T. (1987) Growth rate estimates for cultured Atlantic salmon and rainbow trout. *Aquaculture*, 60, 157–160. https://doi.org/10.1016/0044-8486(87)90307-3.

Avadí, A., Fréon, P. & Tam, J. (2014) Coupled ecosystem/supply chain modelling of fish products from sea to shelf: The Peruvian anchoveta case. *PLOS ONE*, 9, e102057. https://doi.org/10.1371/journal.pone.0102057.

Bacenetti, J., Cavaliere, A., Falcone, G., Giovenzana, V., Banterle, A. & Guidetti, R. (2018) Shelf life extension as solution for environmental impact mitigation: A case study for bakery products. *Science of the Total Environment*, 627, 997–1007. https://doi.org/10.1016/j.scitotenv.2018.01.301.

Baldwin, D.E. (2012) Sous vide cooking: A review. *International Journal of Gastronomy and Food Science*, 1, 15–30. https://doi.org/10.1016/j.ijgfs.2011.11.002.

Basurco, B. et al. (2011) Current status of Sparidae aquaculture. In: *Sparidae: Biology and Aquaculture of Gilthead Sea Bream and Other Species* (edited by M. Pavlidis & C. Mylonas). Wiley-Blackwell.

Bhujel, R.C. (2014). *A Manual for Tilapia Business Management*. CABI Publishing: Wallingford, UK.

Björnsson, B.T., Stefansson, S.O. & McCormick, S.D. (2011) Environmental endocrinology of salmon smoltification. *General and Comparative Endocrinology*, 170, 290–298. https://doi.org/10.1016/j.ygcen.2010.07.003.

Boeuf, G. & Le Bail, P.-Y. (1999) Does light have an influence on fish growth? *Aquaculture*, 177, 129–152. https://doi.org/10.1016/S0044-8486(99)00074-5.

Borges, A. et al. (2014) Quality parameters of pacu (Piaractus mesopotamicus) and tambaqui (*Colossoma macropomum*) gutted and stored on ice for different periods. *International Food Research Journal*, 21, 589–596.

Botta, J.R., Squires, B.E. & Johnson, J. (1986) Effect of bleeding/gutting procedures on the sensory quality of fresh raw Atlantic cod (Gadus morhua). *Canadian Institute of Food Science and Technology Journal*, 19, 186–190. https://doi.org/10.1016/S0315-5463(86)71629-5.

Boziaris, I.S. & Parlapani, F.F. (2017) Specific spoilage organisms (SSOs) in fish. In: *The Microbiological Quality of Food* (edited by A. Bevilacqua, M. R. Corbo & M. Sinigaglia). Woodhead Publishing: Cambridge, pp. 61–98.

Boziaris, I.S., Kordila, A. & Neofitou, C. (2011) Microbial spoilage analysis and its effect on chemical changes and shelf-life of Norway lobster (Nephrops norvegicus) stored in air at various temperatures. *International Journal of Food Science and Technology*, 46, 887–895. https://doi.org/10.1111/j.1365-2621.2011.02568.x.

Brander, K.M. (2007) Global fish production and climate change. *Proceedings of the National Academy of Sciences of the United States of America*, 104, 19709–19714. https://doi.org/10.1073/pnas.0702059104.

Brett, J.R. (1979) Environmental factors and growth. In: *Fish Physiology* (edited by W. S. Hoar, D. J. R. & J. R. Brett). Academic Press: Cambridge, MA, USA, pp. 599–675.

Brinkhof, J., Larsen, R.B., Herrmann, B. & Olsen, S.H. (2018a) Assessing the impact of buffer towing on the quality of northeast Atlantic cod (Gadus morhua) caught with a bottom trawl. *Fisheries Research*, 206, 209–219. https://doi.org/10.1016/j.fishres.2018.05.021.

Brinkhof, J., Olsen, S.H., Ingólfsson, ÓA., Herrmann, B. & Larsen, R.B. (2018b) Sequential codend improves quality of trawl-caught cod. *PLOS ONE*, 13, e0204328. https://doi.org/10.1371/journal.pone.0204328.

Buchanan, R.L., Gorris, L.G.M., Hayman, M.M., Jackson, T.C. & Whiting, R.C. (2017) A review of Listeria monocytogenes: An update on outbreaks, virulence, dose–response, ecology, and risk assessments. *Food Control*, 75, 1–13. https://doi.org/10.1016/j.foodcont.2016.12.016.

Cakli, S., Kilinc, B., Cadun, A., Dincer, T. & Tolasa, S. (2007) Quality differences of whole ungutted sea bream (Sparus aurata) and sea bass (Dicentrarchus labrax) while stored in ice. *Food Control*, 18, 391–397. https://doi.org/10.1016/j.foodcont.2005.11.005.

Cando, D., Moreno, H.M., Borderías, A.J. & Skåra, T. (2016) Combined effect of high hydrostatic pressure and lysine or cystine addition in low-grade surimi gelation with low salt content. *Food and Bioprocess Technology*, 9, 1391–1398. https://doi.org/10.1007/s11947-016-1728-8.

Cao, H., Fan, D., Jiao, X., Huang, J., Zhao, J., Yan, B., Zhou, W., Zhang, W. & Zhang, H. (2018) Effects of microwave combined with conduction heating on surimi quality and morphology. *Journal of Food Engineering*, 228, 1–11. https://doi.org/10.1016/j.jfoodeng.2018.01.021.

Carlucci, D., Nocella, G., De Devitiis, B., Viscecchia, R., Bimbo, F. & Nardone, G. (2015) Consumer purchasing behaviour towards fish and seafood products. Patterns and insights from a sample of international studies. *Appetite*, 84, 212–227. https://doi.org/10.1016/j.appet.2014.10.008.

Catchpole, T.L., Frid, C.L.J. & Gray, T.S. (2005) Discards in North Sea fisheries: Causes, consequences and solutions. *Marine Policy*, 29, 421–430. https://doi.org/10.1016/j.marpol.2004.07.001.

Celik, M. (2008) Seasonal changes in the proximate chemical compositions and fatty acids of chub mackerel (Scomber japonicus) and horse mackerel (Trachurus trachurus) from the north eastern Mediterranean Sea. *International Journal of Food Science and Technology*, 43, 933–938. https://doi.org/10.1111/j.1365-2621.2007.01549.x.

Chan, S.S., Roth, B., Jessen, F., Løvdal, T., Jakobsen, A.N. & Lerfall, J. (2020a) A comparative study of Atlantic salmon chilled in refrigerated seawater versus on ice: From whole fish to cold-smoked fillets. *Scientific Reports*, 10, 17160. https://doi.org/10.1038/s41598-020-73302-x.

Chan, S.S., Roth, B., Skare, M., Hernar, M., Jessen, F., Løvdal, T., Jakobsen, A.N. & Lerfall, J. (2020b) Effect of chilling technologies on water holding properties and other quality parameters throughout the whole value chain: From whole fish to cold-smoked fillets of Atlantic salmon (Salmo salar). *Aquaculture*, 526, 735381. https://doi.org/10.1016/j.aquaculture.2020.735381.

Chavez, F.P., Bertrand, A., Guevara-Carrasco, R., Soler, P. & Csirke, J. (2008) The northern Humboldt Current System: Brief history, present status and a view towards the future. *Progress in Oceanography*, 79, 95–105. https://doi.org/10.1016/j.pocean.2008.10.012.

Cheftel, J.C. & Culioli, J. (1997) Effects of high pressure on meat: A review. *Meat Science*, 46, 211–236. https://doi.org/10.1016/s0309-1740(97)00017-x.

Cheung, W.W.L., Lam, V.W.Y., Sarmiento, J.L., Kearney, K., Watson, R., Zeller, D. & Pauly, D. (2010) Large-scale redistribution of maximum fisheries catch potential in the global ocean under climate change. *Global Change Biology*, 16, 24–35. https://doi.org/10.1111/j.1365-2486.2009.01995.x.

Chevalier, D., Sequeira-Munoz, A., Le Bail, A., Simpson, B.K. & Ghoul, M. (2000) Effect of freezing conditions and storage on ice crystal and drip volume in turbot (Scophthalmus maximus). *Innovative Food Science and Emerging Technologies*, 1, 193–201. https://doi.org/10.1016/S1466-8564(00)00024-2.

Chintagari, S., Hazard, N., Edwards, G., Jadeja, R. & Janes, M. (2017) Risks associated with fish and seafood. *Microbiology Spectrum*, 5. https://doi.org/10.1128/microbiolspec.PFS-0013-2016.

Church, I.J. & Parsons, A.L. (1995) Modified atmosphere packaging technology: A review. *Journal of the Science of Food and Agriculture*, 67, 143–152. https://doi.org/10.1002/jsfa.2740670202.

Cocolin, L., Mataragas, M., Bourdichon, F., Doulgeraki, A., Pilet, M.F., Jagadeesan, B., Rantsiou, K. & Phister, T. (2018) Next generation microbiological risk assessment meta-omics: The next need for integration. *International Journal of Food Microbiology*, 287, 10–17. https://doi.org/10.1016/j.ijfoodmicro.2017.11.008.

Casini, L., Contini, C., Romano, C. & Scozzafava, G. (2015) Trends in food consumptions: What is happening to generation x? *British Food Journal*, 117, 705–718. https://doi.org/10.1108/BFJ-10-2013-0283.

Cortesi, M.L., Panebianco, A., Giuffrida, A. & Anastasio, A. (2009) Innovations in seafood preservation and storage. *Veterinary Research Communications*, 33 (Supplement 1), 15–23. https://doi.org/10.1007/s11259-009-9241-4.

Cosgrove, R., Browne, D., Minto, C., Tyndall, P., Oliver, M., Montgomerie, M. & McHugh, M. (2019) A game of two halves: Bycatch reduction in Nephrops mixed fisheries. *Fisheries Research*, 210, 31–40. https://doi.org/10.1016/j.fishres.2018.09.019.

Dalgaard, P., Mejlholm, O. & Huss, H.H. (1997a) Application of an iterative approach for development of a microbial model predicting the shelf-life of packed fish. *International Journal of Food Microbiology*, 38, 169–179. https://doi.org/10.1016/s0168-1605(97)00101-3.

Dalgaard, P., Mejlholm, O., Christiansen, T.J. & Huss, H.H. (1997b) Importance of Photobacterium phosphoreum in relation to spoilage of modified atmosphere-packed fish products. *Letters in Applied Microbiology*, 24, 373–378. https://doi.org/10.1046/j.1472-765X.1997.00152.x.

Das, O. & Kannuchamy, N. (2014) Processing of frozen seafood products. *INFOFISH International*, 4.

Daskalova, A. (2019) Farmed fish welfare: Stress, post-mortem muscle metabolism, and stress-related meat quality changes. *International Aquatic Research*, 11, 113–124. https://doi.org/10.1007/s40071-019-0230-0.

Dawson-Coates, J.A., Chase, J.C., Funk, V., Booy, M.H., Haines, L.R., Falkenberg, C.L., Whitaker, D.J., Olafson, R.W. & Pearson, T.W. (2003) The relationship between flesh quality and numbers of Kudoa thyrsites plasmodia and spores in farmed Atlantic salmon, Salmo salar L. *Journal of Fish Diseases*, 26, 451–459. https://doi.org/10.1046/j.1365-2761.2003.00477.x.

den Besten, H.M.W., Amézquita, A., Bover-Cid, S., Dagnas, S., Ellouze, M., Guillou, S., Nychas, G., O'Mahony, C., Pérez-Rodriguez, F. & Membré, J.M. (2018) Next generation of microbiological risk assessment: Potential of omics data for exposure assessment. *International Journal of Food Microbiology*, 287, 18–27. https://doi.org/10.1016/j.ijfoodmicro.2017.10.006.

Devlieghere, F. & Debevere, J. (2000) Influence of dissolved carbon dioxide on the growth of spoilage bacteria. *LWT – Food Science and Technology*, 33, 531–537. https://doi.org/10.1006/fstl.2000.0705.

Digre, H., Rosten, C., Erikson, U., Mathiassen, J.R. & Aursand, I.G. (2017) The on-board live storage of Atlantic cod (Gadus morhua) and haddock (Melanogrammus aeglefinus) caught by trawl: Fish behaviour, stress and fillet quality. *Fisheries Research*, 189, 42–54. https://doi.org/10.1016/j.fishres.2017.01.004.

dos Santos, H.M.C., Nishiyama, M.F., Bonafe, E.G., de Oliveira, C.A.L., Matsushita, M., Visentainer, J.V. & Ribeiro, R.P. (2014) Influence of a diet enriched with Perilla Seed Bran on the composition of omega-3 fatty acid in Nile tilapia. *Journal of the American Oil Chemists' Society*, 91, 1939–1948. https://doi.org/10.1007/s11746-014-2545-8.

Drosinos, E.H. & Nychas, G.-J.E. (1996) Brochothrix thermosphacta, a dominant microorganism in Mediterranean fresh fish (*Sparus aurata*) stored under modified atmosphere. *Italian Journal of Food Science*, 4, 323–330.

Duranton, F. et al. (2014) High-pressure processing of meats and seafood. *Sun*. In: (edited by D-W), Emerging Technologies for Food Processing. Academic Press: San Diego, USA.

EC (2009) Commission Regulation (EC) no 710/2009, 5 August 2009 amending Regulation (EC). *Official Journal of the European Union*, 889/2008 laying down de¬tailed rules for the implementation of Council Regulation (EC) No 834/2007, as regards laying down detailed rules on organic aquaculture animal and seaweed production.

EC (2011) Regulation (EU), 1169/2011 of the European Parliament and of the Council of on the provision of food information to consumers, amending Regulations (EC). *No 1924/2006 and (EC) No 1925/2006 of the European Parliament and of the Council, and Repealing Commission* Directive 87/250/EEC, Council Directive 90/496/EEC, Commission directive 1999/10/EC, Directive 2000/13/EC of the European Parliament and of the Council, Commission Directives 2002/67/EC and 2008/5/EC and Commission Regulation (EC) No 608/2004.

Einen, O. & Roem, A.J. (1997) Dietary protein/energy ratios for Atlantic salmon in relation to fish size: Growth, feed utilization and slaughter quality. *Aquaculture Nutrition*, 3, 115–126. https://doi.org/10.1046/j.1365-2095.1997.00084.x.

Einen, O. & Thomassen, M.S. (1998) Starvation prior to slaughter in Atlantic salmon (Salmo salar): II. *Aquaculture*, 169, 37–53. https://doi.org/10.1016/S0044-8486(98)00332-9.

Elbashir, S., Parveen, S., Schwarz, J., Rippen, T., Jahncke, M. & DePaola, A. (2018) Seafood pathogens and information on antimicrobial resistance: A review. *Food Microbiology*, 70, 85–93. https://doi.org/10.1016/j.fm.2017.09.011.

Enders, E.C. & Boisclair, D. (2016) Effects of environmental fluctuations on fish metabolism: Atlantic salmon Salmo salar as a case study. *Journal of Fish Biology*, 88, 344–358. https://doi.org/10.1111/jfb.12786.

Erikson, U., Gansel, L., Frank, K., Svendsen, E. & Digre, H. (2016) Crowding of Atlantic salmon in net-pen before slaughter. *Aquaculture*, 465, 395–400. https://doi.org/10.1016/j.aquaculture.2016.09.018.

Erikson, U., Truong, H.T.M., Le, D.V., Pham, P.D., Svennevig, N. & Phan, V.T. (2019) Harvesting procedures, welfare and shelf life of ungutted and gutted shortfin pompano (Trachinotus falcatus) stored in ice. *Aquaculture*, 498, 236–245. https://doi.org/10.1016/j.aquaculture.2018.06.085.

Erkan, N. & Özden, Ö. (2008) Quality assessment of whole and gutted sardines (Sardina pilchardus) stored in ice. *International Journal of Food Science and Technology*, 43, 1549–1559. https://doi.org/10.1111/j.1365-2621.2007.01579.x.

Esaiassen, M., Nilsen, H., Joensen, S., Skjerdal, T., Carlehög, M., Eilertsen, G., Gundersen, B. & Elvevoll, E. (2004) Effects of catching methods on quality changes during storage of cod (Gadus morhua). *LWT – Food Science and Technology*, 37, 643–648. https://doi.org/10.1016/j.lwt.2004.02.002.

Esmaeilian, S., Rotabakk, B.T., Lerfall, J., Jakobsen, A.N., Sivertsvik, M., Olsen, A. (2021) The use of soluble gas stabilization technology on food – a review. *Trends in Food Science and Technology*, 118(A), 154-166. https://doi.org/10.1016/j.tifs.2021.09.015.

FAO (2005a) Dicentrarchus labrax. Cultured Aquatic Species Information Programme. Text by Bagni, M.. Fisheries and Aquaculture Division. https://www.fao.org/fishery/en/culturedspecies/dicentrarchus_labrax/en.

FAO (2005b) Sparus aurata. Cultured Aquatic Species Information Programme. Text by Colloca, F.; Cerasi, S.. Fisheries and Aquaculture Division. https://www.fao.org/fishery/en/culturedspecies/sparus_aurata?lang=en.

FAO (2007) *General Fisheries Commission for the Mediterranean. Studies and Reviews No 82. Marketing of Aquacultured Sea Bass and Sea Bream from the Mediterranean Basin*. Food and Agriculture Organization: Rome.

FAO (2011). *Global Food Losses and Food Waste*. Food and Agriculture Organization of the United Nations: Rome.

FAO (2015) Climate change and food systems: Global assessments and implications for food security and trade. In: (edited by A. Elbehri). Food and Agriculture Organization: Rome.

FAO (2018a) *Globefish Highlights A Quarterly Update on World Seafood Markets*. Food and Agriculture Organization: Rome.

FAO (2018b) *The State of World Fisheries and Aquaculture, 2018 – Meeting the sustainable development goals*. Food and Agriculture Organization: Rome.

FAO (2019a) *Globefish Highlights A Quarterly Update on World Seafood Markets*. Food and Agriculture Organization: Rome.

FAO (2019b) Bass and bream prices hit multi-year lows as oversupply continues. https://www.fao.org/in-action/globefish/market-reports/resource-detail/en/c/1189941.

FAO (2022) *The State of Word Fisheries and Aquaculture 2022 – Towards Blue Transformation*. Food and Agriculture Organisation: Rome.

Feekings, J., Bartolino, V., Madsen, N. & Catchpole, T. (2012) Fishery discards: Factors affecting their variability within a demersal trawl fishery. *PLOS ONE*, 7, e36409. https://doi.org/10.1371/journal.pone.0036409.

Feng, X., Bansal, N. & Yang, H. (2016) Fish gelatin combined with chitosan coating inhibits myofibril degradation of golden pomfret (Trachinotus blochii) fillet during cold storage. *Food Chemistry*, 200, 283–292. https://doi.org/10.1016/j.foodchem.2016.01.030.

Fogarty, C., Whyte, P., Brunton, N., Lyng, J., Smyth, C., Fagan, J. & Bolton, D. (2019) Spoilage indicator bacteria in farmed Atlantic salmon (Salmo salar) stored on ice for 10 days. *Food Microbiology*, 77, 38–42. https://doi.org/10.1016/j.fm.2018.08.001.

Françoise, L. (2010) Occurrence and role of lactic acid bacteria in seafood products. *Food Microbiology*, 27, 698–709. https://doi.org/10.1016/j.fm.2010.05.016.

Galié, S., García-Gutiérrez, C., Miguélez, E.M., Villar, C.J. & Lombó, F. (2018) Biofilms in the food industry: Health aspects and control methods. *Frontiers in Microbiology*, 9, 898–898. https://doi.org/10.3389/fmicb.2018.00898.

Gambuteanu, C. & Alexe, P. (2015) Comparison of thawing assisted by low-intensity ultrasound on technological properties of pork longissimus dorsi muscle. *Journal of Food Science and Technology*, 52, 2130–2138. https://doi.org/10.1007/s13197-013-1204-7.

Gatlin, III, D.M., Barrows, F.T., Brown, P., Dabrowski, K., Gaylord, T.G., Hardy, R.W., Herman, E., Hu, G., Krogdahl, Å, Nelson, R., Overturf, K., Rust, M., Sealey, W., Skonberg, D., J Souza, E., Stone, D., Wilson, R. & Wurtele, E. (2007) Expanding the utilization of sustainable plant products in aquafeeds: A review. *Aquaculture Research*, 38, 551–579. https://doi.org/10.1111/j.1365-2109.2007.01704.x.

Ghanbari, M., Jami, M., Domig, K.J. & Kneifel, W. (2013) Seafood biopreservation by lactic acid bacteria – A review. *LWT – Food Science and Technology*, 54, 315–324. https://doi.org/10.1016/j.lwt.2013.05.039.

Gjedrem, T., Gjøen, H.M. & Gjerde, B. (1991) Genetic origin of Norwegian farmed Atlantic salmon. *Aquaculture*, 98, 41–50. https://doi.org/10.1016/0044-8486(91)90369-I.

Gjerde, B., Simianer, H. & Refstie, T. (1994) Estimates of genetic and phenotypic parameters for body weight, growth rate and sexual maturity in Atlantic salmon. *Livestock Production Science*, 38, 133–143. https://doi.org/10.1016/0301-6226(94)90057-4.

Gjerde, B., Ødegård, J. & Thorland, I. (2011) Estimates of genetic variation in the susceptibility of Atlantic salmon (Salmo salar) to the salmon louse Lepeophtheirus salmonis. *Aquaculture*, 314, 66–72. https://doi.org/10.1016/j.aquaculture.2011.01.026.

González, S., Flick, G.J., O'Keefe, S.F., Duncan, S.E., McLean, E. & Craig, S.R. (2006) Composition of farmed and wild yellow perch (Perca flavescens). *Journal of Food Composition and Analysis*, 19, 720–726. https://doi.org/10.1016/j.jfca.2006.01.007.

Gram, L. & Huss, H.H. (1996) Microbiological spoilage of fish and fish products. *International Journal of Food Microbiology*, 33, 121–137. https://doi.org/10.1016/0168-1605(96)01134-8.

Gram, L. & Dalgaard, P. (2002) Fish spoilage bacteria – Problems and solutions. *Current Opinion in Biotechnology*, 13, 262–266. https://doi.org/10.1016/s0958-1669(02)00309-9.

Gutierrez, M.M., Meleddu, M. & Piga, A. (2017) Food losses, shelf life extension and environmental impact of a packaged cheesecake: A life cycle assessment. *Food Research International*, 91, 124–132. https://doi.org/10.1016/j.foodres.2016.11.031.

Hansen, A.Å., Mørkøre, T., Rudi, K., Langsrud, Ø & Eie, T. (2009) The combined effect of superchilling and modified atmosphere packaging using CO2 emitter on quality during chilled storage of pre-rigor salmon fillets (Salmo salar). *Journal of the Science of Food and Agriculture*, 89, 1625–1633. https://doi.org/10.1002/jsfa.3599.

Hansen, A.Å. et al. (2012) Advances in bulk packaging for the transport of fresh fish. In: *Advances in Meat, Poultry and Seafood Packaging*, Vol. 8 (edited by J. P. Kerry). Woodhead Publishing: Cambridge, UK, pp. 248–260.

Harmon, T.S. (2009) Methods for reducing stressors and maintaining water quality associated with live fish transport in tanks: A review of the basics. *Reviews in Aquaculture*, 1, 58–66. https://doi.org/10.1111/j.1753-5131.2008.01003.x.

Hazen, E.L., Scales, K.L., Maxwell, S.M., Briscoe, D.K., Welch, H., Bograd, S.J., Bailey, H., Benson, S.R., Eguchi, T., Dewar, H., Kohin, S., Costa, D.P., Crowder, L.B. & Lewison, R.L. (2018) A dynamic ocean management tool to reduce bycatch and support sustainable fisheries. *Science Advances*, 4, eaar3001. https://doi.org/10.1126/sciadv.aar3001.

Henriques, J., Dick, J.R., Tocher, D.R. & Bell, J.G. (2014) Nutritional quality of salmon products available from major retailers in the UK: Content and composition of n-3 long-chain PUFA. *British Journal of Nutrition*, 112, 964–975. https://doi.org/10.1017/S0007114514001603.

Hoang, K.V., Stern, N.J. & Lin, J. (2011) Development and stability of bacteriocin resistance in Campylobacter spp. *Journal of Applied Microbiology*, 111, 1544–1550. https://doi.org/10.1111/j.1365-2672.2011.05163.x.

Hoar, W.S. (1988) 4 the physiology of smolting Salmonids. In: *Fish Physiology* (edited by W. S. Hoar & D. J. Randall). Academic Press: Cambridge, MA, USA, pp. 275–343.

Hoel, S., Mehli, L., Bruheim, T., Vadstein, O. & Jakobsen, A.N. (2015) Assessment of microbiological quality of retail fresh sushi from selected sources in Norway. *Journal of Food Protection*, 78(5), 977–982. https://doi.org/10.4315/0362-028X.JFP-14-480.

Hoel, S., Jakobsen, A.N. & Vadstein, O. (2017) Effects of storage temperature on bacterial growth rates and community structure in fresh retail sushi. *Journal of Applied Microbiology*, 123, 698–709. https://doi.org/10.1111/jam.13527.

Hoel, S., Vadstein, O. & Jakobsen, A.N. (2018) Growth of mesophilic Aeromonas salmonicida in an experimental model of nigiri sushi during cold storage. *International Journal of Food Microbiology*, 285, 1–6. https://doi.org/10.1016/j.ijfoodmicro.2018.07.008.

Hoel, S., Vadstein, O. & Jakobsen, A.N. (2019) The significance of mesophilic Aeromonas spp. in minimally processed ready-to-eat seafood. *Microorganisms*, 7. https://doi.org/10.3390/microorganisms7030091.

Hong, H., Regenstein, J.M. & Luo, Y. (2017) The importance of ATP-related compounds for the freshness and flavor of post-mortem fish and shellfish muscle: A review. *Critical Reviews in Food Science and Nutrition*, 57, 1787–1798. https://doi.org/10.1080/10408398.2014.1001489.

Hoolohan, C., Berners-Lee, M., McKinstry-West, J. & Hewitt, C.N. (2013) Mitigating the greenhouse gas emissions embodied in food through realistic consumer choices. *Energy Policy*, 63, 1065–1074. https://doi.org/10.1016/j.enpol.2013.09.046.

Horner, W.F.A. (1997) Preservation of fish by curing (drying, salting and smoking). In: *Fish Processing Technology* (edited by G. M. Hall). Blackie Academic & Professional: London, pp. 32–73.

Huss, H.H. (1995) *Quality and Quality Changes in Fresh Fish, FAO FISHERIES TECHNICAL PAPER*, Vol. 348. Food and Agriculture Organization: Rome.

Huss, H.H. et al. (2014) Assessment and management of seafood safety and quality: Current practices and issues. FAO Fisheries and Aquaculture Technical Paper no. 574 (edited by J. Ryder, I. Karunasagar & L. Ababouch). Food and Agriculture Organization: Rome.

Hyldig, G., Jørgensen, B.M., Undeland, I., Olsen, R.E., Jónsson, A. & Nielsen, H.H. (2012) Sensory properties of frozen herring (Clupea harengus) from different catch seasons and locations. *Journal of Food Science*, 77, S288–S293. https://doi.org/10.1111/j.1750-3841.2012.02838.x.

ISFA (2018) *Salmon Farming – Sustainnig Communities and Feeding the World*. ISFA. https://www.salmonfarming.org/cms/wp-content/uploads/2018/06/ISFA-JUNE-2018-REVISED-FOR-FOR-WEB1.pdf.

Jakobsen, A.N., Shumilina, E., Lied, H. & Hoel, S. (2020) Growth and spoilage metabolites production of a mesophilic Aeromonas salmonicida strain in Atlantic salmon (Salmo salar L.) during cold storage in modified atmosphere. *Journal of Applied Microbiology*, 129, 935–946. https://doi.org/10.1111/jam.14680.

James, C., Purnell, G. & James, S.J. (2015) A review of Novel and innovative food freezing technologies. *Food and Bioprocess Technology*, 8, 1616–1634. https://doi.org/10.1007/s11947-015-1542-8.

Jessen, F. et al. (2014) Chilling and freezing of fish. In: *Seafood Processing: Technology, Quality and Safety* (edited by I. S. Boziaris). Wiley-Blackwell: Chichester, UK.

Jha, D.K., Bhujel, R.C. & Anal, A.K. (2015) Dietary supplementation of probiotics improves survival and growth of rohu (Labeo rohita Ham.) hatchlings and fry in outdoor tanks. *Aquaculture*, 435, 475–479. https://doi.org/10.1016/j.aquaculture.2014.10.026.

Jhingran, V.G. & Pullin, R.S.V. (1985). *A Hatchery Manual of the Common*, Chinese, and Indian major carsp. Asian Development Bank: Manila and International Center for Living Aquatic Resources Management: Manila.

Kaale, L.D. & Eikevik, T.M. (2016) Changes of proteins during superchilled storage of Atlantic salmon muscle (Salmo salar). *Journal of Food Science and Technology*, 53, 441–450. https://doi.org/10.1007/s13197-015-1979-9.

Kaale, L.D., Eikevik, T.M., Rustad, T. & Kolsaker, K. (2011) Superchilling of food: A review. *Journal of Food Engineering*, 107, 141–146. https://doi.org/10.1016/j.jfoodeng.2011.06.004.

Kaale, L.D., Eikevik, T.M., Rustad, T. & Nordtvedt, T.S. (2014) Changes in water holding capacity and drip loss of Atlantic salmon (Salmo salar) muscle during superchilled storage. *LWT – Food Science and Technology*, 55, 528–535. https://doi.org/10.1016/j.lwt.2013.10.021.

Kapetanakou, A.E. & Manios, S.G. (2014) Application of edible films and coatings on food. In: *Novel Food Preservation and Microbial Assessment Techniques* (edited by I. S. Boziaris). CRC Press: Boca Raton, FL, USA.

Kapetanakou, A.E. & Skandamis, P.N. (2016) Applications of active packaging for increasing microbial stability in foods: Natural volatile antimicrobial compounds. *Current Opinion in Food Science*, 12, 1–12. https://doi.org/10.1016/j.cofs.2016.06.001.

Karapanagiotidis, I.T., Bell, M.V., Little, D.C., Yakupitiyage, A. & Rakshit, S.K. (2006) Polyunsaturated fatty acid content of wild and farmed tilapias in Thailand: Effect of aquaculture practices and implications for human nutrition. *Journal of Agricultural and Food Chemistry*, 54, 4304–4310. https://doi.org/10.1021/jf0581877.

Kassemsarn, B.-O., Perez, B.S., Murray, J. & Jones, N.R. (1963) Nucleotide degradation in the muscle of iced haddock (Gadus aeglefinus), lemon sole (Pleuronectes microcephalus), and plaice (Pleuronectes platessa). *Journal of Food Science*, 28, 28–37. https://doi.org/10.1111/j.1365-2621.1963.tb00155.x.

Kelleher, K. (2005) Discards in the world's marine fisheries. FAO Fisheries Technical Paper.

Khan, M.Y. & Mittal, A. (2017) Freezing of food products: A review. *International Journal of Advance Research in Science and Engineering*, no. 06. http://ijarse.com/images/fullpdf/1504952112_229ijarse.pdf.

Khatun, N., Islam, M.T., Sultana, N., Mrong, S. & Huq, M.A. (2017) Present status of carp hatchery and breeding operations in Bangladesh: A review. *Research in Agriculture Livestock and Fisheries*, 4, 123–129. https://doi.org/10.3329/ralf.v4i2.33724.

Khoa, T.T.L. et al. (2013) Effects of salinity on selected physiological and biochemical parameters of striped catfish (Pangasianodon hypophthalmus) reared in ponds. In: *Positioning for Profits* (edited W.A. Society). HCM City, Vietnam.

Kolstad, K., Grisdale-Helland, B. & Gjerde, B. (2004) Family differences in feed efficiency in Atlantic salmon (Salmo salar). *Aquaculture*, 241, 169–177. https://doi.org/10.1016/j.aquaculture.2004.09.001.

Kolstad, K., Heuch, P.A., Gjerde, B., Gjedrem, T. & Salte, R. (2005) Genetic variation in resistance of Atlantic salmon (Salmo salar) to the salmon louse Lepeophtheirus salmonis. *Aquaculture*, 247, 145–151. https://doi.org/10.1016/j.aquaculture.2005.02.009.

Kostaki, M., Giatrakou, V., Savvaidis, I.N. & Kontominas, M.G. (2009) Combined effect of MAP and thyme essential oil on the microbiological, chemical and sensory attributes of organically aquacultured sea bass (Dicentrarchus labrax) fillets. *Food Microbiology*, 26, 475–482. https://doi.org/10.1016/j.fm.2009.02.008] [PubMed: 19465243.

Koutsoumanis, K.P., Taoukis, P.S., Drosinos, E.H. & Nychas, G.J. (2000) Applicability of an Arrhenius model for the combined effect of temperature and CO(2) packaging on the spoilage microflora of fish. *Applied and Environmental Microbiology*, 66, 3528–3534. https://doi.org/10.1128/AEM.66.8.3528-3534.2000.

Kurt Kaya, G. & Baştürk, Ö. (2014) Organoleptic and chemical changes during storage of sea bass marinades (Dicentrarchus labrax L., 1758). *Journal of Food Processing and Preservation*, 38, 1072–1079. https://doi.org/10.1111/jfpp.12065.

Kuuliala, L., Al Hage, Y., Ioannidis, A.G., Sader, M., Kerckhof, F.M., Vanderroost, M., Boon, N., De Baets, B., De Meulenaer, B., Ragaert, P. & Devlieghere, F. (2018) Microbiological, chemical and sensory spoilage analysis of raw Atlantic cod (Gadus morhua) stored under modified atmospheres. *Food Microbiology*, 70, 232–244. https://doi.org/10.1016/j.fm.2017.10.011.

Lakshmisha, I.P., Ravishankar, C.N., Ninan, G., Mohan, C.O. & Gopal, T.K. (2008) Effect of freezing time on the quality of Indian mackerel (Rastrelliger kanagurta) during frozen storage. *Journal of Food Science*, 73, S345–S353. https://doi.org/10.1111/j.1750-3841.2008.00876.x.

Langsrud, S., Moen, B., Møretrø, T., Løype, M. & Heir, E. (2016) Microbial dynamics in mixed culture biofilms of bacteria surviving sanitation of conveyor belts in salmon-processing plants. *Journal of Applied Microbiology*, 120, 366–378. https://doi.org/10.1111/jam.13013.

Lee, J., Fong, Q. & Park, J.W. (2016) Effect of pre-freezing treatments on the quality of Alaska pollock fillets subjected to freezing/thawing. *Food Bioscience*, 16, 50–55. https://doi.org/10.1016/j.fbio.2016.09.003.

Lerfall, J., Larsson, T., Birkeland, S., Taksdal, T., Dalgaard, P., Afanasyev, S., Bjerke, M.T. & Mørkøre, T. (2012) Effect of pancreas disease (PD) on quality attributes of raw and smoked fillets of Atlantic salmon (Salmo salar L.). *Aquaculture*, 324–325, 209–217. https://doi.org/10.1016/j.aquaculture.2011.11.003.

Lerfall, J., Roth, B., Skare, E.F., Henriksen, A., Betten, T., Dziatkowiak-Stefaniak, M.A. & Rotabakk, B.T. (2015) Pre-mortem stress and the subsequent effect on flesh quality of pre-rigor filleted Atlantic salmon (Salmo salar L.) during ice storage. *Food Chemistry*, 175, 157–165. https://doi.org/10.1016/j.foodchem.2014.11.111.

Lerfall, J. & Rotabakk, B.T. (2015) *Muscle Temperature at the Point of Filleting-Subsequent Effect on Storage Quality of Prerigor Filleted Raw- and Cold-Smoked Atlantic Salmon*. Food Science and Technology International.

Lerfall, J., Bendiksen, E.Å, Olsen, J.V., Morrice, D. & Østerlie, M. (2016) A comparative study of organic- versus conventional farmed Atlantic salmon. I. Pigment and lipid content and composition, and carotenoid stability in ice-stored fillets. *Aquaculture*, 451, 170–177. https://doi.org/10.1016/j.aquaculture.2015.09.013.

Lerfall, J., Bjørge Thomassen, G.M. & Jakobsen, A.N. (2018a) Quality of fresh saithe (Pollachius virens) in modified atmosphere packages as affected by the gas composition. *Food Packaging and Shelf Life*, 18, 147–156. https://doi.org/10.1016/j.fpsl.2018.11.006.

Lerfall, J., Jakobsen, A.N., Skipnes, D., Waldenstrøm, L., Hoel, S. & Rotabakk, B.T. (2018b) Comparative evaluation on the quality and shelf life of Atlantic salmon (Salmo salar L.) filets using microwave and conventional pasteurization in combination with novel packaging methods. *Journal of Food Science*, 83, 3099–3109. https://doi.org/10.1111/1750-3841.14384.

Lerfall, J., Hasli, P.R., Skare, E.F., Olsen, R.E., Rotabakk, B.T., Roth, B., Slinde, E. & Egelandsdal, B. (2017) A comparative study of diploid versus triploid Atlantic salmon (Salmo salar L.). The effects of rearing temperatures (5, 10 and 15°C) on raw material characteristics and storage quality. The effects of rearing temperatures. *Food Chemistry*, 225, 37–44. https://doi.org/10.1016/j.foodchem.2017.01.012.

Li, B. & Sun, D.-W. (2002) Novel methods for rapid freezing and thawing of foods – A review. *Journal of Food Engineering*, 54, 175–182. https://doi.org/10.1016/S0260-8774(01)00209-6.

López de Lacey, A.M., López-Caballero, M.E. & Montero, P. (2014) Agar films containing green tea extract and probiotic bacteria for extending fish shelf-life. *LWT – Food Science and Technology*, 55, 559–564. https://doi.org/10.1016/j.lwt.2013.09.028.

Løvdal, T. (2015) The microbiology of cold smoked salmon. *Food Control*, 54, 360–373. https://doi.org/10.1016/j.foodcont.2015.02.025.

Macé, S., Cardinal, M., Jaffrès, E., Cornet, J., Lalanne, V., Chevalier, F., Sérot, T., Pilet, M.F., Dousset, X. & Joffraud, J.J. (2014) Evaluation of the spoilage potential of bacteria isolated from spoiled cooked whole tropical shrimp (Penaeus vannamei) stored under modified atmosphere packaging. *Food Microbiology*, 40, 9–17. https://doi.org/10.1016/j.fm.2013.11.018.

Maeda, E.E., Mäntyniemi, S., Despoti, S., Musumeci, C., Vassilopoulou, V., Stergiou, K.I., Giannoulaki, M., Ligas, A. & Kuikka, S. (2017) A Bayesian model of fisheries discards with flexible structure and priors defined by experts. *Ecological Modelling*, 366, 1–14. https://doi.org/10.1016/j.ecolmodel.2017.10.007.

Magnussen, O.M., Haugland, A., Torstveit Hemmingsen, A.K., Johansen, S. & Nordtvedt, T.S. (2008) Advances in superchilling of food – Process characteristics and product quality. *Trends in Food Science and Technology*, 19, 418–424. https://doi.org/10.1016/j.tifs.2008.04.005.

Mai, N.T.T., Margeirsson, B., Margeirsson, S., Bogason, S.G., Sigurgísladóttir, S. & Arason, S. (2012) Temperature mapping of fresh fish supply chains – Air and sea transport. *Journal of Food Process Engineering*, 35, 622–656. https://doi.org/10.1111/j.1745-4530.2010.00611.x.

Margeirsson, B., Pálsson, H., Gospavic, R., Popov, V., Jónsson, M.Þ & Arason, S. (2012) Numerical modelling of temperature fluctuations of chilled and superchilled cod fillets packaged in expanded polystyrene boxes stored on pallets under dynamic temperature conditions. *Journal of Food Engineering*, 113, 87–99. https://doi.org/10.1016/j.jfoodeng.2012.05.017.

Maria Poli, B. (2009) Farmed fish welfare-suffering assessment and impact on product quality. *Italian Journal of Animal Science*, 8, 139–160. https://doi.org/10.4081/ijas.2009.s1.139.

McCluney, J.K., Anderson, C.M. & Anderson, J.L. (2019) The fishery performance indicators for global tuna fisheries. *Nature Communications*, 10, 1641. https://doi.org/10.1038/s41467-019-09466-6.

Mente, E., Stratakos, A., Boziaris, I.S., Kormas, K.A., Karalazos, V., Karapanagiotidis, I.T., Catsiki, V.A. & Leondiadis, L. (2012) The effect of organic and conventional production methods on sea bream growth, health and body composition: A field experiment. *Scientia Marina*, 76, 549–560. https://doi.org/10.3989/scimar.03411.07C.

Mente, E., Karalazos, V., Karapanagiotidis, I.T. & Pita, C. (2011) Nutrition in organic aquaculture: An inquiry and a discourse. *Aquaculture Nutrition*, 17, e798–e817. https://doi.org/10.1111/j.1365-2095.2010.00846.x.

Mercier, S., Villeneuve, S., Mondor, M. & Uysal, I. (2017) Time–temperature management along the food cold chain: A review of recent developments. *Comprehensive Reviews in Food Science and Food Safety*, 16, 647–667. https://doi.org/10.1111/1541-4337.12269.

Mochizuki, S., Norita, Y. & Maeno, K. (1998) Effects of bleeding on post-mortem changes in the muscle of horse mackerel. *Nippon Suisan Gakkaishi*, 64, 276–279. https://doi.org/10.2331/suisan.64.276.

Mokrani, D., Oumouna, M. & Cuesta, A. (2018) Fish farming conditions affect to European sea bass (Dicentrarchus labrax L.) quality and shelf life during storage in ice. *Aquaculture*, 490, 120–124. https://doi.org/10.1016/j.aquaculture.2018.02.032.

Mor-Mur, M. et al. (2014) Microbiological aspects of high-pressure processing. In: *Emerging Technologies for Food Processing* (edited by D.-W. Sun). Academic Press: San Diego, CA, USA.

Moran, J.D.W., Whitaker, D.J. & Kent, M.L. (1999) A review of the myxosporean genus Kudoa Meglitsch, 1947, and its impact on the international aquaculture industry and commercial fisheries. *Aquaculture*, 172, 163–196. https://doi.org/10.1016/S0044-8486(98)00437-2.

Murchie, L.W., Cruz-Romero, M., Kerry, J.P., Linton, M., Patterson, M.F., Smiddy, M. & Kelly, A.L. (2005) High pressure processing of shellfish: A review of microbiological and other quality aspects. *Innovative Food Science and Emerging Technologies*, 6, 257–270. https://doi.org/10.1016/j.ifset.2005.04.001.

Møretrø, T., Moen, B., Heir, E., Hansen, A.Å & Langsrud, S. (2016) Contamination of salmon fillets and processing plants with spoilage bacteria. *International Journal of Food Microbiology*, 237, 98–108. https://doi.org/10.1016/j.ijfoodmicro.2016.08.016.

Mørkøre, T. & Rørvik, K.-A. (2001) Seasonal variations in growth, feed utilization and product quality of farmed Atlantic salmon (Salmo salar) transferred to seawater as 0+smolts or 1+smolts. *Aquaculture*, 199, 145–157. https://doi.org/10.1016/S0044-8486(01)00524-5.

Mørkøre, T., Mazo T., P.I., Tahirovic, V. & Einen, O. (2008) Impact of starvation and handling stress on rigor development and quality of Atlantic salmon (Salmon salar L). *Aquaculture*, 277, 231–238. https://doi.org/10.1016/j.aquaculture.2008.02.036.

Nagarajarao, R.C. (2016) Recent advances in processing and packaging of fishery products: A review. *Aquatic Procedia*, 7, 201–213. https://doi.org/10.1016/j.aqpro.2016.07.028.

Naylor, R.L., Hardy, R.W., Bureau, D.P., Chiu, A., Elliott, M., Farrell, A.P., Forster, I., Gatlin, D.M., Goldburg, R.J., Hua, K. & Nichols, P.D. (2009) Feeding aquaculture in an era of finite resources. *Proceedings of the National Academy of Sciences of the United States of America*, 106, 15103–15110. https://doi.org/10.1073/pnas.0905235106.

Newman, M. (2001) *Cryogenic Impingement Freezing Utilizing Atomized Liquid Nitrogen for the Rapid Freezing of Food Products, Rapid Cooling of Food.* Meeting of IIR Commission, Vol. C2. Bristol, UK.

NFBD (2015) Annual report. National Fisheries Development Board, India.

Nortvedt, R. et al. (2007) High-quality seafood products based on ethical and sustainable production. In: *Aquaculture Research: From Cage to Consumption* (edited by M. Thomassen, R. Gudding, B. Norberg & L. Jørgensen). Research Council of Norway: Oslo.

Nowzari, F., Shábanpour, B. & Ojagh, S.M. (2013) Comparison of chitosan–gelatin composite and bilayer coating and film effect on the quality of refrigerated rainbow trout. *Food Chemistry*, 141, 1667–1672. https://doi.org/10.1016/j.foodchem.2013.03.022.

Ohlson, T. & Bengtsson, N. (2000) Introduction. In: *Minimal Processing Technologies in the Food Industries* (edited by T. Ohlson & N. Bengtsson). Woodhead Publishing: Cambridge, UK.

Ohlsson, T. (1994) Minimal processing-preservation methods of the future: An overview. *Trends in Food Science and Technology*, 5, 341–344. https://doi.org/10.1016/0924-2244(94)90210-0.

Olatunde, O.O. & Benjakul, S. (2018) Natural preservatives for extending the shelf-life of seafood: A revisit. *Comprehensive Reviews in Food Science and Food Safety*, 17, 1595–1612. https://doi.org/10.1111/1541-4337.12390.

Olsen, S.H., Tobiassen, T., Akse, L., Evensen, T.H. & Midling, K.Ø (2013) Capture induced stress and live storage of Atlantic cod (Gadus morhua) caught by trawl: Consequences for the flesh quality. *Fisheries Research*, 147, 446–453. https://doi.org/10.1016/j.fishres.2013.03.009.

Olsen, S.H., Joensen, S., Tobiassen, T., Heia, K., Akse, L. & Nilsen, H. (2014) Quality consequences of bleeding fish after capture. *Fisheries Research*, 153, 103–107. https://doi.org/10.1016/j.fishres.2014.01.011.

Osman, G.A. & El-Khateeb, M.A. (2016) Impact of water contamination on tilapia (Oreochromis niloticus) fish yield. *International Journal of ChemTech Research*, 9, 166–181.

Panguila, E.A. et al. (2018) Evaluation of the freshness and shelf-life of fresh and chilled mackerel (Scomber spp.) and horsemackerel (Trachurus spp.), marketed in Luanda (Angola) and in Faro (Portugal). In: *INCREaSE* (edited by A. Mortal, J. Aníbal, J. Monteiro, C. Sequeira, J. Semião, M. Moreira da Silva & M. Oliveira). Springer International Publishing: Cham, Germany, pp. 168–181.

Parfitt, J., Barthel, M. & Macnaughton, S. (2010) Food waste within food supply chains: Quantification and potential for change to 2050. *Philosophical Transactions of the Royal Society of London. Series B, Biological Sciences*, 365, 3065–3081. https://doi.org/10.1098/rstb.2010.0126.

Parlapani, F.F. & Boziaris, I.S. (2016) Monitoring of spoilage and determination of microbial communities based on 16S rRNA gene sequence analysis of whole sea bream stored at various temperatures. *LWT – Food Science and Technology*, 66, 553–559. https://doi.org/10.1016/j.lwt.2015.11.007.

Parlapani, F.F., Mallouchos, A., Haroutounian, S.A. & Boziaris, I.S. (2014) Microbiological spoilage and investigation of volatile profile during storage of sea bream fillets under various conditions. *International Journal of Food Microbiology*, 189, 153–163. https://doi.org/10.1016/j.ijfoodmicro.2014.08.006.

Parlapani, F.F., Kormas, K.A. & Boziaris, I.S. (2015a) Microbiological changes, shelf life and identification of initial and spoilage microbiota of sea bream fillets stored under various conditions using 16S rRNA gene analysis. *Journal of the Science of Food and Agriculture*, 95, 2386–2394. https://doi.org/10.1002/jsfa.6957.

Parlapani, F.F., Haroutounian, S.A., Nychas, G.J. & Boziaris, I.S. (2015b) Microbiological spoilage and volatiles production of gutted European sea bass stored under air and commercial modified atmosphere package at 2°C. *Food Microbiology*, 50, 44–53. https://doi.org/10.1016/j.fm.2015.03.006.

Parlapani, F.F., Mallouchos, A., Haroutounian, S.A. & Boziaris, I.S. (2017) Volatile organic compounds of microbial and non-microbial origin produced on model fish substrate un-inoculated and inoculated with gilt-head sea bream spoilage bacteria. *LWT*, 78, 54–62. https://doi.org/10.1016/j.lwt.2016.12.020.

Parlapani, F.F., Michailidou, S., Pasentsis, K., Argiriou, A., Krey, G. & Boziaris, I.S. (2018a) A meta-barcoding approach to assess and compare the storage temperature-dependent bacterial diversity of gilt-head sea bream (Sparus aurata) originating from fish farms from two geographically distinct areas of Greece. *International Journal of Food Microbiology*, 278, 36–43. https://doi.org/10.1016/j.ijfoodmicro.2018.04.027.

Parlapani, F.F., Michailidou, S., Anagnostopoulos, D.A., Sakellariou, A.K., Pasentsis, K., Psomopoulos, F., Argiriou, A., Haroutounian, S.A. & Boziaris, I.S. (2018b) Microbial spoilage investigation of thawed common cuttlefish (Sepia officinalis) stored at 2°C using next generation sequencing and volatilome analysis. *Food Microbiology*, 76, 518–525. https://doi.org/10.1016/j.fm.2018.08.004.

Parlapani, F.F., Boziaris, I.S., Meziti, A., Michailidou, S., Haroutounian, S.A., Argiriou, A. & Karapanagiotidis, I.T. (2019a) Microbiological status based on 454-Pyrosequencing and volatilome analysis of gilthead seabream (Sparus aurata) fed on diets with hydrolyzed feather meal and poultry by-product meal as fishmeal replacers. *European Food Research and Technology*, 245, 1409–1420. https://doi.org/10.1007/s00217-019-03270-8.

Parlapani, F.F., Michailidou, S., Anagnostopoulos, D.A., Koromilas, S., Kios, K., Pasentsis, K., Psomopoulos, F., Argiriou, A., Haroutounian, S.A. & Boziaris, I.S. (2019b) Bacterial communities and potential spoilage markers of whole blue crab (Callinectes sapidus) stored under commercial simulated conditions. *Food Microbiology*, 82, 325–333. https://doi.org/10.1016/j.fm.2019.03.011.

Peng, J., Tang, J., Barrett, D.M., Sablani, S.S., Anderson, N. & Powers, J.R. (2017) Thermal pasteurization of ready-to-eat foods and vegetables: Critical factors for process design and effects on quality. *Critical Reviews in Food Science and Nutrition*, 57, 2970–2995. https://doi.org/10.1080/10408398.2015.1082126.

Petter Johnsen, J.P. & Eliasen, S. (2011) Solving complex fisheries management problems: What the EU can learn from the Nordic experiences of reduction of discards. *Marine Policy*, 35, 130–139. https://doi.org/10.1016/j.marpol.2010.08.011.

Peterman, M.A. & Phelps, R.P. (2012) Fillet yields from four strains of Nile tilapia (Oreochromis niloticus) and a red variety. *Journal of Applied Aquaculture*, 24, 342–348. https://doi.org/10.1080/10454438.2012.731480.

Pham, T.T.H., Rossi, P., Dinh, H.D.K., Pham, N.T.A., Tran, P.A., Ho, T.T.K.M., Dinh, Q.T. & De Alencastro, L.F. (2018) Analysis of antibiotic multi-resistant bacteria and resistance genes in the effluent of an intensive shrimp farm (Long An, Vietnam). *Journal of Environmental Management*, 214, 149–156. https://doi.org/10.1016/j.jenvman.2018.02.089.

Piñeiro, C., Barros-Velázquez, J. & Aubourg, S.P. (2004) Effects of newer slurry ice systems on the quality of aquatic food products: A comparative review versus flake-ice chilling methods. *Trends in Food Science and Technology*, 15, 575–582. https://doi.org/10.1016/j.tifs.2004.09.005.

Poli, B.M., Messini, A., Parisi, G., Scappini, F., Vigiani, V., Giorgi, G. & Vincenzini, M. (2006) Sensory, physical, chemical and microbiological changes in European sea bass (Dicentrarchus labrax) fillets packed under modified atmosphere/air or prepared from whole fish stored in ice. *International Journal of Food Science and Technology*, 41, 444–454. https://doi.org/10.1111/j.1365-2621.2005.01094.x.

Ponte, S., Kelling, I., Jespersen, K.S. & Kruijssen, F. (2014) The blue revolution in Asia: Upgrading and governance in aquaculture value chains. *World Development*, 64, 52–64. https://doi.org/10.1016/j.worlddev.2014.05.022.

Pournis, N., Papavergou, A., Badeka, A., Kontominas, M.G. & Savvaidis, I.N. (2005) Shelf-life extension of refrigerated Mediterranean mullet (Mullus surmuletus) using modified atmosphere packaging. *Journal of Food Protection*, 68, 2201–2207. https://doi.org/10.4315/0362-028x-68.10.2201.

Powell, J., White, I., Guy, D. & Brotherstone, S. (2008) Genetic parameters of production traits in Atlantic salmon (Salmo salar). *Aquaculture*, 274, 225–231. https://doi.org/10.1016/j.aquaculture.2007.11.036.

Refaey, M.M., Tian, X., Tang, R. & Li, D. (2017) Changes in physiological responses, muscular composition and flesh quality of channel catfish Ictalurus punctatus suffering from transport stress. *Aquaculture*, 478, 9–15. https://doi.org/10.1016/j.aquaculture.2017.01.026.

Rico, A., Oliveira, R., McDonough, S., Matser, A., Khatikarn, J., Satapornvanit, K., Nogueira, A.J., Soares, A.M., Domingues, I. & Van den Brink, P.J. (2014) Use, fate and ecological risks of antibiotics applied in tilapia cage farming in Thailand. *Environmental Pollution*. Barking, Essex, UK, 191, 8–16. https://doi.org/10.1016/j.envpol.2014.04.002.

Robson, A.A., Kelly, M.S. & Latchford, J.W. (2007) Effect of temperature on the spoilage rate of whole, unprocessed crabs: Carcinus maenas, Necora puber and Cancer pagurus. *Food Microbiology*, 24, 419–424. https://doi.org/10.1016/j.fm.2006.07.016.

Rosnes, J.T., Skåra, T. & Skipnes, D. (2011) Recent advances in minimal heat processing of fish: Effects on microbiological activity and safety. *Food and Bioprocess Technology*, 4, 833–848. https://doi.org/10.1007/s11947-011-0517-7.

Rosnes, J.T. & Skipnes, D. (2018) Minimal heat processing applied in fish processing. In: *Trends in Fish Processing Technologies* (edited by D. N. Borda, A. I. & P. Raspor). CRC Press: Boca Raton, FL, USA, pp. 28–69.

Rotabakk, B.T., Birkeland, S., Lekang, O.I. & Sivertsvik, M. (2008) Enhancement of modified atmosphere packaged farmed Atlantic halibut (Hippoglossus hippoglossus) fillet quality by soluble gas stabilization. *Food Science and Technology International*, 14, 179–186. https://doi.org/10.1177/1082013208092051.

Rotabakk, B.T., Skipnes, D., Akse, L. & Birkeland, S. (2011) Quality assessment of Atlantic cod (Gadus morhua) caught by longlining and trawling at the same time and location. *Fisheries Research*, 112, 44–51. https://doi.org/10.1016/j.fishres.2011.08.009.

Rotabakk, B.T., Melberg, G.L. & Lerfall, J. (2018) Effect of season, location, filleting method and storage on the quality of farmed Atlantic salmon (Salmo salar L.). *Food Technology and Biotechnology*, 56. https://doi.org/10.17113/ftb.56.02.18.5346.

Roth, B., Kramer, L., Skuland, A.V., Løvdal, T., Øines, S., Foss, A. & Imsland, A.K. (2014) The shelf life of farmed turbot (Scophthalmus maximus). *Journal of Food Science*, 79, S1568–S1574. https://doi.org/10.1111/1750-3841.12541.

Ruize, V. (2012) Analysis of existing refrigeration plants on-board fishing vessels and improvement opportunities. Second International Symposium on Fishing Vessel Energy Efficiency E-Fishing, Vigo, Spain.

Ryder, J. et al. (2014) Assessment and Management of Seafood Safety and Quality: Current Practices and Emerging Issues. FAO Fisheries and Aquaculture Technical Paper.

Saito, T., Arai, K. & Matsuyoshi, M. (1959) A new method for estimating the freshness of fish. *Nippon Suisan Gakkaishi*, 24, 749–750. https://doi.org/10.2331/suisan.24.749.

Sampaio, F.D.F. & Freire, C.A. (2016) An overview of stress physiology of fish transport: Changes in water quality as a function of transport duration. *Fish and Fisheries*, 17, 1055–1072. https://doi.org/10.1111/faf.12158.

Sarkar, A. & Singh, R.P. (2004) Air impingement technology for food processing: Visualization studies. *LWT – Food Science and Technology*, 37, 873–879. https://doi.org/10.1016/j.lwt.2004.04.005.

Sarkar, A., Nitin, N., Karwe, M.V. & Singh, R.P. (2004) Fluid flow and heat transfer in air jet impingement in food processing. *Journal of Food Science*, 69, CRH113–CRH122. https://doi.org/10.1111/j.1365-2621.2004.tb06315.x.

Shearer, K.D. (1994) Factors affecting the proximate composition of cultured fishes with emphasis on salmonids. *Aquaculture*, 119, 63–88. https://doi.org/10.1016/0044-8486(94)90444-8.

Shearer, K.D., Asgard, T., Andorsdottir, G. & Aas, G.H. (1994) Whole body elemental and proximate composition of Atlantic salmon (Salmo salar) during the life cycle. *Journal of Fish Biology*, 44, 785–797. https://doi.org/10.1111/j.1095-8649.1994.tb01255.x.

Shumilina, E. et al. (2017) Monitoring of quality changes in salmon and salmon rest raw materials by NMR. In: *Modern Magnetic Resonance* (edited by G. A. Webb). Springer International Publishing: Cham, Germany.

Silbande, A., Adenet, S., Chopin, C., Cornet, J., Smith-Ravin, J., Rochefort, K. & Leroi, F. (2018) Effect of vacuum and modified atmosphere packaging on the microbiological, chemical and sensory properties of tropical red drum (Sciaenops ocellatus) fillets stored at 4 degrees C. *International Journal of Food Microbiology*, 266, 31–41. https://doi.org/10.1016/j.ijfoodmicro.2017.10.015.

Silvenius, F., Grönman, K., Katajajuuri, J., Soukka, R., Koivupuro, H. & Virtanen, Y. (2014) The role of household food waste in comparing environmental impacts of packaging alternatives. *Packaging Technology and Science*, 27, 277–292. https://doi.org/10.1002/pts.2032.

Simpson, B. (2000) Digestive proteinases from marine animals. In: *Seafood Enzymes: Utilization and Influence on Postharvest Seafood Quality* (edited by N. Harrd & B. Simposon). Marcel Dekker, New York, pp. 191–214.

Singh, S. & Shalini, R. (2016) Effect of hurdle technology in food preservation: A review. *Critical Reviews in Food Science and Nutrition*, 56, 641–649. https://doi.org/10.1080/10408398.2012.761594.

Singh, V.P. (2018) Recent approaches in food bio-preservation – A review. *Open Veterinary Journal*, 8, 104–111. https://doi.org/10.4314/ovj.v8i1.16.

Sivertsvik, M. (2003) Solubility and diffusion of CO2 into fish: And the effect of CO2 and temperature on the storage quality of fresh Atlantic salmon. *Norges Landbrukshøgskole Ås*.

Sivertsvik, M. (2007) The optimized modified atmosphere for packaging of pre-rigor filleted farmed cod (Gadus morhua) is 63-ml/100ml oxygen and 37ml/100ml carbon dioxide. *LWT – Food Science and Technology*, 40, 430–438. https://doi.org/10.1016/j.lwt.2005.12.010.

Sivertsvik, M. & Birkeland, S. (2006) Effects of soluble gas stabilization, modified atmosphere, gas to product volume ratio and storage on the microbiological and sensory characteristics of ready-to-eat shrimp (Pandalus borealis). *Revista de Agaroquimica y Tecnologia de Alimentos*, 12, 445–454.

Sivertsvik, M., Jeksrud, W.K. & Rosnes, J.T. (2002) A review of modified atmosphere packaging of fish and fishery products – Significance of microbial growth, activities and safety. *International Journal of Food Science and Technology*, 37, 107–127. https://doi.org/10.1046/j.1365-2621.2002.00548.x.

Sivertsvik, M., Rosnes, J.T. & Kleiberg, G.H. (2003) Effect of modified atmosphere packaging and superchilled storage on the microbial and sensory quality of Atlantic salmon (Salmo salar) fillets. *Journal of Food Science*, 68, 1467–1472. https://doi.org/10.1111/j.1365-2621.2003.tb09668.x.

Skare, M., Chan, S.S., Handeland, S.O., Løvdal, T., Lerfall, J. & Roth, B. (2021) A comparative study on quality, shelf life and sensory attributes of Atlantic salmon slaughtered on board slaughter vessels against traditional land-based facilities. *Aquaculture*, 540, 736681. https://doi.org/10.1016/j.aquaculture.2021.736681.

Smith, I.P., Metcalfe, N.B., Huntingford, F.A. & Kadri, S. (1993) Daily and seasonal patterns in the feeding behaviour of Atlantic salmon (Salmo salar L.) in a sea cage. *Aquaculture*, 117, 165–178. https://doi.org/10.1016/0044-8486(93)90133-J.

Solbakken, V.A., Hansen, T. & Stefansson, S.O. (1994) Effects of photoperiod and temperature on growth and parr-smolt transformation in Atlantic salmon (Salmo salar L.) and subsequent performance in seawater. *Aquaculture*, 121, 13–27. https://doi.org/10.1016/0044-8486(94)90004-3.

Speranza, B., Corbo, M.R., Conte, A., Sinigaglia, M. & Del Nobile, M.A. (2009) Microbiological and sensorial quality assessment of ready-to-cook seafood products packaged under modified atmosphere. *Journal of Food Science*, 74, M473–M478. https://doi.org/10.1111/j.1750-3841.2009.01369.x] [PubMed: 20492117.

Stannard, C. (1997) Development and use of microbiological criteria for foods. *Food Science and Technology Today*, 1, 137–176.

Steel, R. (2004) *Understanding and Measuring the Shelf-Life of Food.* Woodhead Publishing: Cambridge, UK.

Stoneham, T.R., Kuhn, D.D., Taylor, D.P., Neilson, A.P., Smith, S.A., Gatlin, D.M., Chu, H.S.S. & O'Keefe, S.F. (2018) Production of omega-3 enriched tilapia through the dietary use of algae meal or fish oil: Improved nutrient value of fillet and offal. *PLOS ONE*, 13, e0194241. https://doi.org/10.1371/journal.pone.0194241] [PubMed: 29641539.

Stormo, S.K., Skåra, T., Skipnes, D., Sone, I., Carlehög, M., Heia, K. & Skjelvareid, M.H. (2018) In-pack surface pasteurization of capture-based, pre-rigor filleted Atlantic cod (Gadus morhua). *Journal of Aquatic Food Product Technology*, 27, 783–794. https://doi.org/10.1080/10498850.2018.1499161.

Strasburg, G. et al. (2007) Physiology and chemistry of edible muscle tissues. In: *Fennema's Food Chemistry* (edited by S. Damodaran, K. L. Parkin & O. R. Fennema). CRC Press: Boca Raton, FL, USA.

Stupar, J., Holøymoen, I.G., Hoel, S., Lerfall, J., Rustad, T. & Jakobsen, A.N. (2021) Diversity and antimicrobial activity towards Listeria spp. and Escherichia coli among lactic acid bacteria isolated from ready-to-eat seafood. *Foods*, 10, 271. https://doi.org/10.3390/foods10020271.

Suresh, V. & Bhujel, R.C. (2019) Tilapias. In: *Aquaculture: Farming Aquatic Animals and Plants* (edited by J. S. Lucas, P. C. Southgate & C. S. Tucker). Wiley-Blackwell: Chichester, UK.

Svanevik, C.S. & Lunestad, B.T. (2011) Characterization of the microbiota of Atlantic mackerel (Scomber scombrus). *International Journal of Food Microbiology*, 151, 164–170. https://doi.org/10.1016/j.ijfoodmicro.2011.08.016.

Svanevik, C.S., Roiha, I.S., Levsen, A. & Lunestad, B.T. (2015) Microbiological assessment along the fish production chain of the Norwegian pelagic fisheries sector – Results from a spot sampling programme. *Food Microbiology*, 51, 144–153. https://doi.org/10.1016/j.fm.2015.05.016.

Sveinsdottir, K., Martinsdottir, E., Hyldig, G., Jorgensen, B. & Kristbergsson, K. (2002) Application of quality index method (QIM) scheme in shelf-life study of farmed Atlantic salmon (Salmo salar). *Journal of Food Science*, 67, 1570–1579. https://doi.org/10.1111/j.1365-2621.2002.tb10324.x.

Thi, A.N.T., Jacxsens, L., Noseda, B., Samapundo, S., Nguyen, B.L., Heyndrickx, M. & Devlieghere, F. (2014) Evaluation of the microbiological safety and quality of Vietnamese Pangasius hypophthalmus during processing by a microbial assessment scheme in combination with a self-assessment questionnaire. *Fisheries Science*, 80, 1117–1128. https://doi.org/10.1007/s12562-014-0786-y.

Thodesen, J., Gjerde, B., Grisdale-Helland, B. & Storebakken, T. (2001a) Genetic variation in feed intake, growth and feed utilization in Atlantic salmon (Salmo salar). *Aquaculture*, 194, 273–281. https://doi.org/10.1016/S0044-8486(00)00527-5.

Thodesen, J., Storebakken, T., Shearer, K.D., Rye, M., Bjerkeng, B. & Gjerde, B. (2001b) Genetic variation in mineral absorption of large Atlantic salmon (Salmo salar) reared in seawater. *Aquaculture*, 194, 263–271. https://doi.org/10.1016/S0044-8486(00)00525-1.

Thostenson, E.T. & Chou, T.-W. (1999) Microwave processing: Fundamentals and applications. *Composites Part A*, 30, 1055–1071. https://doi.org/10.1016/S1359-835X(99)00020-2.

Tintchev, F., Kuhlmann, U., Wackerbarth, H., Töpfl, S., Heinz, V., Knorr, D. & Hildebrandt, P. (2009) Redox processes in pressurized smoked salmon studied by resonance Raman spectroscopy. *Food Chemistry*, 112, 482–486. https://doi.org/10.1016/j.foodchem.2008.05.088.

Tintchev, F., Wackerbarth, H., Kuhlmann, U., Toepfl, S., Knorr, D., Hildebrandt, P. & Heinz, V. (2010) Molecular effects of high-pressure processing on food studied by resonance Raman. *Annals of the New York Academy of Sciences*, 1189, 34–42. https://doi.org/10.1111/j.1749-6632.2009.05204.x.

Torstensen, B.E., Bell, J.G., Rosenlund, G., Henderson, R.J., Graff, I.E., Tocher, D.R., Lie, Ø & Sargent, J.R. (2005) Tailoring of Atlantic salmon (Salmo salar L.) flesh lipid composition and sensory quality by replacing fish oil with a vegetable oil blend. *Journal of Agricultural and Food Chemistry*, 53, 10166–10178. https://doi.org/10.1021/jf051308i.

Tzikas, Z., Ambrosiadis, I., Soultos, N. & Georgakis, S. (2007) Quality assessment of Mediterranean horse mackerel (Trachurus mediterraneus) and blue jack mackerel (Trachurus picturatus) during storage in ice. *Food Control*, 18, 1172–1179. https://doi.org/10.1016/j.foodcont.2006.07.014.

UNEP (2016) *TEAP TF-XXVII-4 Report*; Annex 2.

UNEP (2017) *Cold Chain Technology Brief – Fishing Vessel Application*. International Institute of Refrigeration: Paris.

Vasiliadou, S., Ambrosiadis, I., Vareltzis, K., Fletouris, D. & Gavriilidou, I. (2005) Effect of smoking on quality parameters of farmed gilthead sea bream (Sparus aurata L.) and sensory attributes of the smoked product. *European Food Research and Technology*, 221, 232–236. https://doi.org/10.1007/s00217-005-1138-z.

Veldhuizen, L.J.L., Berentsen, P.B.M., de Boer, I.J.M., van de Vis, J.W. & Bokkers, E.A.M. (2018) Fish welfare in capture fisheries: A review of injuries and mortality. *Fisheries Research*, 204, 41–48. https://doi.org/10.1016/j.fishres.2018.02.001.

Verbeke, W., Sioen, I., Brunsø, K., De Henauw, S. & Van Camp, J. (2007) Consumer perception versus scientific evidence of farmed and wild fish: Exploratory insights from Belgium. *Aquaculture International*, 15, 121–136. https://doi.org/10.1007/s10499-007-9072-7.

Vermeulen, S.J., Campbell, B.M. & Ingram, J.S.I. (2012) Climate change and food systems. *Annual Review of Environment and Resources*, 37, 195–222. https://doi.org/10.1146/annurev-environ-020411-130608.

Vidal, N.P., Goicoechea, E., Manzanos, M.J. & Guillén, M.D. (2017) Effect of smoking using smoke flavorings on several characteristics of farmed sea bass (Dicentrarchus labrax) fillets and on their evolution during vacuum-packed storage at refrigeration temperature. *Journal of Food Processing and Preservation*, 41, e12800. https://doi.org/10.1111/jfpp.12800.

Vieira, V.L.A., Norris, A. & Johnston, I.A. (2007) Heritability of fibre number and size parameters and their genetic relationship to flesh quality traits in Atlantic salmon (Salmo salar L.). *Aquaculture*, 272, S100–S109. https://doi.org/10.1016/j.aquaculture.2007.08.028.

Viji, P., Tanuja, S., Ninan, G., Lalitha, K.V., Zynudheen, A.A., Binsi, P.K. & Srinivasagopal, T.K. (2015) Biochemical, textural, microbiological and sensory attributes of gutted and ungutted sutchi catfish (Pangasianodon hypophthalmus) stored in ice. *Journal of Food Science and Technology*, 52, 3312–3321. https://doi.org/10.1007/s13197-014-1358-y.

Wetten, M., Aasmundstad, T., Kjøglum, S. & Storset, A. (2007) Genetic analysis of resistance to infectious pancreatic necrosis in Atlantic salmon (Salmo salar L.). *Aquaculture*, 272, 111–117. https://doi.org/10.1016/j.aquaculture.2007.08.046.

Whitaker, D.J. & Kent, M.L. (1991) Myxosporean Kudoa thyrsites: A cause of soft flesh in farm-reared Atlantic salmon. *Journal of Aquatic Animal Health*, 3, 291–294. https://doi.org/10.1577/1548-8667(1991)003<0291:MKTACO>2.3.CO;2.

Wiernasz, N., Cornet, J., Cardinal, M., Pilet, M., Passerini, D. & Leroi, F. (2017) Lactic acid bacteria selection for biopreservation as a part of hurdle technology approach applied on seafood. *Frontiers in Marine Science*, 4. https://doi.org/10.3389/fmars.2017.00119.

Woynarovich, A. et al. (2011) Better Management Practices for Carp Production in Central and Eastern Europe, the Caucasus and Central Asia. FAO Fisheries and Aquaculture Technical Paper. Food and Agriculture Organization of the United Nations: Rome.

Wu, C.-h., Yuan, C., Ye, X., Hu, Y., Chen, S. & Liu, D. (2014) A critical review on superchilling preservation technology in aquatic product. *Journal of Integrative Agriculture*, 13, 2788–2806. https://doi.org/10.1016/S2095-3119(14)60841-8.

Ytrestøyl, T., Aas, T.S. & Åsgård, T. (2015) Utilisation of feed resources in production of Atlantic salmon (Salmo salar) in Norway. *Aquaculture*, 448, 365–374. https://doi.org/10.1016/j.aquaculture.2015.06.023.

Zajdband, A. (2012) *Final Seafood Recommendation – Tilapia (Oreochromis niloticus, O. mossambicus, O. Aureus, O. spp.)*. Monterey Bay Aquarium – Seafood Watch.

Zhang, H., Zhu, Y., Chen, S., Xu, C., Yu, Y., Wang, X. & Shi, W. (2018) Determination of the effects of different high-temperature treatments on texture and aroma characteristics in Alaska pollock surimi. *Food Science and Nutrition*, 6, 2079–2091. https://doi.org/10.1002/fsn3.763.

Zotta, T., Parente, E., Ianniello, R.G., De Filippis, F. & Ricciardi, A. (2019) Dynamics of bacterial communities and interaction networks in thawed fish fillets during chilled storage in air. *International Journal of Food Microbiology*, 293, 102–113. https://doi.org/10.1016/j.ijfoodmicro.2019.01.008.

Ørnholt-Johansson, G., Frosch, S. & Munk Jørgensen, B. (2017) Variation in some quality attributes of Atlantic salmon fillets from aquaculture related to geographic origin and water temperature. *Aquaculture*, 479, 378–383. https://doi.org/10.1016/j.aquaculture.2017.06.016.

12

Horizon scanning for aquatic food security

Rachel Norman and Margaret Crumlish

Introduction

Throughout this book we have looked at the role of seafood in the three pillars of food security which are 'availability', 'access' and 'utilization'. A fourth pillar of 'stability' has been added (FAO, 2008) and we have looked at different aspects of these pillars on a global scale. In this chapter we will consider what future challenges and opportunities are coming our way.

There are some large and clearly articulated challenges, such as population growth and climate change. There are political drivers that vary geographically but will surely converge on a demand for net-zero greenhouse gas emissions within the next 20–30 years. That target has to be met through a combination of reduced emissions and carbon sequestration to compensate for those areas where emissions cannot be eradicated. This is obviously a huge challenge but also provides an opportunity to re-prioritize our lives, restructure the food system and develop innovative techniques that will allow us to solve these problems.

It seems clear that we are also going to be dealing with increased global political uncertainty with trade deals and sanctions meaning that supply chains are likely to be disrupted and food prices will fluctuate unpredictably. It is highly likely that emerging diseases in both crops, animals and humans and conflicts will continue to impact on our ability to achieve food security at one scale or another. COVID-19, for example, has had a fundamental impact on how we live our lives and affected livelihoods, food production and supply chains globally.

One way to deal with these uncertainties and the potential lack of stability is to build resilient systems and that is where the opportunities lie. In this chapter we will explore in more detail both the challenges facing us over the next few decades and some of the technological opportunities which are arising throughout the food system including within aquatic food security.

Margaret Crumlish and Rachel Norman (eds) **Aquatic Food Security**
DOI: 10.52517/9781789183368.012,

Challenges

Population growth

One of the major challenges for the food system is the growing population and the fact that the growth is not distributed evenly geographically. The current world population is 7.8 billion (World data) and this is expected to reach 9.7 billion in 2050 (UN, 2019). Predictions from the UN in June 2019 state that nine countries will make up more than one-half the projected growth of the global population between now and 2050: India, Nigeria, Pakistan, the Democratic Republic of the Congo, Ethiopia, the United Republic of Tanzania, Indonesia, Egypt and the USA (in descending order of the expected increase) (UN, 2019). In addition, as some growing countries become richer then the demand for different food types are also predicted to increase both for more expensive food (e.g. more red meat) or for improved nutrition for individuals.

This growth in populations and changes in demand are predicted to require an increase in food production. Some of the shortfall could be achieved by reducing waste or losses of the food that we currently produce. Looking at increased production we find that much of the currently productive land is already in use, and indeed may be in demand for housing or re-forestation to mitigate climate change. Wild capture fisheries production will, at best, be maintained. However, as discussed in previous chapters, seafood production through aquaculture is an area in which we could increase production and therefore supply of a healthy protein source as well as micronutrients essential for human health and development.

Climate change

Climate change is predicted to cause a number of significant changes to aquatic systems including sea level and temperature rises, change in monsoonal rain patterns, extreme climatic events and water stress (Ruby and Ahilan, 2018). These forecasted changes will impact the availability of aquatic food sourced through capture fisheries and aquaculture sectors.

Changing water temperature and quality/chemistry will lead to ecosystem changes, for example, higher temperatures will alter the dissolved oxygen availability for a wide range of aquatic seafood species. The effects of which have already been seen in the range of species observed in capture fisheries (Hughes et al., 2015). Fish species have particular oxygen requirements to ensure their health and welfare, so alterations in the availability of oxygen within the aquatic environments will impact the availability of the species in the wild as well as the production stocking levels that can be maintained in aquaculture systems. Technical changes in fish production, such as moving towards more controlled land-based recirculation systems, would negate this impact, however, these types of farming systems require a high financial investment not available to all sectors.

A 2018 report by the FAO discusses in detail the impact of climate change on fisheries and aquaculture in which they look at the impacts of climate change for different geographical regions (FAO, 2018). Although the impact is expected to vary, particularly in terms of the species affected, there are several overarching issues which need to be anticipated and planned for.

As mentioned above, changes in water temperature and quality will have an effect both on species ranges in wild capture fisheries but also which species it is possible to produce in aquaculture systems. There are three main reasons for this, the first is that species have climate envelopes within which they are able to grow and thrive. Second, these changes will mean a change to the pests and diseases that can survive alongside the target species and will lead to more emerging infections. Finally, climate change is expected to increase the chances of toxic algal blooms.

When considering European seafood, Blanchet et al. (2019) concluded that marine systems are more vulnerable to negative impacts of climate change than freshwater ones, this is because of the higher temperature sensitivity tolerances of marine species. Most species perform poorly outside their optimal temperature range (Blanchett et al., 2019, and references therein). If species are not in their optimal range, then this can affect growth rates and therefore harvest or catch sizes (Cochrane et al., 2009). In capture fisheries where species are free to move, the changing climate is therefore expected to have a significant impact on fish species distribution and model predictions show it will lead to numerous local extinctions within 40 years (UN, 2019b).

While changing water temperature and quality will impact which seafood species can thrive in a given location, they will also change what pests and pathogens can survive, altering the disease outcomes and supporting increases in emerging diseases. The prevalence and virulence of pathogens might also be impacted by these changes (Harvell et al., 1999). Some of the pathogens that are most sensitive to thermal changes are listed in a report from the International Centre for Tropical Agriculture (UNFCCC, 2014). For example, they identify koi herpes virus disease as a key OIE listed disease that is sensitive to climate change because the disease pattern is influenced by temperature – occurring between 16°C and 25°C.

Finally, toxic algal blooms are expected to increase with, among other things, higher temperatures. Blooms can have a devastating impact on the aquaculture industry. For example, in 2016 in Chile high ocean temperatures helped cause a harmful bloom that wiped out up to 20% of Chile's total production for the year and which caused the deaths of 27 million fish and cost £565 million (FAO, 2016). Another study showed that an increase in surface water temperature and reduced freshwater input ultimately resulted in an active harmful algal bloom in coastal southern Chile (León-Muñoz et al., 2018)

While population growth and climate change are significant problems, they are at least widely recognized by the scientific and political sectors. It should theoretically be possible to take actions to prepare in advance to reduce the negative impacts. The increase in extreme weather events is the most difficult to deal with but the year on year changes should be more manageable since they are more predictable and therefore appropriate mitigations can be put in place (e.g. switching to more temperature tolerant species). However, the next few challenges are much more difficult to plan for since they are likely to be quite localized and occur quickly. They are challenges that will affect the whole food system and not just seafood.

Political uncertainty

It is now 2023, Brexit has happened, however there is still an enormous level of uncertainty around how we will manage trade deals, what happens to the Common Fisheries Policy and what the impact of all of that will be on food price and availability. At the same time the war in Ukraine means that trade sanctions are being imposed on Russia. While these are all recent events, they demonstrate a pattern in which trade deals are becoming more politically charged and there is a lack of stability that makes forward planning difficult for producers and consumers alike. This problem is wider than seafood but it can have significant negative impact due to the perishable nature of much of the seafood traded globally.

Social instability

There are a range of events that can cause social instability, the two most significant ones being conflict and disease. According to the FAO reporting in September 2019, 41 countries need external aid with food (FAO, 2019a). In one-half of those cases, aid was needed because of civil unrest or conflicts and in others it is due to an increase in the number of refugees fleeing from conflicts in neighbouring countries.

Considering the example of Syria in order to highlight the specific issues that are associated with conflict, there have been a series of reports from the FAO with one being published in September 2019 (FAO, 2019b). In a report in November 2018, it was stated that there are 6.7 million food insecure people in Syria with 4.5 million at risk (55% of population), there are 6.18 million internally displaced with 5.64 million having sought refuge in neighbouring countries (FAO, 2018b). One-quarter of households suffer poor food quality and insufficient diets. The FAO reports show that year on year wheat production fluctuates depending on the weather, with a poor harvest in 2014, better in 2015 but a 29 year low in 2018 due to a long dry season followed by heavy rainfall (FAO, 2018b). Owing to a decrease in conflict, the harvested wheat area has doubled in 2019 compared with 2018 but is still 25% below the pre-conflict level. Livestock numbers have stabilized, but at low levels and feed prices are high although there is improved access to pasture in 2019 compared with 2018. There is restricted access to vets and medicines resulting in poor control of animal diseases. In particular, lumpy skin disease in cattle has increased, potentially because high temperatures and humidity have increased insect vector densities. Transportation also causes a problem as it becomes unsafe to move from one region to another. Conflicts can mean that markets have fragmented, and food produced in one area cannot easily be transported to another. This has improved in Syria although the costs of transportation mean that in some regions which produce a surplus of fruit and vegetables there is a high level of waste because traders cannot sell their stock (FAO 2019b). Although this is a specific example the issues highlighted will be consistent for any region experiencing conflict and any production system including seafood. Labour shortages, disrupted transportation and lack of veterinary support (personnel and medicines) and other infrastructure means that both food production and the supply chain remain vulnerable and that may be exacerbated by climate changes, which would have made things challenging even without additional impact of local conflicts.

These interactions between the collapse of infrastructure and climate are not unique to areas experiencing conflict. In 2014 there was an Ebola outbreak in West Africa which was both a serious health concern in the region and impacted food security. In a 2014 special alert for Guinea, Liberia and Sierra Leone, the FAO highlighted that movement restrictions due to the disease meant that harvesting that year's crop was going to be difficult, disruption to cross border trade meant price increases and it was anticipated that 1.3 million people would require food aid over a 3 month period (FAO, 2014). The areas with the highest incidences of disease were among the most agriculturally productive. In addition, cash crop production (e.g. cocoa and rubber) would also decline which reduces household income and hence access to food.

While the examples given here do not discuss seafood explicitly, the effects of conflict and disease outbreak are common to all supply chains and production systems.

More recently we have experienced COVID-19 pandemic, which has had significant impact socially and economically around the globe. This impact of this global pandemic varies geographically, but in countries where wide-ranging and lengthy lockdowns took place then the lack of access to restaurants and other prepared food outlets meant more meals were made in the home and hence a higher demand fell on supermarkets resulting in empty shelves and supply issues, at least in the short term. This included rationing of purchasing in some instances. More globally there has been an increase in food insecurity. Those countries who did not provide economic support for people who could not attend work due to lockdown caused significant economic hardship. In this case a number of studies and reports have looked specifically at the impact of COVID-19 on seafood supply chains. Love et al. (2020) found that in the first 5 months of the pandemic some supply chains were more resilient than others. For example, frozen Ecuadorian shrimp were diverted to alternative markets while live supply chains were more heavily impacted. Worldfish have produced a number of reports specifically on the impact of COVID-19 on seafood supply chains in different countries (Worldfish, 2020a) as well as on gender issues, which are exacerbated by the pandemic. For example the mitigations put in place to control COVID-19 resulted in issues with demand and supply chains which severely impacted women traders and processors in sub Saharan Africa play a significant role in the supply chain (WorldFish, 2020b).

Zero carbon agenda

The zero carbon agenda is very much on the social and political radar at the moment but with mixed response globally. There have recently been a series of climate strikes in which millions of people took to the streets in order to convince global leaders to take more action to reduce emissions and work towards a carbon neutral world (BBC, 2019); a world in which most emissions are eliminated and those which remain are offset through carbon capture mechanisms including tree planting. Seafood can have a significant role to play here since it is a production system which is, on average, lower in emissions than red meat production, for example. Capture fisheries use no land and no feed so the emissions are mostly focused on fuel/transportation costs, so in that case a carbon sustainable fishery would be a local, well stocked fishery. However, sustainability in that case is about maintaining stocks so that there are fish in the future.

In terms of aquaculture globally there are many areas where low impact extensive aquaculture is practised, in this case natural lakes are often used and there is little input in terms of additional food, for example. While these systems can, in some cases damage natural environments, they are close to zero carbon excluding transport or supply chain costs. However, more intensive aquaculture can result in emissions, in particular the feed that goes into these systems. In 2017, Robb et al. produced a report for the FAO on greenhouse gas emissions from aquaculture which focused on three Asian systems. Asia was chosen because that is where growth of aquaculture has been greatest. They used life cycle assessment models. Although the emissions varied between systems, the production of feed was the largest source for all three systems (Robb et al., 2017). These came from the production of the raw materials for the feed as well as transportation. Other sources of emissions included energy for the pumps, lighting and transport on the farm. Sustainable (i.e. long lasting) sources of ingredients in feed is one of the major barriers to aquatic food security (as discussed in Chapter 3).

Stability of supply of seafood

All of the issues discussed above may lead to issues in the stability of supply of seafood. The international trade for seafood is growing (Asche et al., 2015). Consumer demands for seafood do not necessarily match what can be produced in country and so trade is a way to meet consumer demand at a local level. However, Asche et al. (2015) showed that developing countries export high-valued species and import low-valued ones with the production amount being exported and imported being similar. Instability therefore has potential benefits economically, disruption to these supply chains that might result in people eating the fish produced in their own country (assuming the economic environment makes that possible) may not affect the amount of seafood they eat and, indeed, in developing countries it might lead to higher quality diets. However, there will be significant economic losses, which will impact on food security.

Opportunities

While there are clearly a range of somewhat dire problems facing us, there are also opportunities and innovations on the horizon that will help to mitigate some of them. Many of them are not yet economically viable but are likely to become so as circumstances change.

These opportunities occur at different stages in the supply chain, they include changes in production system, high-tech approaches to our current systems and changes in the consumer and supply chain stage of the system.

Production systems

As we have discussed already, production systems for aquaculture are impacted by climate change and pests and diseases and those impacts are likely to increase over time. One solution is to reconsider the way in which we produce fish and recirculating aquaculture systems or RAS are proposed as one possible solution. These are closed aquaculture systems in which water is recycled and mechanical and biological filters are used to remove nutrients excreted by the fish and limit disease risk (FAO, 2015). It is also possible to control water temperature and flow in a way that is not possible in natural systems and allows a consistent environment for the fish. Overall this means that the water requirements are much lower than for traditional methods of production and the excreted nutrients can be used as fertilizer. This is therefore an environmentally friendly way of producing fish. However, there are significant costs in setting the system up and energy costs in keeping it running. It is therefore likely that these will become more economically viable as demand for seafood increases and other production methods become less viable, for example, as the climate becomes more unpredictable.

An alternative way to improve nutrient use and move towards a circular economy are integrated aquaculture systems (Little and Edwards, 2003). There are different definitions of integrated framing, but we are interested in the integration of livestock and fish, often within a larger farming system. It is defined by Little and Edwards (2003) as the concurrent or sequential linkage between two or more activities of which at least one is aquaculture. The main potential linkages concern the use of nutrients, particularly the use of livestock manure for fish production. Several studies in Bangladesh, India, Thailand and Vietnam indicate that livestock wastes are the most commonly used input (Little and Edwards 2003). Integrated agriculture is most commonly studied on small-scale farms and has been shown to have a positive benefit on those farms in a number of countries (Dey et al., 2010; Murshed-E-Jahan and Pemsl, 2011).

Offshore aquaculture is another production method that could be a mechanism to meet the growing demand for seafood (Froehlich et al., 2017). The idea being that offshore production might reduce some of the environmental impacts of aquaculture with the requirement for less freshwater, greater depth and distance from shore. Faster currents could reduce disease risks and pollution and cause less conflict with tourism. However, future changes in water temperature and chemistry due to climate change could reduce the suitability of sites. In addition, offshore sites are often exposed to more extreme environmental conditions. Buck et al. (2018) looked at how we could combine the last two production ideas with offshore integrated multi-trophic aquaculture (IMTA). IMTA is the co-cultivation of fed species (such as finfish and shrimp) with extractive species such as suspension-feeding (e.g. mussels and oysters) invertebrates which would feed on the effluents generated by the fed species (Buck et al., 2018). This is an area in which research is being conducted and experimental systems are currently being trialled in order to determine suitable species and the pros and cons of these types of system.

Big data and the internet of things

We are in the midst of a boom in technology with the use of sensors and other forms of data collection meaning we have access to more data than ever before, that can be, for example, data on the environment or on the quality and quantity of food being eaten. The sources are varied and are often a mix of qualitative and quantitative data. How we bring those data sources together and use them effectively in production is an important question (Endo et al., 2019).

In addition, satellite imagery and sensors can be used to guide fishing boats towards fish or as an early warning system for toxic algal blooms, which would allow early harvesting of fish before they are damaged by the bloom (Torres Palenzuela et al., 2019; Tyler et al., 2016).

It is not just at the production level though, we are now able to record and access more data on supply chains and use technology to allow us to track individual fish or batches of fish. This can be an attractive selling point for consumers. Blockchain has been suggested as a technology that would be an independent way to allow traceability (FAO, 2020). At the same time improvements in technology mean that packaging can be used to increase shelf-life and improve food safety.

These data sources are available for a wide range of food products, it is not specific to seafood, but it is technology that seafood producers could take advantage of. In particular, sensors and underwater cameras, for example, can get around the fact that it is more challenging to observe fish than terrestrial livestock (Hung et al., 2016).

Recent laboratory-based advances have driven the research behind the wonderful world of microbiome. These are basically comprised of complex mixtures of microbial organisms (viruses, yeast, fungi but predominantly bacteria) that inhabit the skin, gills, mucosal membranes and gastrointestinal tract of the animals. Current research has identified the composition of many microbiome communities in farmed and wild seafood, predominantly fish, with a view to understanding how these microbial communities function to promote the health of the animals. The DNA-based sequencing technologies have provided these platforms from which we can map the microbial changes in relation to the environmental system, genetic modifications and nutritional/husbandry changes when farming food fish species. Yet the scientific community has yet to understand the function and the communication between these complex microbial environments. As our understanding of these microbial communities increases then we expect to be able to utilize them to improve health and welfare of farmed fish and the surrounding aquatic environments.

Alt-seafood

As well as finding sustainable, plant-based sources to feed to fish, in order to replace fishmeal and fish oil, there is also slow progress towards plant and cell-based alternatives to seafood. These are similar in principle to meat replacement sources (i.e. they mimic seafood but should be more sustainable and are suitable for vegetarians or vegans to eat). Worldfish have recently produced a report on so-called Alt-Seafood (Worldfish, 2020c). They found that plant-based seafood alternatives were already having some success with companies producing products mimicking shrimp, both canned and raw tuna, fish fillets and burgers. Cell-based seafood is less well developed but is gaining momentum, at least scientifically. In this case we are producing a genuine animal seafood by growing cells in a nutrient rich environment. This means that these products more closely resemble seafood because they are made up of the same cell types. However, the technology for scaling this production up is not yet available. However, there is definitely potential for alt-seafood to have an impact on the food system.

Conclusion

In summary, there are plenty of challenges for aquaculture to overcome and opportunities to exploit. Of course we do not know what we do not know, but hopefully lessons have been learned from the unexpected events of the last few years and we are in a stronger place to develop resilient production and supply chains that will protect food security for everyone with seafood playing a, perhaps the most, significant role.

References

Asche, F., Bellemare, M.F., Roheim, C., Smith, M.D. & Tveteras, S. (2015) Fair enough? Food Security and the international trade of seafood. *World Development*, 67, 151–160. https://www.sciencedirect.com/science/article/pii/S0305750X14003246. https://doi.org/10.1016/j.worlddev.2014.10.013.

BBC (2019) Available at https://www.bbc.co.uk/news/world-49777279.

Blanchet, M., Primicerio, R., Smalås, A., Arias-Hansen, J. & Aschan, M. (2019) How vulnerable is the European seafood production to climate warming? *Fisheries Research*, 209, 251–258. https://doi.org/10.1016/j.fishres.2018.09.004.

Buck, B.H., Troell, M.F., Krause, G., Angel, D.L., Grote, B. & Chopin, T. (2018) State of the art and challenges for offshore integrated multi-trophic aquaculture (IMTA). *Frontiers in Marine Science*, 5, 165. https://doi.org/10.3389/fmars.2018.00165.

Cochrane, K., Young, D.C., Soto, D. & Bahri, T. (2009) Climate Change Implications for Fisheries and Aquaculture: Overview of Current Scientific Knowledge. FAO Fisheries and Aquaculture Technical Paper 530. www.fao.org/fileadmin/user_upload/newsroom/docs/FTP530.pdf, Vol. 212.

Dey, M.M., Paraguas, F.J., Kambewa, P. & Pemsl, D.E. (2010) The impact of integrated aquaculture–agriculture on small-scale farms in Southern Malawi. *Agricultural Economics*, 41, 67–79. https://doi.org/10.1111/j.1574-0862.2009.00426.x.

Endo, H. & Wu, H. (2019) Biosensors for the assessment of fish health: A review. *Fisheries Science*, 85, 641–654. https://doi.org/10.1007/s12562-019-01318-y.

FAO (2008) An introduction to the basic concepts of food security. www.fao.org/3/a-al936e.pdf.

FAO (2014) Grave food security concerns following the Ebola outbreak in Liberia, Sierra Leone and Guinea. www.fao.org/3/a-i4003e.pdf.

FAO (2015) A guide to recirculation aquaculture: An introduction to the new environmentally friendly and highly productive closed fish farming systems. www.fao.org/3/a-i4626e.pdf.

FAO (2016) Algal bloom mortalities in Chile have salmon prices climbing even higher. www.fao.org/in-action/globefish/market-reports/resource-detail/en/c/415527/.

FAO (2018a) Impacts of climate change on fisheries and aquaculture: Synthesis of current knowledge, adaptation and mitigation options. www.fao.org/3/I9705EN/i9705en.pdf.

FAO (2018b) Syrian Arab Republic Situation report November 2018. https://www.fao.org/fileadmin/user_upload/emergencies/docs/FAOSyrianArabRepublic_SitReport_November2018.pdf.

FAO (2019a) FAO report cites 41 countries needing external assistance for food. www.fao.org/news/story/en/item/1208508/icode/.

FAO (2019b) FAO/WFP crop and food security assessment mission to the Syrian Arab Republic. www.fao.org/3/ca5934en/CA5934EN.pdf.

FAO (2020) Blockchain application in seafood value chains. www.fao.org/3/ca8751en/ca8751en.pdf.

Froehlich, H.E., Smith, A., Gentry, R.R. & Halpern, B.S. (2017) Offshore aquaculture: I know it when I see it. *Frontiers in Marine Science*, 4, 154. (https://www.frontiersin.org/article/10.3389/fmars.2017.00154). https://doi.org/10.3389/fmars.2017.00154.

Harvell, C.D., Kim, K., Burkholder, J.M., Colwell, R.R., Epstein, P.R., Grimes, D.J., Hofmann, E.E., Lipp, E.K., Osterhaus, A.D.M.E., Overstreet, R.M., Porter, J.W., Smith, G.W. & Vasta, G.R. (1999) Emerging marine diseases-climate links and anthropogenic factors. *Science*, 285, 1505–1510. https://doi.org/10.1126/science.285.5433.1505.

Hughes, K.M., Dransfeld, L. & Johnson, M.P. (2015) Climate and stock influences on the spread and locations of catches in the northeast Atlantic mackerel fishery. *Fisheries Oceanography*, 24, 540–552. https://doi.org/10.1111/fog.12128.

Hung, C.C., Tsao, S.C., Huang, K.H., Jang, J.P., Chang, H.K. & Dobbs, F.C. (2016) A highly sensitive underwater video system for use in turbid aquaculture ponds. *Scientific Reports*, 6, 31810. https://doi.org/10.1038/srep31810.

León-Muñoz, J., Urbina, M.A., Garreaud, R. & Iriarte, J.L (2018). Hydroclimatic conditions trigger record harmful algal bloom in western Patagonia (summer 2016). *Scientific Reports* **8**, 1330 https://doi.org/10.1038/s41598-018-19461-4.

Little, D.C. & Edwards, P. (2003) Integrated livestock-fish farming systems. FAO Inland Water Resources and Aquaculture Service. www.fao.org/3/Y5098E/y5098e00.htm.

Love, D., Allison, E.H., Asche, F., Belton, B., Cottrell, R., Froelich, H., Gephart, J., Hicks, C.C., Little, D.C., Nussbaumer, E.M., da Silva, P., Poulain, F., Rubio, A., Stoll, J., Tlusty, M.F., Thorne-Lyman, A.L., Troell, M. & Zhang, W. (2020) Emerging COVID-19 impacts, responses, and lessons for building resilience in the seafood system. SocArXiv preprint. https://osf.io/preprints/socarxiv/x8aew/.

Murshed-e-Jahan, K. & Pemsl, D.E. (2011) The impact of integrated aquaculture–agriculture on small-scale farm sustainability and farmers' livelihoods: Experience from Bangladesh. *Agricultural Systems*, 104, 392–402. https://doi.org/10.1016/j.agsy.2011.01.003.

Robb, D.H.F., MacLeod, M., Hasan, M.R. & Soto, D. (2017) Greenhouse Gas Emissions from Aquaculture: A Life Cycle Assessment of Three Asian Systems. FAO Fisheries and Aquaculture Technical Paper no. 609. Food and Agriculture Organization: Rome.

Ruby, P. & Ahilan, B. (2018) An overview of climate change impact in fisheries and aquaculture. *Climate Change*, 49, 87–94. www.discoveryjournals.org/climate_change/current_issue/v4/n13/A8.pdf.

Torres Palenzuela, J.M., González Vilas, L., Bellas, F.M., Garet, E., González-Fernández, Á. & Spyrakos, E. (2019) Pseudo-Nitzschia blooms in a coastal upwelling system: Remote sensing detection, toxicity and environmental variables. *Water*, 11, Art. no.: 1954. https://doi.org/10.3390/w11091954.

Tyler, A.N., Hunter, P.D., Spyrakos, E., Groom, S., Constantinescu, A.M. & Kitchen, J. (2016) Developments in Earth observation for the assessment and monitoring of inland, transitional, coastal and shelf-sea waters. *Science of the Total Environment*, 572, 1307–1321. https://doi.org/10.1016/j.scitotenv.2016.01.020.

UN (2019a) Growing at a slower pace, world population is expected to reach 9.7 billion in 2050 and could peak at nearly 11 billion around 2100. https://www.un.org/development/desa/en/news/population/world-population-prospects-2019.html.

UN (2019b) UN Report: nature's dangerous decline 'unprecedented'; species extinction rates 'accelerating'. https://www.un.org/sustainabledevelopment/blog/2019/05/nature-decline-unprecedented-report/.

UNFCCC (2014) Climate change and aquatic animal disease. https://unfccc.int/files/documentation/submissions_from_non-party_stakeholders/application/pdf/517.pdf.

World Data (n.d.) Current world population. https://www.worldometers.info/world-population/.

Worldfish (2020a) Emerging COVID-19 impacts, responses, and lessons for building resilience in the seafood system. https://doi.org/10.31235/osf.io/x8aew.

Worldfish (2020b) COVID-19 impacts on women fish processors and traders in sub-Saharan Africa: Insights and recommendations for building forward better. https://worldfishcenter.org/publication/covid-19-impacts-women-fish-processors-and-traders-sub-saharan-africa-insights-and.

Worldfish (2020c) Alternative seafood: Assessing food, nutrition and livelihood futures of plant-based and cell-based seafood. https://digitalarchive.worldfishcenter.org/handle/20.500.12348/4428.

Index